Mobile Commerce: Technology, Theory, and Applications

Brian E. Mennecke
Iowa State University, USA

Troy J. Strader
Iowa State University, USA

IDEA GROUP PUBLISHING
Hershey • London • Melbourne • Singapore • Beijing

Acquisition Editor: Mehdi Khosrowpour
Senior Managing Editor: Jan Travers
Managing Editor: Amanda Appicello
Development Editor: Michele Rossi
Copy Editor: Maria Boyer
Typesetter: Tamara Gillis
Cover Design: Integrated Book Technology
Printed at: Integrated Book Technology

Published in the United States of America by
     Idea Group Publishing (an imprint of Idea Group Inc.)
     701 E. Chocolate Avenue
     Hershey PA 17033
     Tel: 717-533-8845
     Fax: 717-533-8661
     E-mail: cust@idea-group.com
     Web site: http://www.idea-group.com

and in the United Kingdom by
     Idea Group Publishing (an imprint of Idea Group Inc.)
     3 Henrietta Street
     Covent Garden
     London WC2E 8LU
     Tel: 44 20 7240 0856
     Fax: 44 20 7379 3313
     Web site: http://www.eurospan.co.uk

Library of Congress Cataloging-in-Publication Data

Mennecke, Brian E. (Brian Ernest), 1960-
   Mobile commerce : technology, theory, and applications /Brian E. Mennecke and Troy J. Strader
     p. cm.
   Includes bibliographical references and index.
   ISBN 1-59140-044-9 (hardcover)
     1. Mobile commerce. I. Strader, Troy J., 1945- II. Title.

HF5548.34 .M46 2002
658.8'4--dc21   2002027607

eISBN 1-59140-090-2

British Cataloguing in Publication Data
A Cataloguing in Publication record for this book is available from the British Library.

NEW from Idea Group Publishing

- **Digital Bridges: Developing Countries in the Knowledge Economy**, John Senyo Afele/ ISBN:1-59140-039-2; eISBN 1-59140-067-8, © 2003
- **Integrative Document & Content Management: Strategies for Exploiting Enterprise Knowledge**, Len Asprey and Michael Middleton/ ISBN: 1-59140-055-4; eISBN 1-59140-068-6, © 2003
- **Critical Reflections on Information Systems: A Systemic Approach**, Jeimy Cano/ ISBN: 1-59140-040-6; eISBN 1-59140-069-4, © 2003
- **Web-Enabled Systems Integration: Practices and Challenges**, Ajantha Dahanayake and Waltraud Gerhardt ISBN: 1-59140-041-4; eISBN 1-59140-070-8, © 2003
- **Public Information Technology: Policy and Management Issues**, G. David Garson/ ISBN: 1-59140-060-0; eISBN 1-59140-071-6, © 2003
- **Knowledge and Information Technology Management: Human and Social Perspectives**, Angappa Gunasekaran, Omar Khalil and Syed Mahbubur Rahman/ ISBN: 1-59140-032-5; eISBN 1-59140-072-4, © 2003
- **Building Knowledge Economies: Opportunities and Challenges**, Liaquat Hossain and Virginia Gibson/ ISBN: 1-59140-059-7; eISBN 1-59140-073-2, © 2003
- **Knowledge and Business Process Management**, Vlatka Hlupic/ISBN: 1-59140-036-8; eISBN 1-59140-074-0, © 2003
- **IT-Based Management: Challenges and Solutions**, Luiz Antonio Joia/ISBN: 1-59140-033-3; eISBN 1-59140-075-9, © 2003
- **Geographic Information Systems and Health Applications**, Omar Khan/ ISBN: 1-59140-042-2; eISBN 1-59140-076-7, © 2003
- **The Economic and Social Impacts of E-Commerce**, Sam Lubbe/ ISBN: 1-59140-043-0; eISBN 1-59140-077-5, © 2003
- **Computational Intelligence in Control,** Masoud Mohammadian, Ruhul Amin Sarker and Xin Yao/ISBN: 1-59140-037-6; eISBN 1-59140-079-1, © 2003
- **Decision-Making Support Systems: Achievements and Challenges for the New Decade**, M.C. Manuel Mora and Guisseppi Forgionne/ISBN: 1-59140-045-7; eISBN 1-59140-080-5, © 2003
- **Architectural Issues of Web-Enabled Electronic Business**, Nansi Shi and V.K. Murthy/ ISBN: 1-59140-049-X; eISBN 1-59140-081-3, © 2003
- **Adaptive Evolutionary Information Systems**, Nandish V. Patel/ISBN: 1-59140-034-1; eISBN 1-59140-082-1, © 2003
- **Managing Data Mining Technologies in Organizations: Techniques and Applications**, Parag Pendharkar/ ISBN: 1-59140-057-0; eISBN 1-59140-083-X, © 2003
- **Intelligent Agent Software Engineering**, Valentina Plekhanova/ ISBN: 1-59140-046-5; eISBN 1-59140-084-8, © 2003
- **Advances in Software Maintenance Management: Technologies and Solutions**, Macario Polo, Mario Piattini and Francisco Ruiz/ ISBN: 1-59140-047-3; eISBN 1-59140-085-6, © 2003
- **Multidimensional Databases: Problems and Solutions**, Maurizio Rafanelli/ISBN: 1-59140-053-8; eISBN 1-59140-086-4, © 2003
- **Information Technology Enabled Global Customer Service**, Tapio Reponen/ISBN: 1-59140-048-1; eISBN 1-59140-087-2, © 2003
- **Creating Business Value with Information Technology: Challenges and Solutions**, Namchul Shin/ISBN: 1-59140-038-4; eISBN 1-59140-088-0, © 2003
- **Advances in Mobile Commerce Technologies**, Ee-Peng Lim and Keng Siau/ ISBN: 1-59140-052-X; eISBN 1-59140-089-9, © 2003
- **Mobile Commerce: Technology, Theory and Applications**, Brian Mennecke and Troy Strader/ ISBN: 1-59140-044-9; eISBN 1-59140-090-2, © 2003
- **Managing Multimedia-Enabled Technologies in Organizations**, S.R. Subramanya/ISBN: 1-59140-054-6; eISBN 1-59140-091-0, © 2003
- **Web-Powered Databases**, David Taniar and Johanna Wenny Rahayu/ISBN: 1-59140-035-X; eISBN 1-59140-092-9, © 2003
- **E-Commerce and Cultural Values**, Theerasak Thanasankit/ISBN: 1-59140-056-2; eISBN 1-59140-093-7, © 2003
- **Information Modeling for Internet Applications**, Patrick van Bommel/ISBN: 1-59140-050-3; eISBN 1-59140-094-5, © 2003
- **Data Mining: Opportunities and Challenges**, John Wang/ISBN: 1-59140-051-1; eISBN 1-59140-095-3, © 2003
- **Annals of Cases on Information Technology** – vol 5, Mehdi Khosrowpour/ ISBN: 1-59140-061-9; eISBN 1-59140-096-1, © 2003
- **Advanced Topics in Database Research** – vol 2, Keng Siau/ ISBN: 1-59140-063-5; eISBN 1-59140-098-8, © 2003
- **Advanced Topics in End User Computing** – vol 2, Mo Adam Mahmood/ISBN: 1-59140-065-1; eISBN 1-59140-100-3, © 2003
- **Advanced Topics in Global Information Management** – vol 2, Felix Tan/ ISBN: 1-59140-064-3; eISBN 1-59140-101-1, © 2003
- **Advanced Topics in Information Resources Management** – vol 2, Mehdi Khosrowpour/ ISBN: 1-59140-062-7; eISBN 1-59140-099-6, © 2003

Excellent additions to your institution's library! Recommend these titles to your Librarian!

Mobile Commerce: Technology, Theory, and Applications

Table of Contents

SECTION II: MOBILE COMMERCE THEORY AND RESEARCH

SECTION III: MOBILE COMMERCE CASES AND APPLICATIONS

Preface

A Framework for the Study of Mobile Commerce

INTRODUCTION

This book began as a result of our recognition that mobile commerce is a topic that is of immense importance for business as well as for individual users and consumers. In addition, it seems pretty clear that the importance and relevance of mobile technologies will only increase at an exponential rate in the next few years. Of course, we hear these types of claims about a variety of technologies all the time. What is it that is different about mobile commerce? Further, an obvious question that one should ask in response to this is, "Even if mobile commerce is important, as a business professional why should I care about this topic and take the time to read a book about it?" The answer to this is, we think, at least part of the story that is, in fact, told in this book; but, to put the answer as succinctly as possible, mobile commerce is important because this technological revolution will directly or indirectly affect in a significant way practically every person in the industrialized world. Whether we consider mobile commerce from the vantage point of the consumer as an individual, the businesses that will be servicing consumers, the organizations that will be servicing other organizations, or any combination of the above, mobile commerce technologies will have a profound impact on the way people search out and conduct transactions, interact and communicate, plan and carry out activities, and entertain themselves and play.

The reason for this is that mobile computing technologies will become not only a normal part of the way people conduct business, they will become as ubiquitous as devices that we now take for granted--the phone, the desktop computer, the fax machine, portable entertainment systems, and a host of other commonplace devices that only a few years ago were considered luxuries. For example, if one were to have written a book in 1985 on the topic mobile phones—or, as they were originally called, *car phones*–what would that book have said about the mobile phone? Certainly, it probably would not have indicated that mobile phones would be as common in 2002 as cordless home phones were at that time. Yet, as documented in many places in our book by many of the experts that we have collected as authors of these chapters, it is likely that the growth in mobile computing technology will far surpass that of the voice-centric mobile phone. This is because of the convergence of a variety of functionalities that are coming together into mobile computing devices that will cause them to be essential tools for individuals and businesses. Thus, mobile commerce will not just be about commerce between a consumer and a business, it will be about ease of access, ubiquity of information, flexibility, and freedom to access electronic resources regardless of time or place.

The primary objective of this book is to provide a single source of up-to-date information about mobile commerce including the enabling technologies (i.e., both the hardware and the software that support mobile commerce), conceptual and empirical research and theory regarding the expected impact of this technology on businesses and consumers, and examples demonstrating state-of-the-art mobile commerce applications and lessons learned. We have purposely tried to design the book to be useful to a wide audience. We think that there is content that will be of interest to managers who want to find out how mobile commerce can be used in their firms. In addition, we have included refereed research articles that should be of interest to academic researchers as well as technologists who develop applications and devices. Finally, the book also contains superb resources such as technology descriptions and case studies that should be useful for educators who teach information systems, electronic commerce, or mobile commerce courses.

The book is divided into three sections: (1) technology, (2) theory and research, and (3) cases and applications. In the remaining sections of this introductory chapter, we discuss the range of issues important to understanding the domain of mobile commerce and follow this with an introduction to the chapters included in each of the three sections.

THE DOMAIN OF MOBILE COMMERCE

To develop and discuss where and how mobile commerce should be used, it is useful to consider the work that has been previously completed to understand the domain of electronic commerce. For example, in what is now almost a *classic* book on the economics of electronic commerce, Choi, Stahl, and Whinston (1997) offered a useful model for understanding the relationship between the products, actors, and processes that exist in both electronic and physical markets (Figure 1). In this model, products are differentiated based on whether they are physical or virtual (i.e., electronic). For example, an electronic product is something like a software product or file containing information, whereas a physical product would be an item like an electronic saw or a bottle of milk. A second dimension in the framework differentiates digital from "physical" processes. According to their framework, a digital process is one associated with using the Internet to access the web while a physical process involves a physical act associated with carrying out a commercial transaction. The third dimension is the nature of the agents involved in the transaction. A web-store would be digital while the corner grocery store would be physical.

What is interesting about Choi, Stahl, and Whinston's model is that the core of electronic commerce—the quadrant defined by digital products, digital services, and digital players—is defined based on the degree to which a product, service, or player is not constrained by the limitations imposed by its physical existence. Of course, what makes mobile commerce unique and powerful is that it unleashes this limitation in commerce; with mobile commerce the locational and physical barriers present in electronic commerce disappear and we are left with the potential for *commerce* to be engaged in anytime, anywhere, and for practically anything.

Of course, this does not mean that mobile commerce will be the all-encompassing tool for all applications. But, it does imply that mobile commerce technology will enable individuals and organizations to extend their reach to the Internet in

Figure 1: A Model of Electronic Commerce Market Areas
(adapted from Choi, Stahl, & Whinston, 1997)

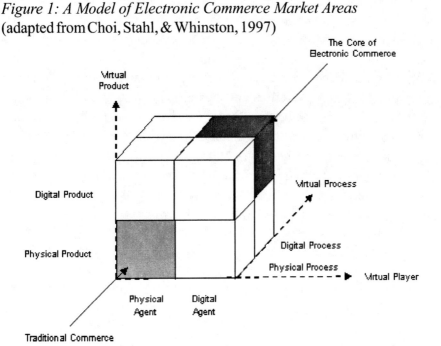

a location-independent manner. Consider for a moment how it is that most people currently think about the Internet and access to it. Although we can use the Internet to break down the barriers presented by location, we often conceptualize of the process of accessing the Internet in a location-dependent manner. In other words, there is a paradox in fixed-line electronic commerce; that is, we must physically go someplace such as an office, a university computer lab, or an Internet café to access the device that frees us from concern about location. Thus, the freedom to access the Internet regardless of location is an important benefit of mobile commerce. What other features that are available via mobile computing technologies add significant value for users relative to traditional, desktop computing forms of electronic commerce? We think that there are three major areas and four interactions between these areas that are important to consider:

1. Location: the relative location of the user when Internet services are needed.
2. Urgency: the relative immediacy of the needed service and the task that is being completed.
3. Utility: the relative importance of the task for the user.
4. Interactions between location, urgency, and utility.
    a. Location-specific urgency: the two-way interaction of urgency and location
    b. Location-specific utility: the two-way interaction of utility and location
    c. Time-dependent utility: the two-way interaction of urgency and utility
    d. Location and time-dependent utility: the three-way interaction of location, urgency, and utility

There are certainly other variables such as task complexity that might be considered, but the three main areas–location, urgency, and utility–represent those variables that either solely or in combination are what most dramatically distinguishes between electronic commerce and mobile commerce.

Let's examine why these variables are so important. Location is obvious; mobile commerce technologies are essentially independent of location when compared to electronic commerce because the user is able to carry the service around as he or she moves from one location to another. However, for urgency and utility it is not quite as obvious why these variables would be important. For example, for tasks that are important and/or immediate, electronic commerce can be a powerful tool to address the task at hand. However, what happens when an electronic commerce terminal is not immediately available? What this question points out is that when we consider the issue of urgency and/or utility in conjunction with location, these variables in combination highlight the importance of mobile commerce technologies. For example, consider a situation where a salesman has a 2 p.m. meeting with an important client, but it is 1:50 p.m. and he cannot find the hotel. It is this type of situation that highlights one of the true benefits of mobile commerce: the ability to leverage computing and communication resources in the

Figure 2: The Domain of Mobile Commerce

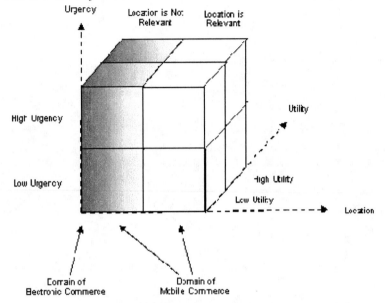

form of location-based services (LBS) to locate the hotel in time to make the meeting. While electronic commerce technology could certainly be used to locate the hotel, traditional electronic commerce services are not always available in the location where the user is situated at the time when he or she needs them. Therefore, it is the interaction of location with urgency and utility that defines the domain of mobile commerce--a domain that is, in many ways, much broader in scope than that of electronic commerce (see Figure 2). While this is obvious, it is nonetheless extremely important in helping to define the role of mobile commerce in supporting users.

With an understanding of the domain of mobile commerce in relation to electronic commerce readily in hand, we are now ready to consider what mobile commerce really means. After all, mobile commerce as a term is broad in scope and therefore lacks limpidity. Thus, to better define the nature of mobile commerce and the content of this book, the next section presents a framework for research and practice associated with the broad concept of mobile commerce.

A FRAMEWORK FOR MOBILE COMMERCE RESEARCH

Technology is neither developed nor used in a vacuum of thought or planning; most technologies of significance are designed, built, researched, and used in the context of a framework that defines what the parts of the technology are, how

Figure 3: Framework for Mobile Commerce Research

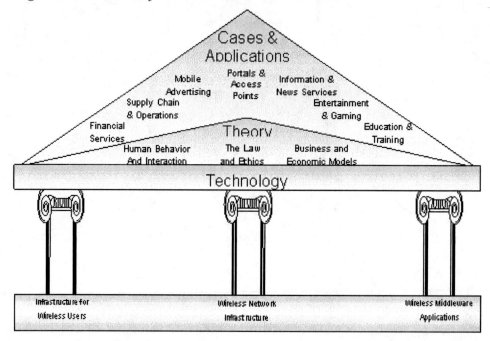

they fit together, and how they should be used. Of course, this does not mean that any such framework is valid or that the framework conceptualized by the designer will be the same framework that will be considered by the user. Nevertheless, frameworks are useful for helping conceptualize topics such as mobile commerce in a way that makes it comprehensible. With this in mind, we propose the framework of mobile commerce research that is summarized conceptually in Figure 3. This research framework is divided into three interrelated categories: technology, theory and research, and cases and applications. As Figure 3 illustrates, technology forms the foundation on which both research and practice are buttressed. On the top of the structure are the applications for which the technology is used. Embedded within these applications is theory, which serves to act as a guide that links the technology with the applications and provides guidance on where, why, when, and how the technology should or shouldn't be used.

This framework is the concept around which this book is organized: technology as the foundation, theory as the guide, and applications as the focus of use and practice. Table 1 outlines these components and elaborates on each by showing the specific categories of research and application areas that are relevant to each grouping. Furthermore, Table 1 also shows a mapping between the framework's components, the book's chapters, and each chapter's related primary (and secondary) research and/or application sub-category. The remainder of this chapter

Table 1: Framework for the Study of Mobile Commerce

Mobile Commerce Research Category	Sub-Categories	Chapter*
1. Mobile Commerce Technology	A. Wireless user infrastructure (browser, hand-held devices)	2, 3
	B. Wireless middleware (connecting applications and wireless network components)	4, 5, (16)
	C. Wireless network infrastructure (LANs, cellular systems, satellites)	6, 7
2. Mobile Commerce Theory and Research	A. Mobile commerce economics, strategy, and business models	(7), 8, 9, 10
	B. Mobile commerce behavioral issues (consumer behavior, technology acceptance, and diffusion)	11, 12, 13
	C. Legal and ethical issues	14
3. Mobile Commerce Cases and Applications	A. Mobile commerce in individual companies or industries	(2), 15, 16, 17
	B. Mobile advertising and retail	18
	C. Mobile portals	(11)
	D. Mobile auctions	
	E. Mobile entertainment and gaming	(12)
	F. Mobile financial services	
	G. Mobile supply chain management	
	H. Mobile service management	
	I. Mobile transportation management	
	J. Mobile education	19
	K. Mobile news and information access	

* Numbers in parentheses represent chapters that have a secondary focus in the selected category

presents an overview of each of the book's chapters in the context of this framework.

MOBILE COMMERCE TECHNOLOGY

The study of mobile commerce technology can be divided into three sub-categories: user interface, middleware, and network infrastructure (Varshney, 2000). In this section, chapters are included that cover each of these three sub-categories.

Chapter 1. If you ask anyone familiar with the mobile commerce industry for an example of an application that has a demonstrated record of success, it is likely that the first words out of his or her mouth would be 'i-mode.' In fact, in Japan i-mode is synonymous with the concept of mobile commerce. This chapter, "Mobile Commerce Reality: NTT DoCoMo's i-mode" by David MacDonald, is the lead contributed chapter in the book for a number of reasons. First, i-mode is a very successful service and MacDonald provides a clearly articulated and compelling discussion about what he sees as the reason for this success. In fact, the success of i-mode is also evidenced by the frequent mention by other chapter authors of this application. In addition, however, this chapter also includes a valuable discussion of not only the i-mode business model, but also the underlying technology, market applications, and partnerships that have made this service successful. Finally, the author also discusses why he believes that this business model can be successful elsewhere. As you will see in other chapters, this viewpoint is not shared by all of the authors; however, this diversity of opinions is useful because it allows you to judge for yourself regarding this fascinating application and its potential for success outside of Japan.

Chapter 2. A well-designed and usable interface is critical for any application. This is particularly true given the interfaces available in most wireless environments, which often makes ease of use a critical factor for the success or failure of mobile commerce applications. The purpose of "Wireless Devices for Mobile Commerce: User Interface Design and Usability," by Tarasewich is to provide the reader with an overview of current wireless device interface technologies. It also provides guidance on designing usable mobile commerce applications and explores the challenges associated with interface design and usability that the mobile commerce environment still creates for users, researchers, and developers.

Chapter 3. As discussed above, location is often important for users of mobile commerce applications. Where you are located may influence what you might do in conjunction with your mobile device, the Internet, or other *normal* behaviors. An application that appears to have great potential for further enabling mobile commerce application is LBS. Yet, there is considerable debate about the business models that might be used to support the development, management, and access to LBS services. Mitchell and Whitmore, the authors of "Location-Based Services: Locating the Money," discuss not only the features of LBS technology, but also these broader concerns about establishing viable models for provisioning LBS. As they point out, the success of LBS is determined by whether and how providers and network operators identify how to locate not only the subscribers, but also the money. As one of the leading LBS firms in several regions of the world, the information presented in this chapter about the LBS industry represents a valuable primer for anyone interested in understanding the factors at play in the LBS space.

Chapter 4. As the Webraska chapter highlights, LBS is an important area in mobile commerce. In fact, as you may have noticed by reviewing the

Table of Contents, a large number of the chapters in this book address this topic directly or indirectly. "Towards a Classification Framework for Mobile Location Services" by Giaglis, Kourouthanassis, and Tsamakos explores a topic related to LBS, Mobile Location Services (MLS), by identifying the most pertinent issues that will determine its future potential and success. The chapter provides a classification of mobile location services that can serve both as an analytical toolkit and an actionable framework that systemizes the author's understanding of MLS applications, underlying technologies, business models, and pricing schemes.

Chapter 5. What makes mobile commerce viable is the wireless network infrastructure that supports the applications and services offered by wireless operators and venders. If you are confused by terms like WLAN, 802.11b, 802.11a, WiFi, Bluetooth, and similar labels and standards, then "Wireless Personal and Local Area Networks" by Tom Zimmerman is a chapter that is a must-read. This chapter discusses terms and concepts concerning wireless technology, wireless standards, and interoperability. It also includes a summary of the important industry associations and standards groups as well as some predictions about which of wireless standards will dominate the marketplace. As a chapter authored by the one of the developers of the concept of the personal area network (PAN), it should be clear that this is an informative chapter that will provide a great deal of useful information about wireless infrastructure and technology.

Chapter 6. Over the next several years, there are expected to be dramatic changes in the capabilities of mobile networks as bandwidth and service offerings increase and expand. Most mobile network operators (MNOs) in the U.S. operate 2nd generation (2G) networks but will soon be moving to 3rd generation (3G) networks. What will the impact of these changes mean for consumers, operators, and other stakeholders? In "The Impact of Technology Advances on Strategy Formulation in Mobile Communications Networks," Constantiou and Polyzos discuss the important industry players and their likely roles as future generations of wireless technologies emerge. The chapter introduces key players in the mobile industry and presents a history of technological innovation in mobile networks (e.g., from 2G to the 4th generation, or 4G) and how this evolution might affect the key industry players. This chapter primarily focuses on the impact of these changes on mobile operators and their strategies. In addition, the relationship that market participants have with other market players is also discussed.

A discussion of wireless technology and infrastructure is certainly important, but it is not sufficient to provide an overall understanding of the impact of mobile commerce on consumers and organizations. As with electronic commerce, information technology is a necessary, but insufficient, condition for successful mobile commerce. In the remaining two sections of the book we have included papers that discuss a broad range of mobile commerce research issues as well as cases and applications of mobile commerce in specific companies and/or industries.

MOBILE COMMERCE THEORY & RESEARCH

Mobile commerce theory and research can be divided into three broad sub-categories: economics, strategy and business models, behavioral issues, and legal and ethical issues. In this section, chapters are included that cover each of these three sub-categories.

Chapter 7. What are the factors that will enable a player to succeed in the mobile commerce marketspace? This is a critical question for many firms that are considering whether and how to build, adjust, adapt, or create a product or service that will be a successful mobile commerce offering. This question is addressed in "The Ecology of Mobile Commerce: Charting a Course for Success Using Value Chain Analysis" by Rülke, Iyer and Chiasson of PRTM consulting. The purpose of the chapter is to present a value chain model for understanding the *ecology* (i.e., the relationship of a firm to its environment) of mobile commerce, including the dynamic relationships among all the elements that are required for a firm to be successful. In doing so, the authors identify the specific technologies, resources, investments, and competencies that firms will need to succeed.

Chapter 8. Stuart Barnes has written an informative chapter entitled "The Wireless Application Protocol: Strategic Implications for Wireless Internet Services." This chapter has the goal of using Porter's model of industry structure to examine the strategic implications of the wireless application protocol (WAP) for enabling wireless Internet services. WAP is a protocol designed to enable mobile phones to display web pages in a manner that is consistent with the small screen size of most mobile handsets. The chapter discusses how WAP was developed and also provides a detailed analysis of the WAP service industry, including the role of customers, suppliers, rivalry, new entrants, and substitutes. The author also discusses the future of WAP in light of its limitations relative to other mobile protocols (e.g., cHTML). This chapter offers a number of useful insights about the WAP protocol, success factors in mobile computing, and research opportunities for studying the strategic aspects of the mobile commerce area.

Chapter 9. "Mobile Business Services: A Strategic Perspective" by Alanen and Autio discusses the strategic and market factors that are expected to be important in the delivery of mobile business services. This is done by discussing the potential benefits of mobile technologies to enterprises in light of both the business and the consumer markets. In addition, the authors also discuss the competitive activity, the value chains, and the starting positions of various types of competitors in the mobile market space. The authors conclude by identifying three key findings: 1) mobile technologies have the potential to revolutionize business processes, 2) opportunities in the mobile space are quite different for businesses when compared to consumers, and 3) the competitive landscape represents a combination of the IT and the telecom value chains. This chapter is informative and insightful and will provide the reader with a great deal of information about the strategic implications of mobile commerce on business and business operations.

Chapter 10. How do you *surf* the Internet from a phone when the screen measures only two inches diagonally and your keyboard is a dial pad? One answer is to use a portal as the launch point for your venture out into cyberspace. It is because of these types of display and input limitations that portals are becoming the preferred starting point for mobile Internet access. The purpose of "Mobile Portals: The Development of Mobile Commerce Gateways" by Clarke and Flaherty is to explore the factors that compose a product mobile portal strategy. In addition, the authors offer several specific recommendations regarding the development and operation of portals, which are particularly useful given the evolving nature of mobile technologies and the associated changes in portal functionality that will be needed to keep a portal relevant and viable.

Chapter 11. As with many technologies, the applications that are most popular in mobile computing are those that designers often did not anticipate. Short messaging services and mobile gaming have turned out to be two of the most popular applications for users of mobile devices. Kleijnen, Ruyter and Wetzels' chapter examines one of these applications, mobile gaming, in an empirical study to identify factors influencing consumers' acceptance and adoption of mobile gaming services in The Netherlands. Their findings indicate that perceived risk, complexity, and compatibility are the three main factors influencing the adoption of mobile gaming applications. In addition to the empirical study, the chapter also identifies several success factors enhancing mobile service adoption that is based on their extensive review of the literature. Although this is an academic research study that focused on gaming services, the authors provide several useful and insightful guidelines for designing a variety of consumer services that have relevance to managers and system designers.

Chapter 12. As new technologies are introduced to a market, there are always risks that the technology may not be accepted by the intended market constituencies. The identification of those factors that cause potential adopters to accept or reject a new technology is important and is the focus of "Mobile Data Technologies and Small Business Adoption and Diffusion: An Empirical Study of

Barriers and Facilitators." This chapter is authored by Van Akkeren and Harker and it presents findings from a two-phase study of the perceptions, needs, and uses of mobile data technologies by various Australian small business owners. The research was conducted in two phases. In Phase I, focus groups were conducted to identify possible uses and applications of Mobile Data Technologies (MDTs) for three types of potential users: those who are either non-adopters of IT and Internet technologies, those who are partial-adopters, and those who are full-adopters. The results of the first phase of the study were applied to the second phase, which involved interviewing 500 small business owner/managers about mobile data technology adoption issues and perceptions of MDT usage. The results offer many insights for managers, developers, and researchers and suggest that technology characteristics, adopter attitudes towards various types of technologies, the relationship of the new technology's benefits to the needs of the organization, and the history of the firm in adopting new technologies all are important factors influencing the adoption of MDTs.

Chapter 13. Location services are expected to be a killer application in mobile commerce because, as discussed above, LBS brings location-specific information about the user into the mix of services that mobile providers can offer customers. However, with the ability to locate users at any given moment, and track movement and activity, the question of user privacy and the ethical issues associated with privacy become a paramount issue to address. In "We Know Where You Are: The Ethics of LBS Advertising," O'Connor and Godar provide an intriguing discussion of many of the issues pertaining to this important topic. The authors begin by elaborating on the three features that differentiate LBS mobile commerce from traditional electronic commerce: mobile location identification, synchronous two-way communication, and provider power. They also point out that there are parallel privacy concerns in other areas such as telemarketing, but that the individualization and location-specific nature of LBS escalates the complexity of the ethical issues associated with location services. The main thesis offered by these authors is that the LBS industry will end up being highly regulated unless the industry develops an effective mechanism for promoting and enforcing self-regulation. The chapter also provides a proposed model for self-regulation that the authors suggest will allow the LBS industry to avoid what they consider to be otherwise inevitable legislative controls.

The chapters in Section 2 provide a broad perspective about a variety of topics relevant to mobile commerce; however, most of these chapters focus on a set of general theoretical or conceptual concerns or research topics. The next section includes chapters that provide us with specific examples and cases that illustrate how mobile commerce will impact or be used in specific companies, industries, and/or institutions.

MOBILE COMMERCE CASES & APPLICATIONS

In our framework for the study of mobile commerce, we have divided the cases and application areas into 11 sub-categories based on our experiences and observations as well as prior published research. The sub-categories include mobile commerce use in individual companies and/or industries, as well as mobile applications for advertising and retail, portals, auctions, entertainment and gaming, financial services, supply chain management, service management, transportation management, education, and news and information services (Cherry Tree & Co., 2000, Durlacher, 1999; Tarasewich, Nickerson & Warkentin, 2002). In this section, we have included chapters that cover a representative sample of five of the 11 sub-categories from our framework.

Chapter 14. How do you use mobile commerce technologies to promote the use of your products or services when you don't sell your products over the Internet? This is a critical question for retailers, manufacturers, and those who produce various consumer goods. In "A Perspective on Mobile Commerce," Mark Lee provides a fascinating account of Coca-Cola's perspective on addressing this question. He points out that an important thing to understand is when and where products would be most likely to be purchased by a mobile consumer. For example, mobile purchases tend to be more of an impulse than a planned expenditure, so one approach for using mobile commerce technologies is to influence the consumer's impulse decision-making process and thereby have a direct impact on their purchasing behavior. To do this, he suggests mobile commerce should be used to build awareness of available products and services, facilitate transactions, develop relationships with consumers, and monitor actual progress and/or results of marketing efforts. The chapter discusses an example of how Coca-Cola North America is currently leveraging the wireless medium in general and mobile commerce in particular, as well as briefly discussing how they use mobile commerce in other parts of the world and in applications such as vending.

Chapter 15. What are the criteria for market adoption of wireless data services? The chapter "Location-Based Services: Criteria for Adoption and Solution Deployment," by Astroth and Horowitz of Autodesk Location Services identifies three key attributes to achieve market adoption: wireless data services must be personalized, localized, and actionable. The chapter provides a case study of Autodesk's experiences in Fiat's Targa Connect project that verifies the success of wireless data services that incorporate the three critical attributes. The authors also suggest that an integrated platform model where location services are packaged into a single software product by network operators provides the best, most consistent service to customers while simultaneously enabling operators to retain these customers. This informative chapter provides a useful overview of the

LBS industry, including revenue and marketing issues for software venders and network operations as well as future industry trends.

Chapter 16. Some of the most advanced applications for mobile commerce exist in the use of mobile technologies in the ultimate mobile device, the automobile. Car-based computing platforms, collectively referred to as "telematics," generally provide a variety of information services such as automatic and manual emergency calls, roadside assistance services, GPS, on-board diagnostics, traffic and dynamic route guidance, Internet communications, and personal concierge services. In "Mobile Commerce in the Automotive Industry – Making a Case for Strategic Partnerships," Solak, Schrauben and Tanniru provide a case study of a telematics application. The chapter provides an analysis of the telematics market and related business opportunities. An important conclusion of this interesting chapter is that there is a risk that divergence in the industry away from common standards and technologies will lead to increased complexity and poor consumer acceptance; therefore, industry partnerships are critical to enable the telematics industry to succeed.

Chapter 17. As noted elsewhere in the book, an important use of mobile commerce is to raise awareness about a product or service in the minds of consumers. The aim of "The Role of Mobile Advertising in Building a Brand" is to discuss how mobile advertising can be used for branding in a cross-media promotion. To do so, Minna Pura presents a case study where mobile advertising was used as an integral part of the cross channel media mix. The case describes how Eera Finland planned and produced a mobile advertising campaign for TUPLA, a Finnish chocolate bar, that was tied to the premier of the movie "Tomb Raider." The promotion focused on activating the target group, the youth segment, into offering information about themselves to the company. The goal, of course, was to identify whether and how this information could be used for future customer relationship management and to identify whether the promotion helped in product branding. The research described in this chapter represents an interesting example of using multiple media and promotional tools to increase consumer awareness and track customers.

Chapter 18. While the bulk of interest in mobile technologies by industry players is focused on business operations and consumer marketing, there is quite a bit of interest in using mobile technologies in support of education. For example, several universities and community colleges have developed fairly extensive wireless networks so that students and faculty can access Internet resources seamlessly across their campuses. This chapter, "Wireless in the Classroom and Beyond" by Jay Dominick, provides an overview of the author's experiences and observations about the use of wireless computing at Wake Forest University. Wireless is used at Wake Forest not only for supporting faculty in the classroom and students in their studies, but also for supporting the management and operation of the entire academic enterprise. Part of the reason for the popularity of mobile computing in education appears to be that this mode of computing seemingly fits well with the mobile lifestyle of today's computer-savvy student population. As with any business operation, however, the implementation of a wireless network necessitates thorough planning, requirements determination, and user involvement. While focused on education, many of the insights presented by this CIO will be applicable to a variety of other organizations and businesses.

CONCLUDING REMARKS

We should note that all of the chapters were reviewed by either the editors or by external reviewers via a blind review process. For chapters submitted by the professionals working for firms in industry, we as editors reviewed and, where appropriate, made recommendations regarding content, scope, and direction. For chapters submitted by academic researchers, papers were submitted to external reviewers who did not know the authors' names or affiliations. In this way, papers were given a thorough scrutiny by experts in the fields of mobile and electronic commerce. In total, we were quite selective regarding actually including a submitted chapter in the book; for example, although we received 34 chapters, we only accepted 18 of the submissions for inclusion in the book. We are delighted to present this book to you and are proud of the many outstanding chapters that are included herein. We are confident that you will find it to be a useful resource to help your business, your students, or your business colleagues to better understand the topic of mobile commerce.

REFERENCES

Cherry Tree & Co. (2000). *Wireless Applications and Professional Services*, November, pp. 1–25, www.cherrytreeco.com.

Choi, S.Y., Stahl, D.O., & Whinston, A.B. (1997). *The Economics of Electronic Commerce.* Indianapolis, IN: Macmillan Publishing.

Durlacher Research Ltd. (1999). *Mobile Commerce Report,* pp. 1–67, www.durlacher.com.

Tarasewich, P., Nickerson, R. C., & Warkentin, M. (2002). Issues in mobile e-commerce, *Communications of the Association for Information Systems*, 8, pp. 41–64.

Varshney, U., Vetter, R. J., & Kalakota, R. (2000), Mobile commerce: A new Frontier. *IEEE Computer*, October, 32–38.

ACKNOWLEDGMENTS

The editors would like to acknowledge the help of all involved in the collation and review process of the book, without whose support the project could not have been satisfactorily completed. Most of the authors of chapters included in this book also served as referees for articles written by other authors. Thanks go to all those who provided constructive and comprehensive reviews.

Support of the Department of Logistics, Operations and Management Information Systems at Iowa State University is acknowledged for use of office personal computers and copiers.

We would also like to thank our graduate assistants, Elif Koc and Pooja Arora, for their assistance.

In closing, we wish to thank all of the authors for their insights and excellent contributions to this book.

Chapter I

NTT DoCoMo's i-mode: Developing Win-Win Relationships for Mobile Commerce

David J. MacDonald
Strategic Alliances, i-mode, Japan

ABSTRACT

In February 1999, Japan's NTT DoCoMo launched the i-mode service, becoming, with over 34 million active subscribers, undoubtedly the world's most successful mobile Internet service. While mobile commerce is an often-discussed topic around the world, it is important to look to the success of i-mode in Japan, to gain real insight into the potential for mobile commerce in other markets. i-mode is a success because of a careful balance of the right technology, the right strategy, the right content, and the right marketing. On this successful platform, many players have developed successful business models, be it premium content, e-commerce, advertising, or others. With the expansion of i-mode, it has now become a "lifestyle infrastructure" and a series of alliances with major players such as Coca-cola has expanded the possibilities. With new i-mode services being launched in Europe and Asia, it is timely to learn, based on the experiences of Japan, what the potential could be.

More than two years have passed since Telecom 99 in Geneva, the industry event for the telecommunications sector. Attendees will remember that at that event we were told that a new era of Internet and commerce was about to begin. "Mobile Internet" and "mobile commerce" became buzzwords overnight. Most industry experts agreed that the new technology before them would lead users into a whole new interactive world beyond their imagination. We were promised that we would surf through multimedia websites using mobile phones and would soon be using those same phones for a multitude of transactions–from online to physical payments. It was an amazing concept.

Over two years have passed and it is still concept. The promised world has not yet developed. Around the world, the uptake by users of Wireless Application Protocol (WAP)-enabled phones and services has been slow. The acceptance of this new technology by industries other than the wireless industry has also been sluggish. When attending the many "Wireless Internet" and "Mobile Commerce" conferences and events, real examples providing real data are difficult to find. In such an environment, it is very easy for skeptics and critics to declare that wireless Internet will never develop beyond concept.

To make such a bold statement is premature. It is a statement that overlooks all of the facts. In truth, in Japan wireless Internet is alive and well, and continuing to grow at an extraordinary rate. As of December 2001, nearly 50 million users in Japan had wireless Internet-enabled handsets of some sort[1], with over 30 million Japanese actively using NTT DoCoMo's i-mode alone. To put that figure into perspective, approximately one in four Japanese are using i-mode. They are using i-mode for a whole range of activities, from sending and receiving e-mail to surfing through over 50,000 websites designed for the small displays of the handsets (see Figure 1). The versatility of the service is great. Having created a solid platform upon which to build and to link with other platforms, as can be seen through numerous projects and services, i-mode proves that wireless Internet and mobile commerce are no longer mere concepts, but reality.

In a book entitled *Mobile Commerce* it is important to examine theory and summarize the results of test projects. It is equally, if not more, important to show actual case studies to explain the realities and to defend the theories. This chapter is intended to be just such a case study. i-mode is an often-used example of wireless Internet, but it is also often misunderstood. Through real examples and real experiences, it is hoped that the real reasons for the success of i-mode will become apparent. This success is not found in some mystic oriental alchemy or in the activities of blond-haired, mini-skirted girls in the entertainment districts of Tokyo, as many propose. i-mode's success is found in a solid strategy, which considered (and continues to consider) the right technology, services and marketing, hand in hand. Most importantly, i-mode seeks to develop win-win relationships to ensure

Figure 1: i-mode Subscriber and Content Partner Growth. Nearly three years after service launch, i-mode users number over 30 million and content providers nearly 2000.

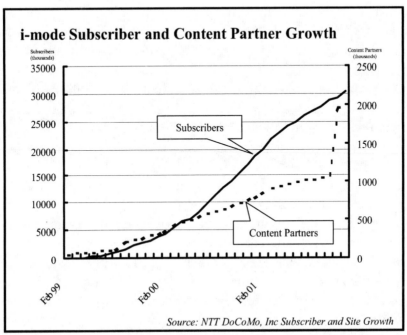

the expansion of the platform. It also shows clear examples of success in "mobile commerce." By looking at the business models of i-mode leaders and the horizontal integration of i-mode with different platforms and new commerce initiatives, it should become clear that the success of "wireless Internet" in general, and "mobile commerce" in particular, is not restricted to the Japanese market alone, and with the right strategy, could be duplicated elsewhere.

WHAT IS I-MODE IN REALITY?

As mentioned, i-mode is often discussed, but is often misunderstood. What really is i-mode and why has it been so popular? What are the secrets to i-mode's wireless Internet success?

Simply put, i-mode is a mobile phone with a larger-than-normal screen, containing both an Internet browser for browsing websites and an e-mail client.[2] Importantly, these phones retain all of the features of a standard mobile phone (battery life, voice quality, size and weight) but are enhanced with the added features of i-mode. The browser allows the user to access well over 50,000 websites specifically designed for the smaller display.[3] These include nearly 3,000 sites (provided by approximately 2,000 companies) within NTT DoCoMo's own portal ("iMenu") and 50,000 other sites, accessible by inputting a URL into the phone. The e-mail client allows communication with not only other mobile phone users, but with e-mail users worldwide, thus expanding the world of communication for the user. In addition to these Internet-type features, the handsets themselves are rich with functionality including polyphonic (multi-voice) ringing tones based on a subset of midi (the latest phones with 16 to 24 voices), and color TFT displays (some with well over 60,000 colors). From the 503i series of handsets, i-mode also incorporates Sun Microsystems' Java technology, which allows the user to download various rich applications to make the users' experience far more dynamic than just a static web page[4] (see Figure 2).

To describe i-mode so simply is to forget the strategy behind the service. It is important to examine in depth how NTT DoCoMo has been able to put this service together and drive its success. When looking at i-mode's success, one must first consider the stakeholders involved. The value chain, or rather value map, for a successful wireless Internet service is composed of a mixture of players. The handset vendor must create a handset that is attractive to the end user. Network and server vendors must create the right infrastructure for high quality of service and scalability. Other vendors must create the enterprise solutions to increase corporate demand. For a mass-market consumer focus, content providers must develop rich services for the end user to use. Finally, the mobile operator is the

Figure 2: 503i Series Handset (NEC). 503i series handsets all contain Sun Micro-systems Java technology and full-color displays. These features, in addition to the 16-voice ringing tones, offer the user an advanced handset for either business or pleasure.

Figure 3: i-mode Basic Value Map. NTT DoCoMo actively coordinates all players in the i-mode value map in an effort to create the best service for the end user.

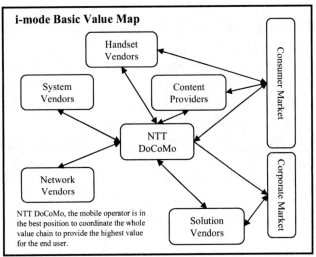

provider of this combined value proposition directly to the end user. In this position, with a focus on the end user's needs, the mobile operator is best placed to coordinate the entire value map to provide the best service. With this in mind, NTT DoCoMo takes an active role coordinating the whole value map, working closely with handset manufacturers, server vendors, content providers and other third-party solution vendors, in the best win-win business model so that all partners in the value map can enjoy success (see Figure 3).

Technology

The aim of this study is to focus on the business success of i-mode, but a brief look at the technology behind the business is essential. There are two key technological features of i-mode that have led to its success. The first is the packet-switch network for Internet services. The second is the right choice of application layer technology for third-party developers.

A packet-switched network (PDC-P) for wireless Internet services is seen as essential. This packet-switch network has several positive effects for the end user. First, what is often called an "always-on" connection allows for quick access to Internet services, by removing the need for a lengthy circuit-switched "dial-up" and "log-on" process. This quick connection promotes greater access to the service overall. In addition, the packet-switch network allows for a different pricing model for the end user. End users now pay per packet (128 bytes) of information downloaded rather than for the number of minutes online. This per-packet pricing scheme is most effective with wireless Internet services where the number of bytes downloaded tend to be very small.[5] With a limited number of kilobytes, the per-packet method is more cost-effective for the average user. This "packet network" is an important technological feature of i-mode, but it is not the only one. With the coming GPRS network of the GSM world, many claim that usage of WAP-based services will increase. This may be the case, but unless the situation for content providers is adjusted, there will still be a lack of content to access. After all, the average user cares little for what is inside the box (that is, the technology) but cares to a great extent about what he or she can see on the screen (the content).

With an underlying layer of the packet network, the top level of the application layer is also important. At the time of planning i-mode, the WAP Forum was just being founded and NTT DoCoMo was playing an active role as a board member. While NTT DoCoMo has always played an active role in the Forum, it was clear early on that content would be the deciding factor, and choosing the right mark-up language would be key. After studying HDML-based services in the US and the lack of content there, it was decided that a browser based on standard HTML would be more effective in attracting content providers to the platform (Natsuno, 2000).

With obvious limitations of the handset, a full HTML specification was not possible, but a subset of standard HTML, already the *defacto* standard of the Internet, with some additional features to take advantage of the mobile handset (such as a "phone to" tag to directly dial a phone number from an HTML page), was developed.[6] For the content provider, unlike the WML of WAP, there would not be a new markup language to learn. Also, unlike WML, HTML is far more forgiving and the presentation of content is far more attractive with a variety of input methods available (radio buttons, drop down lists, check boxes). Finally, basing i-mode on HTML and standard Internet HTTP allowed content providers to interact with existing systems in a far smoother way. While many argue that WML is not too difficult to learn for an experienced programmer and it is in fact a far cleaner markup language, it can be argued that HTML, as the accepted standard of the Internet, promoted the growth of i-mode for non-partner sites (currently numbering over 50,000).[7] In addition, it has allowed companies to concentrate less on the technological adaptation of content and more on the creative side of content development, that is, the development of the right business model and the right content targeting the users' mobile needs.

The Win-Win Strategic Model

The right technology for third-party developers is a key to i-mode's success, but it is not the only one. i-mode is, after all, a business and not a technology, and so the right strategy is also important. As far as possible, the business model of i-mode is designed to create a win-win relationship between all members of the value map. When considering the reality of the PC-based Internet, hardware vendors enjoyed greater sales, software vendors followed suit, and ISPs attracted more users and greater usage. In the days of MOSAIC and then the early Netscape Navigator, it was the explosion of online content that really sparked the positive feedback process that has given us the Internet of today. It is often these same content providers, however, who have the greatest struggle to find the right business model to bring in revenue and profit. While the technology sector overall is suffering today, it is perhaps the content provider who has suffered the most, with many dot com companies vanishing in the last year. i-mode has provided these content providers, who have rich and attractive content, with the missing link. As NTT DoCoMo is a mobile phone operator with a pre-existing billing relationship with the customer, it is possible to collect payment on behalf of content providers for their services. These payments are added to the user's phone bill every month and all but a 9% commission is passed back to the content provider. While this payment method is not for every content provider (in fact, only 30% of i-mode sites take advantage of this option), it has offered a chance, if their content is competitive, to some content providers to receive revenue directly from the end user.[8]

With this being the case, NTT DoCoMo and the content provider work together to create a win-win relationship for both parties. Therefore, NTT DoCoMo does not have to pay for content, unlike some other operators in different markets. Similarly, content providers do not have to pay NTT DoCoMo for any placement fees within the iMenu portal. They do, as it is not possible for NTT DoCoMo to partner with every site on the Internet, have to conform to guidelines for content creation and go through a partnership approval process. However, through this process the content provider also benefits from the experience of a dedicated content development team, who having worked with thousands of content ideas and have insight into what really works, can give advice to support the growth of the content provider's business.

The Right Services

The importance of content should already be clear, but it needs to be stressed again. It is the selection of the best content for the end user, by a dedicated team of specialists, which makes i-mode so strong. Developing the right "portfolio" of content is essential to attracting both the right partners and the right users. With the right kind of portfolio, i-mode has been able to attract a variety of users from young teenagers to the elderly, with the simple philosophy of "something for everyone."

Figure 4: i-mode Content Portfolio. In order to offer the best selection of content for the end user, the content of i-mode is divided into a well-balanced portfolio.

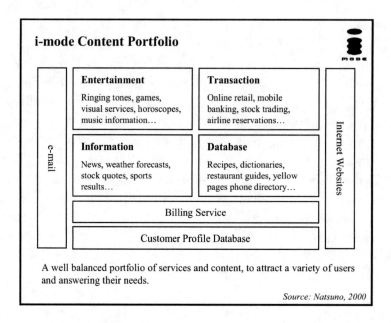

The basic portfolio itself covers four content areas: entertainment, transaction, information and database. By way of "entertainment," i-mode offers a variety of services, from games and horoscopes to visual images and downloadable ringing tones. "Information" is an obvious category for the mobile phone, with news, sports results, financial information and weather. "Transaction" content includes online sales of books, CDs and other goods, in addition to airline and hotel reservations, ticketing, mobile banking and trading and more. The "database" category refers to services linked into a database of content, including restaurant guides, recipes, job information and real estate listings (see Figure 4).

While the types of categories are important, what is more important is the type of content within each category. With the right strategy and relationship in place, it is far easier to work on developing the content concept rather than closing the business deal. One of the strengths of i-mode has been its content team, who, before the service was launched, carefully considered the user needs, paying particular attention to the mobility aspect of the service. This thought process has continued up to today, as the content team works with over 1,000 partners providing content on NTT DoCoMo's portal. The work of these team members is very much like that of a magazine editor, and by understanding the environment, layout and other issues, they are able to suggest improvements to the content, in terms of both user needs and usability[9] (Matsunaga, 2001). This editorial process, while often long and arduous, has given results, which can be seen in the success of a number of i-mode content providers and the service overall.

The Right Marketing

The right technology, the right strategy and the right services are still not enough to make a great service. The right marketing also plays an important role. Another key success factor of i-mode has been a marketing strategy that sells the services rather than the box. That is, i-mode marketing focuses on what the user can do rather than what technology is being used. Posters and commercials never used the word "Internet" or "web" but rather focused on services such as transferring money with mobile banking or airline reservations. Every effort is taken to show how easy i-mode is for anyone to use, through media campaigns and sales tools for users.

I-MODE AND MOBILE COMMERCE

With a better understanding of what i-mode is and how the technology, the strategy, the content and the marketing have come together to create its success, it is time to turn to the real focus of this book, "mobile commerce." This chapter is designed to give real examples of mobile commerce, and not to delve into theory. For the sake of simplicity, let's begin by defining mobile commerce as "making money through the phone." By simplifying the definition in this way, it becomes easier to review the business models of a variety of participants in the i-mode value map, and to show how money can be and is being made on the wireless Internet.

SUCCESSFUL BUSINESS MODELS

There are many different examples of successful business models on i-mode. In reality, these do not vary much from the fixed-line Internet, with a couple of exceptions. In order to make money, it is essential to design the right model, and so it is important to remember that i-mode, while it has provided an excellent environment for most businesses is not a license to print money. The models most often constructed can be divided into seven broad categories: brand building (or "media mix"), customer relationship management, online retail, premium content, aggregation, business-to-business, and advertising. There are a host of variations on these models and different levels of sophistication. Also, the benefits of each of these models must be calculated in different ways. Some of the benefits are direct and quantifiable, such as online retail and premium content, and some are harder to quantify, such as brand building or CRM, but their significance should not be forgotten.

Brand Building and Media Mix

The oldest use of the Internet is to build one's brand. Almost every company has a corporate website which describes its products and services. This type of website is a sales brochure, a billboard and a corporate profile all in one. Not everyone wants, however, to see a commercial. A more subtle way of building one's brand is to offer useful content to the end user. Many consumer-oriented websites offer not only product information, but also useful content. On i-mode, too, this type of model can be successful. An even more successful derivative of this model is "media mix," that is, blending the content of i-mode with content in other media to help drive that business too. i-mode is not intended to replace other media, and is not an appropriate medium to present certain types of content. It can, however, be a useful access point for other content or media. This could take the form of an interactive television program or information such as anchor diaries, presented in a more interactive way as "content" to the user.

An example of this model is the recipe database provided by Ajinomoto. Ajinomoto is one of Japan's largest producers of food products. On their fixed-line website they offer a large database of recipes, "A-Dish," as a service to their customers. Certainly, the hope is that the user will be more inclined to buy Ajinomoto products. On i-mode, the same database of recipes is being provided and building the brand of Ajinomoto on the wireless Internet too. Of course, the look and feel of the service on i-mode is different from the PC interface, and the mobility needs of the user are stressed.

Two issues arise when attempting this type of brand building exercise. The first is the perception of the user. Does the user see this as content or rather a commercial for the company? The second is the mobile needs of the user: what type of content, in what way do they want to interact with the content, and the freshness of that content. NTT DoCoMo's content team advises partners based on these needs, and in fact the idea of mobile recipes was first developed by the DoCoMo team, giving considerable thought to how, when and where the user would use the content on the phone. Accordingly, the content team can advise the company on the best way to present the content to the user, with the best functions for the mobile device. In this way, attracting useful content to the i-mode service, and giving the partner company the opportunity to expand their brand into a new medium, a win-win relationship is developed.

Customer Relationship Management

Customer relationship management (CRM) has a variety of meanings and can be used to express a multitude of services. In the case of i-mode, CRM can be described simply as providing support for one's customers, with useful online services in the mobile environment. These can be as simple as a shipment tracking service, operated by several courier services on the i-mode platform. More sophistication is presented by a variety of financial institutions. Over 350 banks currently offer mobile banking services. Most major credit card companies offer online services to their customers through i-mode. Consumer-oriented stockbrokers, such as DLJ Direct or Nomura Securities, offer real-time trading functionality to their account holders, while offering other useful services to non-clients. Insurance companies, from automobile and travel to life, provide clients with information and the ability to make transactions online. CRM is always a wise investment, as it increases both the satisfaction and the loyalty of customers. CRM applications on i-mode can be far more immediate, effective and timely.

Banking applications are perhaps one of the best examples of this timeliness and effectiveness. As already indicated, a great number of banks offer mobile banking services through i-mode. The first content partner of i-mode was Sumitomo Bank (now Mitsui Sumitomo). Sumitomo was already a leader in Internet banking when the i-mode service was first being planned in 1997/98. Today, users are able to access account information from a variety of different devices, including i-mode, the PC and the telephone, all with the same password. Not only can the user check account balances at any time in any place, they can also transfer money between accounts and to another party's account in order to pay bills.[10] The freedom that these services offer the user is enormous. The user is no longer restricted by time, and unlike an ATM or PC banking service, is no longer restricted by location. Customers who don't have time to get to a bank can easily check account balances and make payments when walking down the street or riding an elevator. From the bank's point of view, like an ATM service, staffing costs can be reduced while at the same time increasing customer satisfaction. With banking, like the other types of CRM, one of the biggest issues to tackle is finding the right mix of applications for the mobile user. Another issue is security, but with end-to-end SSL from the 503i series of phones, and server-to-server SSL or dedicated line connections with the first two series of handsets, the banking industry has been willing and able to take the lead in wireless Internet and enjoy the benefits of mobile commerce.

Online Retail

Online retail has been one of the most talked about business models of the Internet revolution. The initial success of online-only retailers, by snatching business away from traditional retailers, grabbed business headlines. While the number of companies who continue online is dwindling (especially those without a traditional retail arm), online sales are still a valuable revenue stream for many businesses. With 30 million subscribers, i-mode is a very attractive environment for these online marketplaces. Today, there are many retailers selling their wares over i-mode, including bookstores, music stores, game software vendors, flower shops, fashion boutiques, ticket agents, airlines and mail-order businesses. Shopping through the phone, once thought impossible, is very much alive and well. The challenge is to find the right mix to make the business successful.

Tsutaya Online is the Internet branch of Tsutaya and their "Culture Convenience Club" chain of video and CD rental stores. Tsutaya has a large chain of stores with a national coverage. These stores are quite large, with rental videos, DVDs and CDs, as well as limited video and DVD sales. The stores are also high-tech, with an Internet-based inventory system which tracks what videos are out at which locations. With its online channel, Tsutaya has expanded into a quality music and video retailer. i-mode, which combines both the interactiveness of the Internet with the aspect of mobility, became an attractive new medium for Tsutaya. They have expanded their retail business onto the phone, but have also been able to mix it with their bricks and mortar business of Culture Convenience Clubs. Users can search for titles on the move and check to see if the video they want is in stock. Using i-mode for coupons, which can be shown directly to the store staff, and permission-based direct mail, they have been able to increase sales and rentals. The benefits of a push e-mail function have also been proven. Tsutaya discovered that the frequency of click-throughs to an online purchase page when using e-mail through i-mode was much higher than compared to the fixed-line Internet. These e-mail messages, like the content online, are meant for entertainment rather than a hard sell, and users return frequently for the latest release information and entertainment news.

A number of issues must be considered when planning online retail in the mobile environment. A number of these issues overlap with the fixed-line Internet, including distribution system, payment method, and brand familiarity and comfort (traditional retailer versus online-only retailer). The payment system for goods on i-mode is up to the retailer, and many use common methods, such as credit cards, bank transfers or cash-on-delivery. Other online methods of payment are slowly developing, but even with a simple credit card, retail opportunities exist. There are a number of issues that, while they may exist in the fixed-line Internet world, are magnified in the wireless Internet. Two key issues are product type and usage frequency. The limitations of the small screen limit the range of products that users feel comfortable buying "sight unseen." With CDs, books or game software, the user generally expects what he gets. With other items, such as clothing, it is more difficult to try something on. For that reason, many retailers have also begun to use the concept of media mix. Mail-order businesses, while they may present a catalog through i-mode, would have more success using i-mode as an "order terminal" for their customers. Once again, i-mode is not seen as a competitor to other media, but rather a complementor. Another key issue to resolve is frequency of use. Many online retailers are just that, a point of sale. Often, users visit only when they want to buy a product. While a shopper might visit a brick and mortar shop once a week, and a PC-based retailer once a day, with the wireless Internet always in your hand, the frequency of access could be many times per day. Therefore, it is important to present information in the form of content to attract the user back frequently, and to keep it fresh so they are satisfied each time they return. Tsutaya accomplishes this by providing new release information, entertainment news, reviews, and more.

Premium Content

Many of the business models and concepts presented so far do not differ greatly from the concepts on the fixed-line Internet. The mobility of the user is the biggest factor that magnifies the need to find the right blend of services. In the Internet world, there are millions of content providers providing a variety of content all for free, primarily because there has never been a satisfactory way to receive payment from users. There are several so-called micro-payment services, but none have been that successful. i-mode has provided the missing link in the premium content model by providing a convenient way for users to pay and content providers to get paid. If content providers wish to charge a monthly premium subscription fee for content (to a maximum of 300 yen, or approximately US$3), then DoCoMo collects that charge from subscribed users as part of their phone bill at the end of the month. For the user, the registration procedure is painless, and by entering only their i-mode PIN number through the phone, they can register for a site immediately. At the same time the content provider can register the individual into their database to control access and provide personalized services. Once NTT DoCoMo has received payment for the user and processed the amount, the content provider has that amount, minus a 9% commission for collection and processing, transferred to their bank account. Premium services today range from news and sports information to entertainment services such as games, visual services and ringing tones. The whole market for premium content on i-mode is estimated to be around $60 million per month.

There are many examples of companies using this service. News companies, who traditionally provide news in other media using advertising revenue, can now charge users small fees for essential information. Entertainment companies, who have not yet found a true moneymaking model on the fixed-line Internet, can transform their properties into mobile content and begin to receive revenue. The

Walt Disney Company has been in the entertainment business for a long time, and has developed a portfolio of content to appeal to a wide variety of users, young and old. Using the same content, but repositioned for a mobile user, Disney has developed a business model that is providing an excellent revenue stream. Disney provides over 10 different types of content, including downloadable ringing tones based on Disney favorites, cartoon character screen savers, games and horoscopes. These are all premium services, ranging from 100 to 300 yen per month per subscription, and they have been able to achieve a very high penetration rate among i-mode users. At the same time, by repositioning content from the Disney Fan Magazine, they can provide information to users and continue to build their brand in a new medium. Recently, Disney has launched content based on Tokyo Disneyland and Disney Sea, teamed with Oriental Land, the majority shareholder in Tokyo Disney Resort. This allows users to enjoy their visit, before, during and after, and this type of "media mix" is as attractive to Disney as it is to other content providers.

Disney is but one example of the premium content providers on i-mode. While this is a very successful model, there are a number of issues to consider. With 30 million subscribers, the i-mode subscriber market is very attractive. With nearly 2,000 partner companies offering content, however, it is also very competitive. To make the premium service model work, the content provided must be of high quality. When the DoCoMo content team works with any content partner, they focus on four principles, for not only premium content, but all content. First, is the content fresh, that is, is it something new that has never been tried and will it quickly attract new users? Second, does the content offer continuity, is there something to attract the user back day after day to use the content and is the rate of refresh sufficient to satisfy the frequent visitor? Third, is the content deep, by looking at, for example, the number of restaurants listed, the number of stories available and the number of ringing tones that can be downloaded? Finally, is clear benefit for the user considered? This last point is evident, but what is stressed are the needs of the mobile user, or rather, the need for this content on the phone. With all four of these criteria measured, it is possible to build the right premium business model for success on i-mode (Natsuno, 2000).

Aggregation

Another business model, found often in the "database" category of content, is that of the aggregator. These content providers include restaurant guides, real estate guides or job search sites. The classic aggregator would be the yellow pages telephone directory. While free to the user, companies pay for listings or extra advertising space. While a call to a directory assistance operator might be premium, to cover the extra staffing charges, an Internet, and by extension wireless Internet, service is free to the end user. The same model is used on i-mode. These aggregators of information develop a number of "backend" models so that costs do not have to be incurred by the users. These include real estate sites, town guides and job sites. Each can develop intricate business models, such is the case with the job site, which might be free to the job seeker, but premium to the job poster. The key to this type of model is to develop the right win-win so that all partners can see the benefit.

Business-to-Business

i-mode is not often seen as a business-to-business solution. Magazines and television news like the image of teenage girls with their phones, but it is important not to forget that i-mode, as an Internet-based system, is an attractive medium for business-to-business specialists to create and sell mobile workforce solutions to enterprise customers. Using the same HTML and HTTP of the Internet world, many third-party solution vendors have developed software packages for companies to allow their employees to access essential groupware packages, such as the scheduling of Microsoft Outlook, while on the move. Others have developed more custom-made enterprise solutions for such tasks as inventory management. The business model is very much like that of any enterprise solution, selling packages and after service care to companies, in addition to any other recurring revenue streams that can be developed. Many trading firms and investment banks have also developed separate applications for i-mode for their corporate clients.

Sagawa Express is a freight transportation service company, offering delivery across the country. Sagawa has been known as a leader in using information technology to enhance performance of its business. In order to streamline communication between call centers and drivers in the field, and to correct mistakes which are often made in taking customer orders, an i-mode CRM application, NEC Corporation's Clarify eFrontOffice, was introduced. Now, all drivers use i-mode phones to respond to customer requests at all times and to make reports and requests. The visual aspect of the system has increased the reliability of service. In addition, the initial investment, compared to a truck-based communication system, was small, enabling rapid implementation.

Advertising

A final model now being developed on i-mode is that of advertising. The success or failure of an advertising model depends greatly upon the number of users watching. When i-mode first launched with no subscribers, there was no way to put a monetary value on banner ad space. Over a year passed before i-mode was ready for advertising, both in terms of number of users and user acceptance. With a much smaller screen, a banner ad takes up much more space than on a PC screen, and so the chance of a negative user reaction is always great. In June 2000, NTT DoCoMo formed a joint venture with Dentsu, Japan's largest advertising agency, and NTT Ad, a member of the NTT Group. This new company, called D2 Communications, was given the task of developing a viable advertising model for i-mode. The primary mission was to create an advertising model with meaning, and to use the mobile phone as more than just a billboard.

Today, advertising on i-mode can be broken down into three schemes. The first is simple banner ads, within content providers' sites and the "Weekly i-Guide" area on NTT DoCoMo's portal. These small banners are kept to a minimum, in order to reduce the negative impact to the user, and in highly trafficked pages to increase the click-through rate. Once a user clicks through a banner, the most effective advertising consists of campaign information for the advertiser or takes the user through to the advertiser's website. This concept is not very different from the fixed-line Internet. D2 Communications is the agency responsible for ad space within DoCoMo's own portal pages, but then competes with other online advertising agencies to represent other content providers. The second advertising pattern is through a dedicated "Specials" menu page on the i-mode portal, with links to special campaign information, coupons and more. Users can also register, so that the information can be more targeted. The third advertising method is the "message free" service, which is a special push mail function of i-mode, in which transmission costs are born not by the user but by the advertiser. This is an opt-in push advertising service, which allows advertisers to target their message to certain demographics to increase the rate of success.

HORIZONTAL INTEGRATION AND THE EXPANSION OF MOBILE COMMERCE OPPORTUNITIES

i-mode is all about strategic alliances and designing the best value map so that numerous players can enjoy success. With this in mind, DoCoMo actively explores new alliances with a variety of complementary companies to expand the world of i-mode. i-mode's vertical integration includes developments with advertising, location-based services and content growth. i-mode's horizontal integration includes linkage with an assortment of other platforms and devices to expand the user's experience. To date, DoCoMo has announced alliances or projects with Sony Computer Entertainment for the PlayStation game console, Lawson Convenience stores for online retail, Coca-Cola for new multimedia vending machines, Sega for game center video games, America Online for fixed-mobile convergence and several vendors of car navigation systems for i-mode compatible Intelligent

Figure 5: i-mode Horizontal Integration. Through a series of alliances with other companies and platforms, NTT DoCoMo expands the world of i-mode.

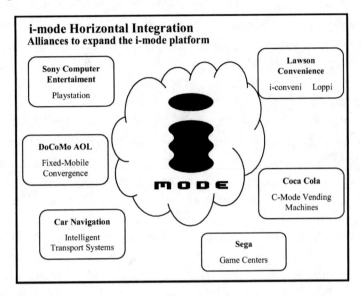

Transportation Systems (ITS). DoCoMo is also an investor in several joint ventures, including Payment First, designed to develop an online payment platform, and Japan Net Bank, an online bank. While there are numerous alliances to expand the i-mode platform; four are especially important for mobile commerce, and will be highlighted here (see Figure 5).

PlayStation

The PlayStation game console, in both its original and PlayStation 2 versions, is an extremely popular game machine. In Japan alone there are approximately 25 million game consoles sold, and about 70 million globally. What the PlayStation lacks, however, is network interactivity. DoCoMo's alliance with Sony Computer Entertainment led to a project to create cables to link i-mode phones to the PlayStation game console. With this linkage, there are several new concepts for both entertainment and commerce that emerge.

For entertainment, linkage with the i-mode phone expands the fun. First, and very simply, content from i-mode can now be seen on the user's TV through the game console, with special browser software sold by Sony. At the same time, the packet network of i-mode can be used to download new game data for old games. This functionality can prolong the life of a game and give game makers the opportunity for extra revenue streams for a game that once only yielded the purchase price. In the same way, game data can be uploaded and stored on a server, to be accessed and played anywhere, through the i-mode phone. This could be done simply with an HTML-based game. Characters could be put through a virtual "training session," and the results can be stored in a server to be downloaded into the PlayStation for continued play at home. The recent addition of Java into the 503i handsets has meant that arcade-style action can also be simulated to a certain extent on the mobile handset. With the introduction of subscription models linked to the i-mode billing service, game makers can continue to grow their businesses in a variety of ways.

The linkage with the PlayStation platform also has implications for other industries, apart from the computer game manufacturers. With 25 million game consoles in homes across Japan, the terminals, when linked to the packet network of i-mode, can become a powerful Internet device. While children may play games, with software provided by mail-order businesses, mothers could be shopping. Some mail order businesses have already begun services on i-mode, linking content online with content from printed and distributed catalogs. As already discussed, the i-mode display is not always the best device to present products for purchase. The PlayStation could offer this better experience, and provide an easy to use interface for ordering. The i-mode phone and packet network becomes the transmission network and also a device for customer relationship management, allowing shoppers to check on the status of orders, answer questionnaires and provide other useful marketing data. In this way, the PlayStation and i-mode can create a new environment for e-commerce (Natsuno, 2000).

Lawson and iConvenience

Japan is a country of convenience stores, and it is hard to travel more than a kilometer or two before finding one. Convenience stores are more than just a shop to pick up some milk, but sell a wide range of products using a carefully managed inventory system. Today, many convenience stores offer bill payment services, ticket purchases and have in-store multimedia terminals for online retail. The large density of these shops and their sophisticated inventory systems make them ideal distribution points for online purchases.

Lawson is one of the largest chains of convenience stores in Japan, with over 7,500 shops nationwide. Lawson stores already have "Loppi" multimedia kiosks that allow shoppers to select products, print out a ticket with barcode and pay at the checkout counter to receive their purchase. DoCoMo and Lawson have aligned for the iLawson online convenience store, and "iConvenience" on i-mode. With iConvenience users can make purchases from a huge inventory of books, CDs, cosmetics and other products. A purchase number is provided and the user then goes to the physical store. At the store, the user inputs their purchase number into the Loppi kiosk (in the future, it is hoped that an infrared link between the i-mode handset and the Loppi kiosk will eliminate this input step). The ticket and barcode can be taken to the counter for payment and exchanged for the actual product. The user first registers online and selects their preferred location, but the pickup store can be selected and changed at time of purchase. This registration procedure also provides valuable information for customer profiling, which is usually carried out by store clerks upon checkout. The customer relationship management aspect of iConvenience is very important for Lawson.

iConvenience does not have to be limited to the products in Lawson's own inventory. As already indicated, convenience stores can serve as a point of distribution for online purchases. Other companies can choose to sell their products through the iLawson online shop or can sell through their own online shop, but use the Lawson chain as an infrastructure provider. Lawson solves the problem of both distribution of goods and purchases. The volume and density of these shops, combined with the mobility aspect, makes Lawson a perfect complementor for i-mode.

Coca-Cola and C-Mode

In the spring of 2001, DoCoMo announced a joint project with Coca-Cola and Itochu to study and implement next-generation vending machines. The shear volume of vending machines, and the range of products that can be purchased through them, have probably surprised anyone who has ever visited Japan. Coca-cola alone has approximately 1 million vending machines around the country. From the first meeting to the launch of a test product, over a year was spent discussing business models and technology. From August of 2001, Coca-Cola and DoCoMo began testing the new-generation vending machine, called C-Mode. Like the previous example of Lawson and the online convenience store, this project aims to expand the world of i-mode and, using a networked approach, provide real business for the players involved.

The C-Mode vending machine includes all of the features of a traditional vending machine, but also includes a high-resolution video monitor, printer and reader. These machines are networked, and when combined with the networked i-mode, create an exciting new arena for mobile commerce. Coca-cola has also launched a new i-mode website, called Coca-cola Moment and C-Mode Club, to round out the mobile commerce environment. Currently, there are only a handful of these high-tech vending machines in a testing phase, but the numbers will be expanded and distributed to high population areas over the next few months and years (see Figure 6).

Users can, of course, buy a Coke with cash. C-Mode also becomes an electronic wallet, and after registering with the service, users can use the vending machine to deposit up to 5,000 yen (approximately US$50) into their virtual

Figure 6: Coca-Cola's C-Mode.   Next generation vending machine with barcode reader and video screen, which, when linked with i-mode, becomes not only a drink dispenser but also a multimedia terminal, capable of ticketing, and more.

account. This money can be used to buy drinks by generating a barcode on the i-mode screen and scanning it with the reader in the vending machine. (Like Lawson and the Loppi kiosk, in the future it is hoped to replace this by infrared.) This alone is not so exciting. After all, as is often mentioned, the mobile users in Finland have been buying Cokes by SMS for years. It is the other functions and features of C-Mode that will lead to bigger mobile commerce opportunities. The device can be used to sample and select content, such as ringing tones or graphics, which, once paid for through the online wallet, can be downloaded to the phone. This introduces event-based billing to i-mode. Also, when combined with the printer, coupons or maps can be provided. A ticketing business, similar to the convenience stores, will also be possible. These purchases are all interlinked with a loyalty reward program.

The vending machine could be used as a point of distribution for small products, in addition to traditional drinks. Payment methods always pose a challenge for online retailers, and so the deposit feature and online wallet offered by C-Mode may help to solve this problem. While buying a Coke through i-mode is mobile commerce, DoCoMo, together with Coca-Cola and Itochu, hope to create a far more exciting environment.

DoCoMo AOL

In the autumn of 2000, NTT DoCoMo announced a partnership with the world's largest ISP, America Online, and a large investment in AOL Japan. Subsequently, the name has been changed to DoCoMo AOL, and the partnership with AOL plays an important part of the concept of fixed-mobile convergence for i-mode. When DoCoMo launched i-mode in February 1999, the Internet penetration in Japan was very low. In fact, there were only around 5 million home PCs, and most Internet users accessed the Internet from work or school.[11] The situation is now changing in Japan, and more homes are hooking up to the Internet with personal computers. In this changing environment, DoCoMo and i-mode have to adjust to bring about convergence and promote the complementary aspects of the PC and mobile Internet.

Fixed-mobile convergence does not mean the same content on both devices. It does not mean multi-access portals. One of the key success factors of i-mode has been the ability to identify what services and content the user wants to use on the mobile device. In the same way there are certain types of content more appropriate on the PC. Together with AOL, DoCoMo is studying areas of overlap. The first is e-mail, and the launch of AOLi on i-mode with access to one's AOL e-mail was the beginning. As the relationship grows, the type of content could be expanded to Instant Messaging. In terms of mobile commerce, like the vision of PlayStation, online commerce sites on the PC could also have i-mode interfaces, using the same infrastructure. While the majority of shopping might take place on the PC through AOL, the i-mode phone could act as an ordering device or a customer relationship management application. i-mode is not designed to replace the PC, but to complement it, so the right mixture of content interlocking and mobile needs must be determined.

CONCLUSIONS AND EXTRAPOLATIONS
The Future

Two and a half years ago, i-mode was a very simple service of black and white handsets, simple HTML browsers and e-mail. There were 67 sites at launch and no users. Today, there are color handsets with more sophisticated browsers, Java applications, 2900 sites (within the portal, over 50,000 outside) and over 30 million subscribers. What is the future of i-mode? Where do we go from here?

The i-mode platform will expand to include more alliances, like those with Coca-cola, Sony and Lawson. These alliances, and similarly the expansion of content, slowly build the user's reliance on i-mode for every aspect of their daily life. As other industries join platforms with i-mode, the opportunities will increase for mobile commerce. The strength of platforms and the interlocking of these platforms have become evident, and have developed win-win relationships for the stakeholders involved.

While the alliance structure and the services increase on i-mode, so does the technology. The first leap was made with Java, and today 34% of i-mode users are using Java-enabled phones. The next leap came on October 21, 001, when DoCoMo launched its third-generation mobile network commercially, based on W-CDMA technology. Branded FOMA, for Freedom of Mobile Multimedia Access, this new network offers download speeds of up to 384 kbps (see Figure 7). This faster network allows not only for faster surfing, but also for even more exciting applications. First, the size of Java applications has been increased to allow

Figure 7: FOMA Visual Phone (by Matsushita). One of several third-generation mobile phones launched by NTT DoCoMo in October 2001, heralding a new era in mobile multimedia.

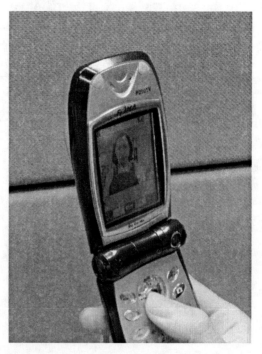

programmers even more freedom to stretch their imaginations. The next application to be launched is i-motion, allowing short video and sound clips to be downloaded and played from i-mode websites. The most important thing to bear in mind when watching the growth of i-mode is the step-by-step process of building upon success after success. 3G is not a whole new world, but a natural succession to the successful 2G service of i-mode. The business models do not necessarily change, but the ability to provide richer content to support those business models is key.

Lessons Learned

Hopefully, while reading these few pages, the important lessons of i-mode have become clear. This chapter was opened with the statement that i-mode success is not based on some mystic oriental alchemy. While it is not alchemy, it is a careful concoction of many factors. First, the right technology has helped to attract a great number of partners, reducing barriers for entry. Second, the alliance structure, which searches for the win-win in every relationship, has attracted not only eager content providers, but also other platform creators, to develop a better environment for all involved. Third, the right services, always considering the mobile needs of the user, have strengthened i-mode overall. Fourth, the right marketing has stressed what can be done, and has not misled the user or focused on aspects not of interest, such as technology. These four factors have been essential, but the most important thing that DoCoMo has done is coordinate all stakeholders in the value map, to design the best proposition for the user. Mobile commerce follows the same philosophy, by developing useful services for users and win-win relationships for the players involved.

Is i-mode a "Japan Only" Phenomenon?

It is time now to return to Telecom 99 in Geneva. More than two years later, Japan is the only market in the world that can truly say that wireless Internet is a success. With this fact, it is all too easy for critics to say that i-mode is a "Japan-only" phenomenon and similar services will never take off elsewhere in the world. "All Japanese ride trains, and we drive," they say. "Our fingers are too big and we want big screens," they say. "i-mode is all about teenagers and games," they say. "The Japanese don't use PCs," they say. Questions such as these are based on misperceptions and stereotypes. If thought through logically, there is a counter argument for every question. WAP-based services, it is true, have not taken off in other markets, but when examined in detail, many of the key factors, including technology, alliances, services and marketing, are different to i-mode.

Over the last two years, NTT DoCoMo has made numerous investments in mobile operators in other markets, in Asia, Europe and the Americas. The primary mission of DoCoMo with these partners is to offer know-how and experience to help them increase their value. Often, this leads to the discussion of i-mode, and how to launch it in these markets. It would be the decision of DoCoMo's overseas partners if they would launch i-mode-like services. But what is "i-mode-like?" The most important lesson of i-mode has been its strategy and the concept of coordination of the value map to provide the best services to the end user. This coordination, combined with the right technology, the right strategy, the right services and the right marketing, is i-mode. Of course, the details would be different for every market, but the essential concept of the win-win relationship does not differ. If this can be accomplished, other i-mode-like services should be a success. With that success, the opportunity for mobile commerce will move out of the realm of theory and concept, and into real business.

ENDNOTES

1   Statistic from the Telecommunications Carriers Association, December 2001.
2   Screens are a minimum of 16 characters by 6 characters, but many are larger and today in color. i-mode phones are manufactured by a number of vendors, including NEC, Matsushita, Fujitsu, Mitsubishi and Sony.
3   Data for number of sites, partners and users is as of time of writing, December 2001.
4   NTT DoCoMo announced its partnership with Sun Microsystems in March 1999, shortly after the launch of the i-mode service, with the intention of bringing Sun's Java technology to the mobile handsets. Over nearly two years of planning and preparation, the now branded "i-appli" was launched in January 2001.
5   With both the limited screen size and the limited cache size of the handset, the average i-mode page is 3 Kb. Many are far less than this.
6   For a complete list of compatible tags of HTML for i-mode, visit NTT DoCoMo's website, www.nttdocomo.co.jp/english/p_s/imode/index.html, which is updated as new functions are added to the browser.
7   Like the PC-based Internet before it, i-mode has also created a whole industry around i-mode site development. A trip to any bookstore or computer store in Japan will reveal literally hundreds of books and developers' packages for creating i-mode homepages. This can be attributed in part to the similarity to Internet standards.
8   More detail about business models for "premium" sites will be discussed later in the chapter.

9   Mari Matsunaga, originally an editor on several magazines at the company Recruit, played an important role in the design of i-mode and of considering the needs of the user.

10 The *furikomi* system (account transfer) in Japan is among one of the most popular ways of paying for goods and services. This is often an option for Internet commerce, and is especially used with larger payments, such as airline tickets.

11 This situation is often used to explain the success of i-mode in Japan. While the relatively low Internet penetration had an effect on the uptake of i-mode, it was not the only factor, and it cannot be said that i-mode-like services will not take off in area of high Internet penetration. As has been described in the chapter, technology, strategy, content and marketing all play an important role in making i-mode successful.

REFERENCES

Matsunaga, M. (2001). i-mode: The Birth of i-mode. Singapore: Chuang Yi Publishing Pte. Ltd.

Natsuno, T. (2000). i-mode Strategy. Tokyo: Nikkei BP Planning.

Telecommunications Carrier Association of Japan. (2001). December. *www.tca.or.jp*

NTT DoCoMo, Inc. (2001). *www.nttdocomo.co.jp/english.htm.*

Chapter II

Wireless Devices for Mobile Commerce: User Interface Design and Usability

Peter Tarasewich
Northeastern University, USA

ABSTRACT

Well-designed and usable interfaces for mobile commerce applications are critical. But given the uniqueness of the wireless environment, usability becomes even harder to ensure. This chapter describes the benefits and limitations of various wireless device interface technologies. It provides guidance on determining the usability of wireless devices, emphasizing the fact that context will factor heavily into the use of mobile applications. Some of the additional challenges that developers face when designing applications for wireless devices, such as infrastructure and software issues, are also discussed.

INTRODUCTION

An increasing number of technologies and applications have begun to focus on mobile computing and the wireless Web. *Mobile commerce* (m-commerce) encompasses all activities related to a (potential) commercial transaction conducted through communications networks that interface with wireless (or mobile) devices (Tarasewich, Nickerson, and Warkentin, 2001). Ultimately, researchers and developers must determine what tasks users really want to perform anytime from anywhere and decide how to ensure that information and functionality to support those tasks are readily available and easily accessible.

A well-designed and usable interface to any application is critical. For example, properly designed Websites help ensure that users can find information that they are looking for, perform transactions, spend time at the site, and return again. Given the uniqueness of the wireless environment, usability becomes even harder to ensure for m-commerce applications. The purpose of this chapter is to provide the reader with an overview of current wireless device interface technologies. It will provide guidance on designing usable m-commerce applications that take advantage of the benefits and respect the limitations of these devices. This chapter will also explore the interface design and usability challenges that the m-commerce environment still presents for users, researchers, and developers.

This chapter is organized as follows. The first section describes the benefits and limitations of various wireless device interfaces. The next section looks at how the usability of wireless devices affects the feasibility and success of m-commerce applications. The third section discusses some of the additional challenges that developers face when designing applications for wireless devices. The final section reiterates the need for good wireless application design, and describes some of the safety and security issues related to wireless device interface design.

WIRELESS DEVICES AND THEIR INTERFACES

The devices currently most important to m-commerce can be classified according to the categories listed in Table 1. There is some feeling that devices will become completely generic, and take the place of items like televisions, pagers, radios, and telephones (Dertouzos, 1999), but the question remains as to what form the devices will ultimately take. This important issue will be investigated further in the section on mobile system developer issues later in the chapter. But first we look at the current interfaces of these devices, their strengths, and their limitations. The discussion is separated into input and output interactions. Research that has been

Table 1: Wireless Device Categories

Laptop Computer
Handheld (e.g., Palm, Pocket PC, Blackberry)
Telephone
Hybrid (e.g., "smartphone" PDA/telephone combination)
Wearable (e.g., jewelry, watches, clothing)
Vehicle Mounted (in automobiles, boats, and airplanes)
Specialty (e.g., the now defunct Modo)

performed with various types of interface devices will be discussed in the next section on usability.

Input Interaction with Wireless Devices

Input interaction concerns the ways in which users enter data or commands. Common technologies used for input interaction with wireless devices include keyboards, keypads, styluses, buttons, cameras, microphones, and scanners. Each of these will be discussed in turn, emphasizing the benefits and limitations of each in the mobile environment.

The keyboard still remains popular as a form of input for many types of computing devices. The QWERTY configuration of keys (named for the sequence of keys at the upper left of the keyboard), while not the most efficient layout possible, remains a standard because of its wide user acceptance. Laptop computers have carried the concept of QWERTY keyboards forward, although keys are usually made smaller to conserve room. Devices such as phones and handhelds, however, have generally foregone the integration of a full keyboard because of the desire to create a device that is as small and light as possible. The exception to this is the Blackberry device, which includes a miniature keyboard. The problem with this keyboard is that a user must adjust to smaller keys, oftentimes learning to type messages with both thumbs. Data entry and error rates can suffer with smaller keys as well.

Smaller mobile devices usually rely on a more limited keypad for input. Most mobile phones use a standard 12-button numeric keypad, sometimes augmented by several special purpose keys (such as "clear" and "ok"). Each of the keys 2 through 9 also corresponds to a set of three or four letters. There are several approaches to entering text using a keypad. In the first, known as the multi-press input method, the user must hit a numeric key that also corresponds to the desired letter. For example, the letter "s" would require that the "7" key (labeled with "pqrs") be depressed four times. A capital "S" would then require eight or more keystrokes. A user must also pause or press an additional key to move onto the next letter. A different method that uses two-key input requires selecting a letter's group with the first key press and the location of the desired key with the second. For example, the letter 'E' (the second character on the "3" key which is labeled "def") requires the key press sequence 3-2. Another approach uses dictionaries of words and linguistic models to "guess" the word intended by a series of keystrokes. For example, the sequence 8-4-3 (corresponding to "tuv"-"ghi"-"def") might produce the word "the" out of all possible letter combinations.

One way to eliminate the use of a keypad for text entry is to attach a temporary keyboard to the device being used. Several vendors have developed miniature and/or full-size folding keyboards for this purpose. A more radically designed alternative

Figure 1: Matias Half Keyboard (taken from halfkeyboard.com)

Figure 2: Fabric Keyboard (taken from electrotextiles.com)

is the Matias Half Keyboard (Figure 1), which contains only those keys from the left-hand side of a traditional keyboard. When the space bar is pressed, the same keys function as the right-hand side. Another alternative is a fabric keyboard, being developed by ElectroTextiles, that can be rolled up for storage (Figure 2). Researchers are also developing "non-keyboards" in the form of gloves (Goldstein et al., 1999) or "FingerRings" (Fukumoto and Tonomura, 1997) that sense finger movements of users typing on a virtual keyboard and use software to interpret the movements. Essential Reality (www.essentialreality.com) is producing a glove called P5 that can be programmed to respond to users' hand gestures with combinations of keystrokes and mouse clicks. A potential problem with these types of devices is the additional training time that might be needed to use the device effectively.

Another way to eliminate the use of a keypad (and keyboard as well) is to use a stylus to write input directly on the screen of the device, a process known as gesture recognition. With this method, the device must recognize each character or symbol that is written, which can take a good deal of processing time and oftentimes suffers from inaccuracy. Palm has developed a proprietary system for character recognition (called Graffiti) that seems more accurate than other recognition systems, but forces the user to conform to a writing style for letters that is somewhat different than normal. Another gesture recognition technique is Jot (often used with

Pocket PC devices). In both cases, the user must learn which pen strokes represent a particular character to the device, rather than the device interpreting the handwriting of the user. As an alternative to keypads, Smart Design (www.smartdesign.com) is developing a system called Thumbscript that replaces a keypad on phones with a nine-point grid. Users tap a keystroke sequence on the grid for each character (Roman letter or Asian character) that they wish to input.

As an alternative to gesture recognition, keyboards (or other key configurations) can be created virtually on a screen, with each key being "pushed" by touching it with a stylus. These so-called "soft-keyboards" are sometimes implemented in sections (e.g., the alphabetic characters separated from numbers and other characters) to save screen space and create larger keys. Styluses can also be used to activate icons, menu choices, or hyperlinks displayed on a screen. Virtual keyboards currently suffer from a lack of tactile feedback often found on keyboards and some keypads, although feedback can be provided through sounds generated as keys are "pressed."

Mobile device input can also be achieved through "mouse buttons," thumbwheels, and other special-purpose buttons. The user interface of the telematics system OnStar consists of just three buttons, labeled "call," "help," and "off." Mobile phones often have dedicated buttons with labels such as "call," "ok," and "clear" in addition to a numeric keypad. Mouse buttons are toggle switches that allow one-dimensional cursor movement. An alternative to a mouse button is the "navi-roller," which allows scrolling by rolling and selection by clicking. Small joysticks, which allow two-dimensional cursor movement, are sometimes found integrated into the keyboards of laptop computers, and more recently on mobile phones. Handheld devices usually have a mouse button and a few other special-purpose buttons, but no keyboard or keypad. CyMouse by Maui Innovative Peripherals (maui-innovative.com) is an eight-ounce headset that acts as a wireless mouse. A version called Miracle Mouse is aimed at providing more control options to people with physical disabilities. The now defunct Modo device (Figure 3), which featured one-handed operation, had a thumbwheel to move between selections and to scroll text

Figure 3: Modo Device (taken from www.useit.com Alertbox 9/17/00)

up and down. Pressing the wheel activated the current selection. Some other handheld devices also feature a similar built-in thumbwheel. However, the location of the thumbwheel limits which hand can hold the device for one-handed operation.

Using human speech as input to mobile devices is also becoming increasingly practical as voice recognition technology continues to improve. Whether or not voice interfaces will ultimately succeed as a primary form of input depends on how well certain limitations of the technology can be overcome. These limitations include the need to train devices to recognize a user's voice, the relative slowness of voice versus other input means, and the difficulty in using visual information (e.g., graphics) with voice input. Benefits of voice input include the ability of users to interact with the device in their natural language. Voice input allows those users who cannot type or use a stylus to interact with a device. It may also be a viable interface alternative for devices too small for buttons or for those without a screen. However, voice input suffers from possible privacy and social issues. For example, users may feel uncomfortable speaking input aloud instead of typing or writing it, and certain places (e.g., libraries) might restrict the use of voice input to maintain a quiet environment. One option that allows a voice interface with mobile devices, but does not require direct Internet access from the mobile device, is Voice Extensible Markup Language (VXML). This standard allows consistent access to Web applications from both the wired and wireless environments.

With the shrinking size of camera lenses and the increasing sophistication of digital photography, video is becoming more common as a form of input with mobile devices. Some laptops, phones, and handheld devices have built-in or attachable cameras. DoCoMo has been developing specialized mobile Internet appliances, some of which are cameras that can take pictures, adorn them with overlays, and send them to users with similar devices or i-mode phones. Video might also be used as input through the recognition of hand gestures or facial expressions.

Similarly, scanners may also become part of the wireless environment. They can be used for reading text, bar codes, or other symbols. Wireless devices that scan UPC symbols as input could be part of in-store mobile commerce applications used for comparison-shopping or for purchasing merchandise without the need of a cash register and sales attendant.

Finally, input can come from technologies that sense location, or from those that can receive information from their environment based on their location. The Global Positioning System (GPS), a set of satellites owned and operated by the U.S. Department of Defense, allows any device equipped with a GPS receiver to determine its geographic location within about 10 meters. All mobile phones sold in the U.S. will be required to have the ability to determine their location. Bluetooth technology, which allows short-range communications, will allow mobile devices to receive information automatically when they are in close proximity of another

Bluetooth-equipped device. As we will discuss later, location is a key factor in designing useable mobile applications. However, privacy issues dealing with the use of location data must also be addressed.

Output Interaction with Wireless Devices

Output interaction concerns the ways in which users receive data, prompts, or the results of a command. Common technologies used for output interaction with wireless devices include video screens and speakers. Both of these will be discussed in turn, emphasizing the benefits and limitations of each in the mobile environment.

The liquid crystal display (LCD) screen is the primary technology used to produce output in the form of images and text on current wireless devices. Screen size varies greatly from one type of device to another. Most mobile phones have small (1' to 2' square) screens that can display 4 to 8 lines of 10 to 20 alphanumeric characters each. Handheld devices have relatively larger screens (about 3' by 4') that are more suitable for graphics as well as text, but are still limited by low screen resolutions (usually 240 by 320 pixels). Most phones and handhelds have monochrome screens, although more are being sold with color screens, which can increase device usability. Laptops have fairly large color screens (up to 15' diagonal) with resolutions that compare favorably to desktop monitors. Vehicle-mounted devices have screens ranging from smaller than the size found on phones to the size found on small laptops, depending on the intended purpose of the device (e.g., displaying song titles versus a map of a city).

The current limitations of screens on wireless devices are their size, resolution, and color capabilities, all of which are usually less than those found on desktop computers. These limitations make it difficult to display large amounts of text and graphic-based output (e.g., maps, charts, or Web pages). There are also tradeoffs in improving the screen characteristics of mobile devices. Increasing screen size will increase the size and weight of a device. Color screens with high resolutions use more power than their monochrome counterparts, resulting in increased battery weight and/or less time before the battery needs to be recharged (although research into better batteries continues).

There are, however, some recent technological developments that may address some of the disadvantages of current wireless device screens. Flexible screens are on the horizon, which may eventually allow screens that can be rolled or folded up. E Ink (www.eink.com) and Gyricon Media (www.gyriconmedia.com) are developing displays with electronic ink technology (e-paper), first in black and white, but possibly in color in the future. The screens hold an image until voltage is applied to produce a new image, using less overall power than LCD screens.

Figure 4: Olympus' Eye-Trek Device (taken from www.eye-trek-olympus.com)

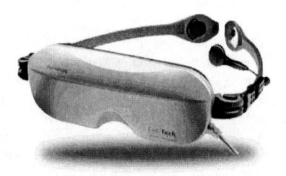

Monocular units or goggles can be used with magnifying glasses to enlarge small displays (less than an inch diagonal) so that they look like an 800 x 600 resolution monitor. Goggle-type products include InViso's eShade (www.inviso.com/products), Sony's Glasstron (www.ita.sel.sony.com/products/av/glasstron), and Olympus' Eye-Trek (www.eye-trek-olympus.com, see Figure 4). Microvision (www.mvis.com) is developing a device that projects an image, pixel by pixel, directly onto the viewer's retina. Heads-up displays, which have seen limited use in automobiles in the past, might also be used for vehicle-mounted devices. These types of devices allow viewing of color images with similar sizes and resolutions as those found on desktop computers. Potential concerns with these technologies include interference with users' other visual inputs, and the social acceptance of wearing and using such technologies.

Sound is the other primary form of output from a wireless device. Forms of this output range from words to music to various beeps, buzzes, and other noises. These can be created through speakers or through headphones. Newer laptops usually have a set of speakers built in for stereo sound production. Most smaller mobile devices have a single speaker at best. Stereo speakers can be used to generate sounds coming from a particular direction, which as we shall see later can be used to enhance usability. This same effect can be achieved through headphones, but at the cost of possible interference with a user's other audio input (i.e., sounds from the environment).

Sound output may be a viable interface alternative for devices without a screen, although there may be difficulties in presenting certain visual information (e.g., graphics). Voice output is also generally produced and comprehended more slowly than visual output. On the positive side, sound allows those users who cannot see a screen to receive output. Ultimately, it may be that multi-modal browsing, where voice and visual output are combined, may be best suited for wireless devices (Nah and Davis, 2001).

WIRELESS DEVICE USABILITY

This section looks at the usability of wireless devices and how usability affects the feasibility and success of m-commerce applications. Some of the recent research on interface design and usability for mobile and wireless devices will be discussed, along with usability issues present with wireless devices. The section will also consider whether or not current HCI standards can be applied to wireless devices, and what further research issues regarding the usability of wireless devices need to be addressed.

Usability can be defined as the quality of a system with respect to ease of learning, ease of use, and user satisfaction (Rosson and Carroll, 2002). It also deals with the potential of a system to accomplish the goals of the user. *Usability testing* asks users to perform certain tasks with a device and application while recording measures such as task time, error rate, and the user's perception of the experience. Methods for evaluating usability include empirical testing, heuristic evaluations, cognitive walkthroughs, and analytic methods such as GOMS (goals, operators, methods, and selection rules).

Many of these same usability methods can be applied successfully to test the usability of a particular application on a device, or to compare usability across different devices or configurations. Chan and Fang (2001) reported on research in progress that is conducting a heuristic evaluation and cognitive walkthrough of 15 m-commerce sites across three different device platforms (Palm, Pocket PC, and WAP phone). Their preliminary results indicate that many Web sites are trying to duplicate their wired Web architecture and design for the wireless Web, resulting in poor navigation and information overload.

Likewise, many of the current principals of interface design can be transferred to newer devices, although soundly applying these principals may be more difficult due to the unique nature of mobile systems and devices. Fundamental rules such as consistency, shortcuts for advanced users, the use of feedback, error prevention, easy reversal of actions, and minimization of short-term memory requirements (Shneiderman, 1998) will undoubtedly transfer to mobile applications. However, as shown in the previous section, the devices that the user might interact with are quite different than the desktop computers used in much of the interface design research to date. While further study is needed, it is likely that much of the specific research on effective screen design and information output cannot be generalized to mobile devices.

Furthermore, *context* will factor heavily into the use of mobile applications and devices, which is something that was not as much (if any) of a concern with stationary desktop applications. Mobile tasks and technology use are significantly different than their stationary counterparts. People can now literally be anywhere at anytime and use a mobile application, which was not true with the traditional (wired) Web since a physical connection was needed to the Internet. Location will need to be factored into the usability of an application and a device, as will the dynamic nature of the environment within which it is used. Conceivably, a mother could be walking down a street in an unfamiliar city trying to use a mobile application to find the location of an office for an appointment, while keeping track of her three children and processing all the other input coming from her environment. Interface design that may be well suited to a relatively stable office or home environment will not necessarily work well in the Amazon rain forest or in an automobile cruising down a highway.

Let us now turn to some of the recent research that specifically addresses the design and usability of mobile applications and devices, first from the viewpoint of input interaction. One usability concern is how well users can perform tasks using the assortment of keypads and keyboards found on many wireless devices. Looking at keypad text entry performance, Silfverberg, MacKenzie, and Korhonsen (2000) created models to predict the entry rates for multi-press, two-key, and linguistic-based keypad text entry methods. Using empirical data, they estimated that expert users could achieve rates of up to 27 words per minute (wpm) using thumb (one-handed) or index-finger (two-handed) input with the multi-press and two-key methods. For the particular linguistic-based method that they investigated, they predicted speeds up to 46 wpm for expert users using two hands and their index finger. A study done by Weiss, Kevil, and Martin (2001) on a particular mobile phone found that users in general had difficulty in using its keypad. Some user frustration came from confusion as to which keys performed what functions, and how the keys were labeled. All subjects had difficulty in entering text. Difficulties in navigating through applications were also encountered, in part due to use of the keypad and in part due to the confusing structure of the applications tested.

There have been many studies on soft keyboard performance. Those by Lewis, LaLomia, and Kennedy (1999) and MacKenzie and Zhang (1999) found that users could achieve speeds of up to 40 words per minute with a QWERTY layout on a soft keyboard, although speed varied with the devices used, the tasks performed, and the amount of practice. Alternate soft keyboard layouts can produce even higher text entry speeds than the QWERTY configuration, but usually after much experience with the alternate layout (e.g., MacKenzie and Zhang, 1999). A study by Zha and Sears (2001) showed that the size of a PDA soft keyboard did not affect data entry or error rates. Additional subjective ratings did not suggest that users preferred larger keyboards, which implies that soft keyboards could be successfully implemented on smaller devices, such as mobile phones.

Looking at virtual keyboards, a study by Goldstein et al. (1999) found that their non-keyboard (i.e., glove) device resulted in fewer errors and higher subjective satisfaction than a soft keyboard and a miniature keyboard on mobile devices (although a full-size keyboard was still the most preferred). The Fukumoto and Tonomura (1997) FingerRing device was tested only with users producing chords (symbols) rather than individual characters on a QWERTY keyboard, so there is no way to compare use of their device to other keyboard types.

If a stylus is used to write input on the screen of a mobile device (using gesture or handwriting recognition), performance is generally much poorer compared to using any type of keyboard. Studies such as MacKenzie and Chang (1999) found that data entry rates of up to 18 words per minute (wpm) can be achieved using various gesture recognition systems. But these studies did not test performance using handheld devices. An exception to this is Lewis (1999), which reported speeds of up to 24 wpm on PDAs, but used simulated "perfect" handwriting recognition where any attempt at creating a letter was considered correct. More recently, Sears and Arora (2001) compared Jot and Graffiti using Pocket PC and Palm devices, respectively. They used tasks that they felt were more realistic than previous studies, and kept track of data entry times and error rates. Novice data entry rates of 7.37 wpm were obtained for Jot and 4.95 wpm for Graffiti. The recognition of "gestures" also covers stylus-made marks other than letters or numbers used for data input or commands. A survey of handheld device users completed by Long, Landay, and Rowe (1997) showed that users generally liked using gestures for device input, although they often found them difficult to remember and became frustrated when a device did not recognize what they wrote. More recent research such as Long et al. (2000) looked at designing gestures that are easier for people to use and remember.

There is also research that looks into assisting the user with the data input or command process. Dunlop and Crossan (1999) proposed a text entry method for mobile phones that anticipates words based on a dictionary of common words stored on the device. The method was tested using a PC-based emulation of a mobile phone. Results showed some success with and a general user preference for the new method, although more testing needs to be done. Masui (1999) developed a dictionary-based text entry method that uses the context of the phrase or document being typed. Given the current input limitations of mobile devices, usability might also be increased by changing the nature of the data or instructions required by the application. Versign is introducing a service called WebNum, which would substitute a telephone number or other numeric string for a standard Web address (e.g., www.neu.edu). Testing still needs to be completed on this method as well.

Voice recognition technology continues to improve, but there is still the question of how well it works for different applications and tasks. De Vet and Buil (1999) listed some general findings from user studies on the use of voice control compared to entering text data on limited-key devices. User operations that favor voice control included: 1) direct addressing of content (e.g., calling out someone's name), 2) menu navigation and option selection, and 3) setting a range (e.g., the starting and stopping times on a VCR). The operation of scrolling through a long list favored the use of cursor keys rather than voice commands for people who were browsing.

Now we look at some of the research concerning design and usability related to output interaction. Output technology has received a fair amount of attention from researchers, with much of the recent focus on small displays. The fundamental question here is: "Can users perform tasks as well using small displays rather than larger ones?" The answer to this will, of course, vary based on the size of the display and the task being performed. A study by Jones et al. (1999) found that users in a "small screen" environment (simulated by setting monitor resolution to 640x480 pixels) were less effective in completing search and retrieval tasks than users with a "large screen" environment (1074x768 pixels).

Reading text on small devices, especially the size found on many mobile phones, can be difficult. There are various options that can be considered for formatting text on small screens and providing navigation. Melchior (2001) developed a method called "wiping" that may make it easier for people to read text on small displays. The method adds a perceptual guide (the graying of text that will be removed from the screen) during scrolling that aids in refocusing the user's attention after the paging of text. A study on wireless application protocol (WAP) interface usability was done by Chittaro and Cin (2001) using novice users. Each screen of material on a WAP device is known as a card. They evaluated: 1) navigation among cards using links versus an action screen, and 2) single-choice lists using a list of links versus a selection screen. Results showed that users performed better using links and a list of links, and perceived greater difficulty in using the action screen and selection screen environments.

Rapid serial visual presentation (RSVP), which serially presents one or more words at a time at a fixed place on a screen, is another option for presenting text on a small screen. There are many studies that investigate the use of RSVP, but overall the results seem inconclusive as to whether the method works better for text presentation than other methods. Bernard, Chaparro, and Russell (2000) compared RSVP against presenting three lines of text at a time and 10 lines at time on a simulated small-screen interface. Overall reading comprehension levels were about the same for the RSVP and 10-line methods, which were marginally higher than the three-line method's comprehension levels. Subjects were equally satisfied with each method of presentation, and did not seem to prefer one method to the others. However, they did prefer a slower text speed and thought that the RSVP method produced more eyestrain. Studies comparing RSVP to sentence-by-sentence presentation were performed by Rahman and Muter (1999). They concluded that RSVP was not liked by subjects but is as efficient (as measured by reading speed and reading comprehension) as sentence-by-sentence and full-page presentations.

Variations of RSVP are also being investigated for use on small-screen devices, and may provide better presentation alternatives. Adaptive RSVP allows the exposure time for each word or group of words to vary, based on word length and familiarity. Sonified RSVP attaches appropriate sounds (such as earcons) to groups of text. Details on the development of these two concepts can be found in Goldstein et al. (2001). The concept of RSVP has also been applied to Web browsing on small screen devices (De Bruin, Spence, and Chong, 2001). The idea behind this concept is to rapidly display navigation choices sequentially when space is limited, allowing users to see the range of alternatives (links) available without a lot of searching. Initial testing of an RSVP browser against a WAP browser showed RSVP browsing to be at least as effective as WAP browsing for experienced users.

Other types of browsers for small-screen devices are being developed and tested as well, all hoping to increase the usability and effectiveness of mobile devices for Web-based tasks. When viewing a Web page on a small screen, most current browsers show a subset of the original page (usually with minimal graphics) after processing it through a proxy server. An application called Power Browser was developed and tested by Buyukkokten et al. (2000) against various other handheld device Web browsers. Their method presented Web pages as text-only summary views based on information collected about link importance. While the Power Browser does require use of a proxy server, its performance seemed to be better than the other browsers tested. Gomes et al. (2001) presented ongoing research into a mobile device interface that does not require a separate server to store and provide Web content to mobile devices (i.e., the system works with existing Web pages). It first uses a clipping filter to get rid of items that users do not want to see on a handheld device (e.g., ads and other content). It then minimizes the text it presents to the user through heuristics that use parsing and abbreviations. However, the user can zoom into greater levels of detail if desired, to the point of seeing the complete original text.

Other small-screen browsers seek to maintain the "look and feel" of Web pages as much as possible in the mobile environment. Instead of transcoding a Web page into a text-based subset of the original page for mobile devices, the ZFrame (www.zframe.com) browser shrinks the Web page down to fit the screen. When a user moves a stylus across the screen, parts of the Web page are enlarged for easier readability. Along this same line of thinking, Rist and Brandmeier (2001) have proposed ways to change graphics into images that are suitable for small displays. These include transforming graphics (either blindly or after an analysis of the source) or generating a new picture from a content description of the current graphic obtained through semantic analysis.

With the increasing use of color displays in mobile devices, color and its manipulation are important considerations for visual interfaces. Issues here include whether or not to allow the user to change colors, how many colors to use, what colors to use, what the colors should represent, and what colors should be adjacent to each other. Shneiderman (1998) gave some interface color use guidelines that can generally be carried over to mobile devices, although some of the effects of color may be different on smaller screens. Research by Deshe and Van Laar (1999) discussed applying a perceptual layers methodology to tabular displays on handheld computing devices. Tables that are too large to fit on the display can force the user to scroll from one part of the table to another, causing frustration and wasting time. Using the perceptual layers methodology, related areas of the table can be color coded, rather than using labels and headings that take up room on the screen.

Usability of mobile applications and devices can also be increased through the use of sound output. Brewster, Leplâtre, and Crease (1998) suggested that non-speech sound might be used to overcome some of the limitations from the lack of screen space on many mobile devices. Going beyond the ubiquitous beeps and ringing tones that many phones use, they suggest that structured audio messages called "earcons" can be used as part of the interface of a wireless device. Walker and Brewster (1999) proposed using three-dimensional audio space surrounding the user in conjunction with graphical user interface techniques to expand the display capabilities of mobile devices. Information is presented in multiple spatially segmented "windows" of sound. However, such an interface requires the user to wear headphones, because most current mobile devices have no more than one speaker.

Sound may be especially useful where the user of a mobile application may not be able to give his or her full attention to an output screen of any size. Holland and Morse (2001) are investigating an audio interface for a GPS system that requires minimal attention from the user so that they can use their eyes and hands for other purposes. This is done through tones projected through headphones at locations relative to the user (e.g., left, right, forward) to indicate direction, along with pulses that increase in rapidity as the user gets closer to the destination.

Many input and output technologies still need to be tested further. There seems to be a lack of research testing and comparing the various buttons and wheels that appear on many mobile devices. Using cameras and scanners with mobile devices should be investigated. There is also the question of what mobile devices should or should not do. Are they meant to have all the capabilities of a desktop machine, or are they meant for a limited set of tasks performed in a certain context? And will the usability engineering methods that work well for "fixed" computer systems meet the demands for evaluating mobile systems?

Lastly we turn to research that concerns the effect of context on factors related to mobile application design and usability. The unique nature of the m-commerce environment requires a focus on usability that goes beyond the device itself. Mobile applications, by definition, can be used in various locations, meaning that the context of the application and the device must be taken into account when looking at usability. Developers need to understand people and how they interact with their surroundings, and design systems that work well in the range of environmental conditions that may exist. Mobile device users may also be much more sensitive to task time than those who are sitting at a desk. Researchers cannot simply design a device, test it in a controlled laboratory setting, and conclude that it is usable.

Johnson (1998) looked at the challenges that researchers and practitioners face in the design of mobile systems. He noted four problems that need to be addressed:

1.   The demands of designing systems for mobile users increase when the context of usage is considered.
2.   A diversity of wireless devices, network services, and mobile applications need to be accommodated and integrated (also see Olsen, 1998; Tarasewich and Warkentin, 2002).
3.   Current human-computer interaction models are limited in their ability to address the demands of mobile systems.
4.   Usability evaluation methods for mobile systems will need to be developed and tested.

Mobile activities can become very complex because of changing interactions between the user and the environment. It will be very difficult to model these interactions. What works with wired systems will not necessarily work on mobile systems, not only due to wireless device differences, but also because of the unpredictability of user priorities and the context in which the application might be used.

Researchers have begun to investigate the additional usability and design issues that result from the use of wireless devices in complex mobile environments. The circumstances under which mobile applications are used can be significantly different than those for desktop machines. Holland and Morse (2001) recently summarized these differences from various research papers. Mobile device use can be characterized by:

•   Limited user attention given to the device and application (interactions with the real world being more important).
•   User's hands being used to manipulate physical objects other than the device.
•   High mobility during the task, with the adoption of a variety of positions and postures.

- Context dependent interactions with the environment.
- High speed interactions with the device, driven by the external environment.

With these differences comes the question of whether or not graphical or Windows-based interfaces are appropriate for mobile devices. Some researchers have formulated alternative interaction methods that begin to address the needs of mobility. Kristoffersen and Ljungberg (1999) developed an interaction method called MOTILE, which requires little visual attention and provides audio feedback to the users. Input is provided using four buttons and structured commands on a handheld device. Pascoe, Ryan, and Morse (1999) discussed a context-aware application called "stick-e notes," which allows users to type messages on a mobile device and virtually attach it to their current location. Contexts other than location can also be used, such as time of day, temperature, and weather conditions. The format of the notes is not limited to plain text, and the notes reappear if the user approaches the same location again. Pascoe, Ryan, and Morse (2000) formulated and discussed two general principles for mobile interface design. The first is Minimal Attention User Interfaces (MAUI), which seek to minimize the user's attention (but not necessarily the number of interactions) required to operate a device. The second is context awareness, in which the mobile device assists the user based on a knowledge of the environment.

Perhaps ethnographic methods are better suited for the design of mobile systems and devices than traditional laboratory usability testing. Väänänen-Vainio-Mattila and Ruuska (1998) discussed the use of contextual inquiry during the requirements analysis for a smartphone device. Contextual inquiry observes potential users of a device as they perform tasks in a real setting (e.g., office workers in their building), and could also be used to observe users performing tasks with mobile devices. Performing this type of study can be very time consuming and challenging, but can add insights not obvious from controlled laboratory testing.

One way to address the issue of usability in a dynamic environment is to design devices that derive input indirectly from the user. Schmidt (1999) discussed a vision of mobile computing where devices can "see, hear, and feel." Devices act according to the situational context in which they are used. Schmidt sees a shift from explicit interaction with devices (e.g., using speech input) to *implicit interaction*, where the actions performed by the user are not necessarily directed at the device but are understood as input by the device. For example, a device might turn on automatically when grasped by a user, and power down after being left alone for a certain length of time.

Devices might also receive input from their surroundings rather than from the user. Addlesee et al. (2001) are investigating systems that react to changes in the environment according to a user's preferences. They use the term *sentient computing* because the applications appear to share the user's perception of the environment. They have created a device called a "Bat" which determines its three-dimensional location within a building in real time. These devices can be carried by users or attached to equipment, and can be used as virtual mice or buttons. They can also be used to augment and/or personalize a user's experience regardless of physical location.

DEVELOPER ISSUES

This section looks at some additional challenges that developers face when designing applications for wireless devices. There are problems with creating applications that work on more than one device, some of these due to the devices themselves and some due to the infrastructure that supports wireless communications. Developers especially need to consider designing applications that work well given the relatively limited bandwidth, processing power, and storage capacity of mobile devices.

Limited bandwidth restricts the amount of material that can be realistically sent across wireless communication pathways. Developers need to carefully consider the amount of data that is sent to the wireless device from a server or from another device. Many current methods for data storage used in mobile applications do not maintain a single source of data for mobile and non-mobile applications, nor do they allow direct sharing of data among devices and applications, which creates an issue of data integrity. Many organizations are transcoding (converting the content of) their current Websites to make them useable with wireless devices. This creates "wireless Web" applications, but ones that are separated from their wired counterparts. Another decision that is affected by bandwidth availability is whether to develop applications that use text, graphics, or a combination of both. Text is very efficient in terms of data transmission requirements, and can be used with almost any device. Navigation, however, may be more difficult with a textual interface. Graphics can often convey information more concisely than text, but at increased transmission costs. Graphics are also limited by the size and capabilities of display screens.

Data must be stored so that it is readily useable and accessible by mobile applications. Extensible Markup Language (XML), which tags data and puts content into context, is one possible solution to this problem. Another is Relational Markup Language (RML), which acts as in intermediate format between languages such as HTML and Wireless Markup Language (WML), and allows the automatic markup of all markup languages (Saha, Jamtgaard, and Villasenor, 2001). With RML, device output is generated without regard to the initial markup language.

Developers must select the best technique(s) for storage of data used by different wireless devices.

The relatively limited processing power and memories of current wireless devices have forced developers to carefully revisit both operating systems and applications software on mobile platforms. Operating systems such as Symbian's EPOC have been created to function using the limited amount of memory available in mobile phones. Other limited function operating systems such as Microsoft's Pocket PC and Palm's PalmOS have been developed for handheld devices. Symbian and Palm have agreed to collaborate on technologies, which could result in the eventual combination of the two operating systems.

Another important building block for this emerging infrastructure landscape may be the Wireless Application Protocol (WAP), which enables wireless devices such as mobile phones and handheld devices to access the Internet (Ralph and Aghvami, 2001). Many WAP-enabled devices have already appeared, although there is doubt as to whether WAP will become a globally accepted standard, especially with the popularity of Japan's i-mode. Developers ultimately face the issue of deciding which set of protocols to accept, or risk the potential problems of working with multiple standards and/or choosing to ignore some.

Wireless Markup Language acts as a page description language within WAP (Herstad, Thanh, and Kristoffersen, 1998). Based on XML, it is not compatible with HTML, although it borrows many of the latter's tags. WML is optimized for displaying information on small-screen form factors, and uses specific tags for text and table representation. Another language that can be used for viewing text portions of Web pages on wireless devices, but is not based on XML, is Handheld Device Markup Language (HDML). Companies are also beginning to explore the use of Java applications with wireless devices. Carriers such as DoCoMo have begun to introduce services that can take advantage of Java-enabled wireless phones. This will allow the development of "push" applications that can initiate contact with users (e.g., alerting someone to a breaking news story) rather than waiting for the user to pull information off of the Web.

Safety will be a critical issue when designing mobile commerce systems to be used in automobiles. Operating the wireless device is not the primary task, for the user needs to concentrate on driving. If car-mounted devices eventually allow regular Internet access, safety issues of "browsing while driving" must be addressed. Companies such as Nokia have done testing with mobile phones and in-car communication systems under simulated driving conditions (Koppinen, 1999). Automobile manufacturers such as Ford are beginning to test the use of telematics devices under simulated driving conditions as well. Graham and Carter (1999) reported results of a comparison of speech versus manual operation of a mobile phone system under simulated driving conditions. While driving performance was significantly better using speech input, task performance was significantly worse. However, users' attitudes were favorable toward the speech interface, with most desiring it over a manual interface for a car phone.

Developers can try to provide applications for all different types of mobile devices, but there is still the big question of what form devices will ultimately take. Users may find specific purpose devices most desirable and usable. Or they may want multi-purpose (e.g., smartphone) or "all-purpose" devices that perform multiple functions. Information viewing can also be personalized for the user across multiple wireless devices, which will allow users to have a consistent and familiar environment when going from one device to another. One issue that arises with personalization, however, is whether or not organizations will want to control personalization, or at least want to limit it when their own content is involved. For example, organizations may not want users reformatting data taken from their Website before displaying it on a wireless device.

DISCUSSION

This chapter has looked at the input and output interfaces available for wireless devices, along with some of the benefits and limitations of each. Usability of these interfaces was then discussed, focusing on current research in the field. The issue of context, which differentiates mobile systems from their wired counterparts, was emphasized. Various challenges that developers face in the design of mobile systems were then summarized.

While often neglected or left as an afterthought by many organizations, proper interface design is necessary to the success of any system. M-commerce application developers must look carefully at potential users, devices, and contexts of use. One usability issue that has not been addressed yet is the need for organizations to determine how people can best use mobile applications and access information through different wireless devices. It may not make sense to perform certain tasks through specific wireless devices, or through any wireless device at all.

To measure the success of mobile applications and devices, researchers need to find the best ways to test their usability. A crucial factor here is taking context into account, including not just location but factors such as the available communications infrastructure, the current physical conditions, the user's social setting, and the user's emotional state (Schmidt, 1999). Current usability testing methods may be generally applicable to mobile devices and applications, but new or improved methods will need to be developed as well. The mobility of devices and applications may require dynamic interfaces that change with the user's changing needs, status, and environment. Interfaces might be more effective if they differ based on the social setting (e.g., a work meeting versus a group of friends) or the emotional state (e.g., anxious versus relaxed) of the user.

Security of wireless information is another important issue in m-commerce (Ghosh and Swaminatha, 2001). The increased use of wireless devices for e-commerce makes the issue of positive identity verification even more important yet more difficult to ensure. One consequence of this need is the increasing importance of biometrics. Future wireless devices may include a thumbprint or retinal scanning ID device, or may use smart cards to store user authentication information. These security requirements, and their effect on wireless device interfaces and usability, will need to be considered during the design of mobile applications.

Another development that may affect user interface design and usability of mobile devices is electronic signatures (Broderick, Gibson, and Tarasewich, 2001). Software recently developed by Brokat (www.brokat.com) allows the use of mobile digital signatures. A user will receive electronic verification of a transaction, and can digitally confirm that it is correct. The software works with existing mobile devices, and does not depend on the implementation of a public key infrastructure specifically for the mobile market. Other implementations of electronic signatures on mobile devices may require access to smart cards or biometric readers.

Aesthetics, along with usability, may also be part of designing an overall enjoyable user experience with mobile devices. Karlsson and Djabri (2001) have begun to investigate "aesthetics in use," which they define as dynamic interaction that invokes a positive affective response from the user. They are investigating whether parameters such as engagement (feedback) and transparency (understanding the interaction flows of an interface) can be used as user interface design principles for small screen devices.

While this chapter dealt primarily with the current wireless environment, the ultimate wireless device may still be far from reality. Promised increases in available wireless bandwidth from third-generation technologies may be useless if people cannot or will not use the devices themselves. Folding screens, a technology that is currently under development, could be the answer--or perhaps a device with multiple physical windows (screens), each showing a different (but associated) view of the application. Maybe wireless devices need screens and keyboards that can be stretched to larger sizes before being used, and then shrunk back to their original size. Or maybe a virtual keyboard will work best, along with an output device that projects an image directly onto the user's retina. The ultimate mobile interface might even be the integration of human and machine, with technology that is implanted under the skin to detect the user's every intent and to automatically receive and process signals from the outside environment. Designing a usable m-commerce application using current technologies is difficult at best. Sometimes the most successful approach will involve waiting for a technology that better fits the application (and user) to be developed.

REFERENCES

Addlesee, M., Curwen, R., Hodges, S., Newman, J., Steggles, P., Ward, A., & Hopper, A. (2001). Implementing a sentient computing system. *Computer, 34*(8), 50–56.

Bernard, M., Chaparro, B., & Russell, M. (2000). Is RSVP a solution for reading from small displays? *Usability News, 2*(2). Available at http://psychology.wichita.edu/surl/usabilitynews/2S/rsvp.htm (accessed on 6/29/02).

Brewster, S., Leplâtre, G., & Crease, M. (1998). Using non-speech sounds in mobile computing devices. In Johnson, C. (Ed.), *Proceedings of the First Workshop on Human Computer Interaction for Mobile Devices*. Scotland: University of Glasgow. Available at http://www.dcs.gla.ac.uk/~johnson/papers/mobile/HCIMD1.html (accessed on 6/29/02).

Broderick, M., Gibson, V., & Tarasewich, P. (2001). Electronic signatures: They're legal, now what? *Internet Research, 11*(5), 423–434.

Buyukkokten, O., Garcia-Molina, H., Paepcke, A., & Winograd, T. (2000). Power browser: Efficient web browsing for PDAs. In *Proceedings of CHI 2000,* 430–437.

Chan, S. S. & Fang, X. (2001). Usability issues in mobile commerce. In Strong and Straub (Eds.), *Proceedings of the Seventh Americas Conference on Information Systems*. Atlanta, Georgia: Association for Information Systems, 439-442.

Chittaro, L. & Cin, P. D. (2001). Evaluating interface design choices on WAP phones: Single-choice list selection and navigation among cards. In Dunlop and Brewster (Eds.), *Proceedings of Mobile HCI 2001: Third International Workshop on Human-Computer Interaction with Mobile Devices*. Available at: http://www.cs.strath.ac.uk/~mdd/mobilehci01/procs/ (checked on 6/29/02).

De Bruun, O., Spence, R., & Chong, M. Y. (2001). RSVP browser: Web browsing on small screen devices. In Dunlop and Brewster (Eds.), *Proceedings of Mobile HCI 2001: Third International Workshop on Human-Computer Interaction with Mobile Devices*. Available at: http://www.cs.strath.ac.uk/~mdd/mobilehci01/procs/ (accessed on 6/29/02).

Dertouzos, M. (1999). The oxygen project: The future of computing. *Scientific American, 281*(2), 52–55.

Deshe, O. & Van Laar, D. (1999). Applying perceptual layers to colour code information in hand-held devices. In Brewster & Dunlop (Eds.), *Proceedings of the Second Workshop on Human Computer Interaction with Mobile Devices*. Available at: http://www.dcs.gla.ac.uk/~mark/research/workshops/mobile99/ (accessed on 6/29/02).

De Vet, J. & Buil, V. (1999). A personal digital assistant as an advanced remote control for audio/video equipment. In Brewster & Dunlop (Eds.), *Proceedings of the Second Workshop on Human Computer Interaction with Mobile Devices*. Available at: http://www.dcs.gla.ac.uk/~mark/research/workshops/mobile99/ (accessed on 6/29/02).

Dunlop, M. & Crossan, A. (1999). Dictionary based text entry method for mobile phones. In Brewster & Dunlop (Eds.), *Proceedings of the Second Workshop on Human Computer Interaction with Mobile Devices*. Available at: http://www.dcs.gla.ac.uk/~mark/research/workshops/mobile99/ (accessed on 6/29/02).

Fukumoto, M. & Tonomura, Y. (1997). Body coupled finger-ring: Wireless wearable keyboard. *Proceedings of the 1997 ACM Conference of Computer–Human Interaction*, 147-154.

Ghosh, A. K. & Swaminatha, T. M. (2001). Software security and privacy risks in mobile e-commerce. *Communications of the ACM, 44*(2), 51–57.

Goldstein, M., Book, R., Alsiö, G., & Tessa, S. (1999). Non-keyboard QWERTY touch typing: A portable input interface for the mobile user. *Proceedings of the 1999 ACM Conference of Computer-Human Interaction, 32*–39.

Goldstein, M., Öqvist, G., Bayat-M, M., Ljungstrand, P., & Björk, S. (2001). Enhancing the reading experience: Using adaptive and sonified RSVP for reading on small displays. In Dunlop and Brewster (Eds.), *Proceedings of Mobile HCI 2001: Third International Workshop on Human-Computer Interaction with Mobile Devices*. Available at: http://www.cs.strath.ac.uk/~mdd/mobilehci01/procs/ (accessed on 6/29/02).

Gomes, P., Tostão, S., Goncalves, D., & Jorge, J. (2001). Web clipping: Compression heuristics for displaying text on a PDA. In Dunlop and Brewster (Eds.), *Proceedings of Mobile HCI 2001: Third International Workshop on Human-Computer Interaction with Mobile Devices*. Available at: http://www.cs.strath.ac.uk/~mdd/mobilehci01/procs/ (accessed on 6/29/02).

Graham, R. & Carter, C. (1999). Comparison of speech input and manual control of in-car devices while on-the-move. In Brewster & Dunlop (Eds.), *Proceedings of the Second Workshop on Human Computer Interaction with Mobile Devices*. Available at: http://www.dcs.gla.ac.uk/~mark/research/workshops/mobile99/ (accessed on 6/29/02).

Herstad, J., Thanh, D. V., & Kristoffersen, S. (1998). Wireless markup language as a framework for interaction with mobile computing communication devices. In Johnson, C. (Ed.), *Proceedings of the First Workshop on Human Computer Interaction for Mobile Devices*. Scotland: University of Glasgow. Available at http://www.dcs.gla.ac.uk/~johnson/papers/mobile/HCIMD1.html (accessed on 6/29/02).

Holland, S. & Morse, D. R. (2001). Audio GPS: Spatial audio in a minimal attention interface. In Dunlop and Brewster (Eds.), *Proceedings of Mobile HCI 2001: Third International Workshop on Human-Computer Interaction with Mobile Devices*. Available at: http://www.cs.strath.ac.uk/~mdd/mobilehci01/procs/ (accessed on 6/29/02).

Johnson, P. (1998). Usability and mobility; interactions on the move. In Johnson, C. (Ed.), *Proceedings of the First Workshop on Human Computer Interaction for Mobile Devices*. Scotland: University of Glasgow. Available at http://www.dcs.gla.ac.uk/~johnson/papers/mobile/HCIMD1.html (accessed on 6/29/02).

Jones, M., Marsden, G., Mohd-Nasir, N., Boone, K., & Buchanan, G. (1999). Improving web interaction on small displays. *Computer Networks, 31*, 1129–1137.

Karlsson, P. & Djabri, F. (2001). Analogue styled user interfaces: An exemplified set of principles intended to improve aesthetic qualities in use. In Dunlop and Brewster (Eds.), *Proceedings of Mobile HCI 2001: Third International Workshop on Human-Computer Interaction with Mobile Devices*. Available at: http://www.cs.strath.ac.uk/~mdd/mobilehci01/procs/ (accessed on 6/29/02).

Koppinen, A. (1999). Design Challenges of an In-Car Communication System UI. In Brewster & Dunlop (Eds.), *Proceedings of the Second Workshop on Human Computer Interaction with Mobile Devices*. Available at: http://www.dcs.gla.ac.uk/~mark/research/workshops/mobile99/ (accessed on 6/29/02).

Kristoffersen, S. & Ljungberg, F. (1999). Designing interaction styles for a mobile use context. In Gellersen, H. W. (Ed.) *Handheld and Ubiquitous Computing, First International Symposium (HUC '99)*. Berlin, Germany: Springer-Verlag, 281-288.

Lewis, J. R. (1999). Input rates and user preferences for three small-screen input methods: Standard keyboard, predictive keyboard, and handwriting. *Proceedings of the Human Factors and Ergonomics Society 43rd Annual Meeting*, 425-428.

Lewis, J. R., LaLomia, M. J., & Kennedy, P. J. (1999). Evaluation of typing key layouts for stylus input. In *Proceedings of the Human Factors and Ergonomics Society 43rd Annual Meeting* (420-424).

Long, Jr., A. C., Landay, J. A., & Rowe, L. A. (1997). PDA and gesture use in practice: Insights for designers of pen-based user interfaces. *Technical report UCB//CSD-97-976*. Berkeley, California: University of California Berkeley.

Long, Jr., A. C., Landay, J. A., Rowe, L. A., & Michiels, J. (2000). Visual similarity of pen gestures. In *Proceedings of CHI 2000* (360-367).

MacKenzie, I. S. & Chang, L. (1999). A performance comparison of two handwriting recognizers. *Interacting with Computers, 11*, 283-297.

MacKenzie, I. S. & Zhang, S. X. (1999). The design and evaluation of a high-performance soft keyboard. *Proceedings of the 1999 ACM Conference of Computer-Human Interaction* (25-31).

Masui, T. (1999). POBox: An Efficient Text Input Method for Handheld and Ubiquitous Computers. In Gellersen, H.-W. (Ed.), *Handheld and Ubiquitous Computing, First International Symposium (HUC '99)* (289-300). Berlin, Germany: Springer-Verlag.

Melchior, M. (2001). Perceptually Guided Scrolling for Reading Continuous Text on Small Screen Devices. In Dunlop and Brewster (Eds.), *Proceedings of Mobile HCI 2001: Third International Workshop on Human-Computer Interaction with Mobile Devices*. Available at: http://www.cs.strath.ac.uk/~mdd/mobilehci01/procs/ (accessed on 6/29/02).

Nah, F. F. & Davis, S. (2001). Research Issues in Human-Computer Interaction in the Web-Based Environment. In Strong and Straub (Eds.), *Proceedings of the Seventh Americas Conference on Information Systems* (1332-1334). Atlanta, Georgia: Association for Information Systems.

Olsen, D. R. (1998). Interacting in Chaos. *Proceedings of the ACM Second International Conference on Intelligent User Interfaces* (97).

Pascoe, J., Ryan, N., & Morse, D. (2000). Using While Moving: HCI Issues in Fieldwork Environments. *ACM Transactions on Human-Computer Interaction, 7*(3), 417-437.

Pascoe, J., Ryan, N., & Morse, D. (1999). Issues in developing context-aware computing. In Gellersen, H.-W. (Ed.), *Handheld and Ubiquitous Computing, First International Symposium (HUC '99)* (208-221). Berlin, Germany: Springer-Verlag.

Rahman, T. & Muter, P. (1999). Designing an interface to optimize reading with small display windows. *Human Factors, 41*(1), 106-117.

Ralph, D. & Aghvami, H. (2001). Wireless application protocol overview. *Wireless Communications and Mobile Computing, 1*(2), 125-140.

Rist, T. & Brandmeier, P. (2001). Customizing graphics for tiny displays of mobile devices. In Dunlop and Brewster (Eds.), *Proceedings of Mobile HCI 2001: Third International Workshop on Human-Computer Interaction with Mobile Devices.* Available at: http://www.cs.strath.ac.uk/~mdd/mobilehci01/procs/ (accessed on 6/29/02).

Rosson, M. B. & Carroll, J. M. (2002). *Usability Engineering: Scenario-Based Development of Human-Computer Interaction.* San Francisco: Academic Press.

Saha, S., Jamtgaard, M., & Villasenor, J. (2001). Bringing the wireless Internet to mobile devices. *Computer, 34*(6), 54–58.

Schmidt, A. (1999). Implicit human computer interaction through context. *Personal Technologies, 4*(2, 3), 191–199.

Sears, A. & Arora, R. (2001). An evaluation of gesture recognition for PDAs. *Proceedings of HCI International 2001.* 1–5.

Shneiderman, B. (1998). *Designing the User Interface: Strategies for Effective Human-Computer Interaction* (3rd ed.). Reading, MA: Addison-Wesley.

Silfverberg, M., MacKenzie, I. S., & Korhonen, P. (2000). Predicting text entry speed on mobile phones. *Proceedings of CHI 2000, 9–16.*

Tarasewich, P., Nickerson, R., & Warkentin, M. (2002). Issues in mobile e-commerce. *Communications of the AIS, 8,* 41–64.

Tarasewich, P., & Warkentin, M. (2002). Information everywhere. *Information Systems Management, 19(1),* 8–13.

Väänänen-Vainio-Mattila, K. & Ruuska, S. (1998). User needs for mobile communication devices: Requirements gathering and analysis through contextual inquiry. In Johnson, C. (Ed.), *Proceedings of the First Workshop on Human Computer Interaction for Mobile Devices.* Scotland: University of Glasgow. Available at http://www.dcs.gla.ac.uk/~johnson/papers/mobile/HCIMD1.html (accessed on 6/29/02).

Walker, A. & Brewster, S. (1999). Spatial audio in small screen device displays. *Personal Technologies, 4*(2&3), 144–154.

Weiss, S., Kevil, D., & Martin, R. (2001). *Wireless Phone Usability Research.* New York: Useable Products Company.

Zha, Y. & Sears, A. (2001). Data entry for mobile devices using soft keyboards: Understanding the effect of keyboard size. In *Proceedings of HCI International 2001,* 16–20.

Chapter III

Location Based Services: Locating the Money

Kirk Mitchell and Mark Whitmore
Webraska Mobile Technologies, Australia

ABSTRACT

Location based services (LBS) are considered by some to be the 'golden child" of wireless data services and one of the few areas where users would be willing to pay a premium for usage. Mobile Operators however are yet to be convinced, and despite acknowledging location services as strategic, have not considered it a priority. Recent LBS deployments however focusing on a holistic view of user behaviour are showing positive signs of success. These deployments focus on providing services that integrate different content from multiple sources to provide users with a coherent and logically connected flow of application options. These applications are called "Find it, Route it, Share it & Buy it. Importantly this model maximises return on investment (ROI) by motivating user to undertake multiple transactions. The challenge for those within the LBS industry is to convince mobile operators that LBS is viable and can deliver a strong ROI. Indeed, the future success of LBS is as much dependant on locating the money as it is about locating the subscribers.

INTRODUCTION

The ability to communicate across a cellular network has had a significant impact on the way both individuals and businesses undertake daily tasks to the point where today many people are totally dependent on the mobile phone. The promise of Location Based Services (LBS) has the potential to further revolutionise consumer and commercial activity, however like many good ideas, the potential may not be realised.

LBS is considered by some to be the 'golden child" of wireless data services and one of the few areas where users would be willing to pay a premium for usage. Schema estimates that location sensitive services could generate US$30 to $40 additional yearly revenue per user by 2005 and US$100 by 2010. Mobile operators however are yet to be convinced, and despite acknowledging location services as strategic, have not considered it a priority.

The challenge for those within the LBS industry is to convince mobile operators that LBS is viable and can deliver a strong Return on Investment (ROI). Indeed, the future success of LBS is as much dependant on locating the money as it is about locating the subscribers.

BACKGROUND

Mobile Phone Operators Are Under Pressure...

An important measure of value within the mobile telecommunications sector is Average Revenue Per User or ARPU. To date, ARPU has almost exclusively been generated from voice-related services. As the mobile telecommunications sector matures, a combination of factors are causing ARPU to steadily decline; these include:

- Increased market penetration of low value market segments,
- Increased market competition
- Fixation by the investor community on customer acquisition not customer spend as a measure of value
- Inability of Mobile operators to unlock new revenue streams

Until recently the decline in ARPU has been masked by double-digit user growth rates within the mobile telecommunications sector. According to the US-based company, Strategic Consulting, global user growth rates will decline from 50 to 15% by 2002 and will continue to decline to single-digit growth rates thereafter.

Seduced by the dot.com euphoria and buoyed by the success of the Japanese operator NTT DoCoMo, many operators believed wireless data was the solution to their ARPU problems. Wireless data services, however, have been slow to take off. Many claim a lack of data-enabled handsets has hindered take up while others believe that operators simply don't have a compelling business model. One thing for certain is that many operators made the mistake of comparing wireless data services with those services available on the web and commenced to migrate readily available fixed content to wireless devices. Experience has since shown that only those services that are uniquely mobile and relevant to wireless users will create value and lead to mass adoption.

Wireless Data Services Need To Be Uniquely Mobile

A PC provides a rich multimedia interface that allows users to efficiently process large amounts of information. This information is generally a combination of both offline and online interactions and typically includes Internet searching or surfing. Access to the Internet via the PC is therefore one dimensional, that is, the "What" of information. Internet access via the mobile phone is uniquely different as it encompasses a distinctly broader set of dimensions. These dimensions include:

- The "Where" of location
- The "When" of time
- The "Who" of personalisation

While multi dimensional, the mobile phone poses a number of limitations that shape its use today. These include:

- A small, low resolution screen
- A monochrome display
- Limited graphics, primarily text-based interface
- Keypad as the only method for data entry
- Limited offline processing and storage capabilities
- Cost of access

The challenge for mobile operators and the industry is to provide users access to the multi-dimensional elements of wireless data through an interface and device which to date has been optimised for voice communication.

LOCATION-BASED SERVICES DEFINED

What is LBS?

Location-based services can simply be defined as *mobility services that exploit the derived location of a user (specified by user, network or handset) to provide services that have a geographic context.*

Why is LBS uniquely mobile?

Location defines and uniquely differentiates the mobile device. As a result, services that exploit location information are by definition uniquely mobile and typically incorporate the following elements:

- *Convenience* in that they are available through portable devices (i.e. mobile phones) which are more than likely carried by users when they are lost or are seeking information about their immediate situation;

- **Personalised** by their very nature in encouraging users to better define their mobile activity (monitor the traffic on my daily route and notify me of any incidents);
- **Relevant** simply through filtering only that information proximal to a users location and simplifying an already complex user experience; and
- **Real Time** in their ability to link dynamic content relevant to a user's context (i.e. closest cheapest petrol station).

What are the Key Drivers for LBS?

The motivation to deploy LBS to date has been:

- **Regulatory:** In the US, the Federal Communication Commission (FCC) has mandated that all operators be able to locate the position of any emergency call placed by a mobile phone with an accuracy of 125 meters (~40 feet). In Europe, the European Commission is considering a recommendation by the EU Communication Review that location for emergency purposes become mandatory by January 2003.
- **Technology:** Location is one of the key enablers for many new services.
- **Competitive Advantage:** Operators and service providers have tried to capture and secure "green field" opportunities by being first to market.
- **Revenue:** Unlock new revenue streams by charging for location information and services.
- **Efficiencies:** Many corporations have deployed LBS in order to reduce costs or run business activities more effectively.

What are the Key Barriers Delaying Deployment?

- **Operator Debt Levels:** Overpayment on spectrum and continued revenue pressure has seen many operators slow down deployment of LBS to cut capital expenditure.
- **UMTS Availability:** Some operators are deliberately delaying investment in location technology preferring instead to wait for UMTS (i.e., 3G) deployments, which are scheduled to start more broadly in 2003.
- **Limited Standards:** To date there are no standards defined for location determination technology, although forums like the Location Interoperability Forum (LIF) are working to remedy this.
- **Pre-occupation with Precision:** Some operators question the value of LBS due to the low accuracy of many mobile device positioning solutions
- **Privacy Concerns:** Operators are vulnerable to public criticism and are sensitive to deploying services that enhance their "Big Brother" image.
- **Business Model Paralysis:** Indecision by operators on whether to keep location for themselves or release to others is causing delays.

LOCATING THE PERFECT BUSINESS MODEL

The industry benchmark for the successful deployment of wireless Internet services is the highly successful DoCoMo i-mode service. Launched on February 22, 1999, i-mode has over 40 million users and is the second largest ISP in the world after AOL Time Warner.

The success of i-mode is due to the following characteristics:

- Intensive marketing campaign targeting the youth segment
- Existing packet-based network and low-cost usage charges
- Availability of good quality handsets
- Strong revenue share to i-mode partners
- Control over handset design and production quantities
- Low, fixed Internet penetration

However, despite the success of the i-mode in Japan, it has been spectacularly unsuccessful when deployed in other countries. This is evident where DoCoMo has attempted to apply its model through equity partners and where operators have attempted to replicate i-mode themselves. This failure has been due to a combination of factors including:

- A lack of data-friendly handsets
- Poor delivery of the service over circuit-based cellular infrastructure
- And last, and very importantly, the lack of a critical mass of multimedia developers who were able to provide sufficient quantities of locally developed content

A New Model Is Emerging

Realising the limitations of i-mode, leading operators have embarked on the development of a model of their own. This model is characterised by the move from a "one service-one content" model to a "multiple content–multiple service" model.

The i-mode model involves the aggregation of individual applications from different partners. Each partner has their own application limited to the content they can provide. The limitation of this model is that, as users move between applications, they must re-enter information to maintain service continuity. To overcome this, operators have moved to a model of providing services that integrate different content from multiple sources. This provides users with a coherent and logically connected flow of services. Importantly this model maximises ROI by motivating the user to undertake multiple transactions.

Commoditize and Destroy

To maximise operator revenue, the i-mode model encouraged the aggregation of large numbers of application partners. The impact of this is that the most popular services can get lost among the less popular services, preventing them from achieving their full potential. Commoditize and Destroy involves promoting the most popular services or content to the top menu system and removing those services not being used. The graph in Figure 1 below shows the distribution of usage across a range of search topics throughout Western Europe. The graph clearly shows the concentration of usage around the top six to seven main categories. Rather than attempting to replicate the Internet, which is about content breadth, operators are finding that the best model is to provide a subset of highly targeted services.

Own Rather Than Share

A shortcoming of the i-mode model is that the operator is sharing revenue on services that are highly profitable. Rather than adopting the i-mode model of "sharing," operators are instead electing to "own" best-of-breed applications. Owning the applications is not meant to imply operators are becoming application developers, but rather licensing and re-branding best of breed applications. Owning applications also gives operators the opportunity to further maximise returns by directly investing in content quality and ensuring consistency in the user interface across the application suite. Operators are accommodating the lack of any particular service by providing powerful search engines.

Functional Rather Than Entertainment

An important distinction between i-mode and this new emerging model is the shift away from the heavy reliance on entertainment applications to functional applications. This has removed the requirement for operators to develop a thriving

Figure 1: October 2001 Webraska Reporting: Usage Patterns in Western Europe

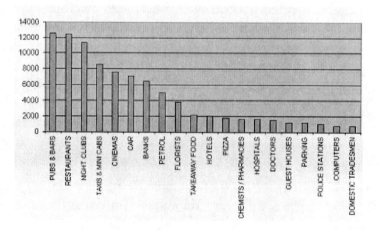

local multimedia industry in order to provide compelling services. Operators can now focus on functional mobile specific applications, an important component of which is location. Functional applications are simple information services that relate to users who are out and about. Functional applications typically enable users to answer the questions. Where is the nearest ATM? What cinema is a particular movie showing? How do I get across town?

REAL-WORLD USER BEHAVIOUR

Wireless Internet users typically do not want to simply find a restaurant—they want to find the nearest restaurant by cuisine type, obtain directions on how to get there, invite their friends and make a reservation. This holistic view of user behaviour relates to four areas: finding the object of interest, determining the route to get to that interest point, sharing information about this point of interest with other friends or associates, and completing a transaction related to the selected location. Webraska has developed (see Figure 2) a platform that integrates these functions and applications and supports the user through each point in the process; these applications are called "Find it, Route it, Share It, Buy it."

Finding

With so much information around, 'Finding' is becoming a commodity. The introduction of location enables users to access spatially relevant and time-sensitive information. Webraska uses its SmartZone Platform to provide this capability. This

Figure 2: Webraska User Interaction Model

platform enables places to be located on a map and subsequently identified in order of accessibility—where accessibility may be nearest place by driving or walking. By relating information to a map, it is possible to take into account both natural and man-made boundaries—so the search will cater for the existence of rivers, highways, one-way streets, dead-ends and so on.

The ability of LBS to intelligently consider real-life elements ensures the delivery of meaningful services and avoids the situation of the user having to scale a mountain or swim a river to get to the nearest Chinese restaurant. Equally important to spatial relevance is time relevance--there is no point being guided to the nearest car park if there are no spaces, or directed to a nearest restaurant only to find it closed or booked out.

Several firms have developed various tools to assist users in using *finding* what they need, when they need it. For example, Webraska has develop a range of finder applications, including *Hotel Finder, Restaurant Finder, ATM/Cash Machine Finder, Event Finder, Movie Finder* and *Flight Finder*. These applications have in common their focus on providing customers with real-time access to information about location-specific resources that are relevant to their business and leisure needs.

Routing

Once the user has selected a particular destination, typically their next requirement is for detailed directions based on their current location. To do this, an application must utilize a cartographic database that comprises tables of information and attributes that together represent the road network. To geo-locate objects, each element or road segment in the database must be given a longitude and latitude coordinate (commonly referred to as X,Y) so that it can be referenced in a standard way. To be useful, a dataset must also include points of interest or landmarks (i.e., railway stations, shopping centres, park, etc.) that are each given an X,Y coordinate.

A location service must be capable of transforming the X,Y coordinates of the user's current location into a street address or a point on a map. A number of mathematical algorithms then calculate the optimum path based on the interconnecting road network between the user's location and nominated destination. An important element of this calculation is the consideration of the prevailing traffic conditions. In this way, the information that would be given to the user represents as closely as possible their real-life experience.

Webraska has developed a range of routing applications, some of which include *Driving Directions (DD),* which allow consumers to retrieve driving directions based on the shortest drive time, *Walking Directions (WD),* which provides pedestrian specific directions (e.g., one-way streets can be utilised for walking, while freeways and highways cannot be used by pedestrians), and *Finder Maps,* which enable users to plot a location or route overlayed on a map.

Sharing

Sharing information is fundamental to the way we live our lives. However, by the very nature of being mobile, it can be difficult to share information particularly to multiple people. Today's WAP or SMS phones currently lack the ability to store (independent of the handset) and/or publish information to either a private group or public group (a private group maybe a set of friends, whereas a public group is akin to a newsgroup on the Internet—a forum for public discussion).

Webraska has developed tools that can be used to publish or store information for friends or associates. This information is cached in the form of an m-note, which offers a structured message format by which users can pull, push, share and search for information. There are two types of m-notes: 1) *m-Vite,* which acts as a structured mobile invitation application that allows users to use predefined invitation templates to post and retrieve mobile invitations, and 2) *m-Classifieds*, which allows users to post and retrieve classified announcements.

Buying

It is important to distinguish between mobile commerce and location commerce. Mobile commerce can be defined as anything purchased through a mobile device, whereas location commerce can be defined as the purchase of goods made as a direct result of information received on the mobile device related to the user's position. Location commerce may occur through a physical store or via a mobile device. For example, a music CD purchased through Amazon's wireless site is mobile commerce, whereas the purchase of a meal due to an LBS advertising coupon announcing a special lunchtime price is Location Commerce. Location is a key factor behind most consumer and business purchasing decisions. Once effective mobile payment platforms are commonplace, LBS will provide an effective stimulant to commerce transactions.

Webraska has developed both mobile commerce and location commerce applications. One of these is *Movie Tickets, which is a mobile commerce application*, that allows users to use their mobile phones to not only search and find the nearest cinema showing a particular movie, but also purchase movie tickets from their mobile phone. A second application is *M-coupon, which is a location commerce application* that allows mobile consumers to be alerted of deals on products and goods across a range of predefined shopping categories, brands and stores.

DEPLOYING LBS SERVICES

The LBS industry can be a confusing place. There are a large number of companies all claiming market leadership, yet it is often unclear exactly what they provide. As companies fight to secure a position in the LBS value chain many are producing product-marketing material that is remarkably similar. This confusion, however, is not unexpected, and while the search for a successful business model continues, companies will continue to blur the line between their activities and those of potential partners.

Despite this confusion a number of core components have emerged as essential for any commercial LBS deployment; these are shown in Figure 3 and include the following;

- Handset Location Measurement Technology
- Location Information Management Platform
- Application/Geo-Spatial Platform
- Applications

Locating the Users

Techniques for positioning the location of the handset and ultimately the user can be broken into three main categories:

- User defined
- Network defined
- Handset defined

Figure 3: Core LBS Components

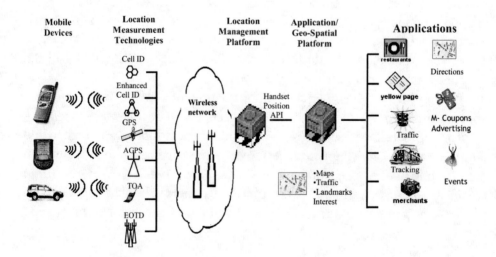

User Defined

The majority of early location service deployments had to make do without the benefit of the handset being locatable by the network operator. In such services, the user had to define their own position. This can be done by the user entering a street name, city or postcode. Webraska also offers the ability for users to store predefined address information on a server and to recall this information through easily remembered names such as "Work," "Girl Friend," "Home," etc.

In Japan, J-Phone's J-Navi service currently utilizes no positioning technology. Users locate themselves by entering an address or landmark within any major Japanese city. The J-Navi service will then locate businesses within a 100m radius of the entered position and provide business address, phone number and product info. The service will also display the information overlayed on a map accessible through the WAP browser of the mobile phone. 30% of J-Navi requests involve a location other than the user's current one, as people use the service to find out information in advance.

Network Defined

Mobile phone networks are composed of thousands of base stations, with each base station covering an area called a cell. Technologies based on Cell ID use proximity to a base station to determine mobile handset position.

Cell ID information is already supplied in the GSM radio network, which therefore makes this method very inexpensive. The positioning accuracy depends on the cell size or the number of base stations in a given area. As urban areas generally have a higher density of base stations, urban areas provide higher accuracy services than in rural areas. In short, accuracy can range from 10m (where micro cells are installed, such as in an airport) up to 2km. In a typical urban area, the achievable accuracy is often sufficient for proximity services, such as finding the nearest restaurant.

To improve the accuracy of handset positioning for Cell ID systems, two different techniques can be utilised, these include Angle of Arrival (AOA) or Time of Arrival (TOA).

Angle of Arrival (AOA)

The AOA method calculates the angles at which a signal arrives at two base stations. Once the angles are obtained, a simple triangulation calculation is performed to determine the coordinate solution.

Time of Arrival (TOA)

TOA method involves listening to the handover access burst across three or more mobile base stations. The base stations measure the difference in the time of arrival of data at each station. This information is then compared with the time reading from a GPS absolute time clock, and a simple calculation is performed to determine the coordinate solution.

Although utilisation of AOA and TOA can increase positioning accuracy by up to 50%, both require base station modifications making them costly to implement. Neither AOA or TOA require handset modifications.

Handset Defined

The best-known method of location, Global Positioning System (GPS), has been used in vehicle navigation systems and dedicated hand held devices for some time. Provided that at least 3 satellites can be tracked, GPS is capable of positioning objects to a resolution of within 1m. However because GPS signals are weak they suffer from attenuation (eg., by buildings, bridges, tunnels, etc.), reducing their suitability as a solution for urban environments. To overcome this, companies are developing enhanced GPS solutions that combine GPS signal information with cellular network handset positioning information.

One of these solution is known as network-assisted GPS (A-GPS). A-GPS involves installing fixed GPS receivers every 200km to 400km to fetch data to complement the readings by the mobile handset. With assistance from these fixed receivers, the mobile handset can make timing measurements without having to decode the GPS information, greatly enhancing calculation time. Measurement results are sent to a location information management platform, there the position of the handset is calculated using differential positioning.

Another emerging solution is Estimated Observed Time Difference (E-OTD). E-OTD relies solely on software in the mobile handset to listen to bursts from multiple base stations. The time difference in the arrival of these bursts is used to triangulate where the mobile handset is located. This method requires the exact location of the base stations to be known, and the sending of data from each base station to be tightly synchronised. Synchronisation is most often done through the use of fixed GPS receivers. Although potentially not as accurate as AGPS, E-OTD does not rely on a clear view of the sky.

Both A-GPS and E-OTD, despite the highly accurate positioning they provide, have not been widely implemented as they require modifications to both the network and handset making them prohibitively expensive to implement.

A Clear Winner Is Yet To Emerge

Whatever the merits of various handset-positioning technologies, the absence of a clear winner has made operators reluctant to invest heavily in higher accuracy solutions. Even if the technology they deploy is suitable, selecting one not picked by other operators, particularly in the same country, could be disastrous. Operators face the following issues if they move early and find they have selected a non standard solution;

- Loss of revenue as it will not be possible to offer LBS to inbound roaming customers
- Customers may not be able to buy handsets that support their LBS services
- Interoperability with other operator services may not be possible
- Limited vendor equipment, leading to higher prices through limited competition

The lack of a definitive business model and the real risk of technology obsolescence have resulted in the majority of global operators limiting their investment in handset positioning technology to Cell ID. In the U.S. operators were ordered by the FCC, for network-based solutions, to be capable of positioning the handset within 125m. It is interesting to note that many operators in the U.S. are prepared to pay the fines for noncompliance, rather than face the potential losses incurred by implementing the wrong solution.

It is often said that if bandwidth is a problem, then you are using the wrong application. Similarly, the same concept holds for LBS. Inadequate location granularity is only an issue for those applications which require accurate positioning (i.e., emergency services, driving directions).

Yet even these services can be implemented successfully using a Cell ID-based solution coupled with simple location refinement techniques (user defined position). Unfortunately too much of the debate today regarding LBS has been focused on the accuracy of different positioning technologies rather than application concepts and market requirements.

Turning Coordinates Into Value

Identifying the X,Y coordinates for the location of a user is only the first step. To create real value requires the interaction of a number of core systems; these include:

1. A Location Management Platform is required to ensure that only authorised applications have access user coordinate information.
2. A Geo-Spatial platform is required to convert the coordinates into location information (street address or point on a map).
3. An Application Platform is required to integrate the location information with content and business logic to create the LBS service.

The operation of the various system components is further illustrated in Figure 4; which shows the flow of commands in the case of a simple LBS application: Find the nearest Petrol station.

Figure 4: The Flow of Commands in the Case of a Simple LBS Application

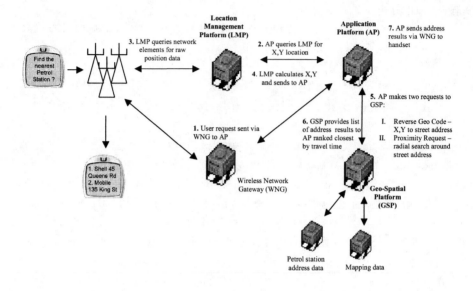

Motorists Will Have The Final Say

Motorists have always demonstrated a strong desire to access mobility services whilst driving. A recent survey by US based IDC suggested that as many as 69% of wireless calls were either placed or received in vehicles. Telematics, which until recently was essentially a niche industry delivering a limited range of RF (Radio Frequency) based commercial services to vehicles (i.e., fleet management, emergency response, taxi dispatchment, etc.), is now poised to enter the mass market and LBS is one of several technologies that will drive this transition.

Many industry players and analysts often find the overlap between LBS and telematics confusing. As a result, considerable 'double counting' has occurred which has resulted in considerably inflated market sizes. Essentially, LBS is just one of many telematics applications (others include vehicle diagnostics, broadband services, vehicle security, etc.). Regardless of which applications are eventually deemed necessary within a telematics portal, many believe that consumers must be able to access applications both within and outside the vehicle to achieve the economies of scale necessary to escalate mass adoption. Embedded devices have challenges with respect to product lead times (3–4 years), long vehicle replacement cycles and scalability. However, the technical challenges of retrofitting after market devices (such as a PDA) remains a significant challenge for non-embedded device bundles.

Telematics has also been known as the ultimate convergence. Automotive manufacturers once had the opportunity to become wireless operators back in the days when the mobile phone was known as the car phone and actually bolted in vehicles. Telematics has presented manufacturers with the opportunity to again consider whether they want to be the provider of mobility services and compete for subscriber ARPU. Will it be automotive manufacturers or wireless carriers who deliver telematics, or will there be a role for independent telematics service providers?

Noncommercial factors may also have a significant bearing on the adoption of telematics. How will the user interfaces be designed to prevent driver distraction and rest the safety concerns of manufacturers, regulatory bodies and consumers?

Whether LBS has a significant role to play within telematics portals will most likely again be determined by whether motorists are prepared to pay. To date, most mass telematics services which incorporate LBS are offered on a monthly subscription basis (General Motors offered their US 'Directions and Connections' Onstar service for $34.95 p/mth). Whether motorists can justify this price over simply purchasing a street directory is a major challenge for the LBS industry, and the requirement to clearly differentiate LBS, such as providing real-time traffic information, is paramount. Whether it is through gaining incremental revenue or through saving costs on such items as insurance or vehicle maintenance, LBS is certain to be a critical factor in the ability of telematics to grow from a niche enterprise offering to a mass market service.

CONCLUSION

LBS is without a doubt one of the most exciting developments to emerge from the mobile telecommunications sector. It will enable the mobile phone to evolve into a true personal co-pilot device. However, before the full benefits of LBS can be realised there remains a number of significant barriers to overcome, the most significant of which is a lack of standards.

In Europe we are seeing early deployment of LBS yielding positive user acceptance and providing operators with a strong ROI. The success of early LBS deployments is incredibly important, not just to those in the LBS business, but also the entire mobile telecommunications sector. The failure of LBS to live up to market expectations, on top of the failure of WAP and GPRS, could provide a knock out blow from which the wireless data industry will take years to recover.

The challenge for those within the LBS industry is to convince mobile operators that LBS provides the key enabler to stimulate usage of wireless data applications and that this in turn provides mobile operators with the potential to locate new revenue streams.

REFERENCES

Lawrence, J. & Leung, M. (2001). *Telematics and Location-Based Services: Collateralization*. Dain Rauscher Wessels Market Research Report, (April).

Green, J., Betti, D. & Davison, J. (2000). *Mobile Location Services, Market Strategies*. Ovum Report.

Glover, T. (2001). Getting through, *Communications International*, (October), 25–31.

Baines, S. (2002). Eyes Wide Shut, *Communications International*, (January), 19–23.

Koh, J. & Kim, YM (2000). *Mobile Internet Applications Primer*, UBS Warburg, Global Equity Research Report. (August).

Scuka, D. (2001). Made in Japan. *Mobile Communications International*, (October), 38–46.

Chapter IV

Towards a Classification Framework for Mobile Location Services

George M. Giaglis
University of the Aegean, Greece

Panos Kourouthanassis and Argirios Tsamakos
Athens University of Economics and Business, Greece

ABSTRACT

The emerging world of mobile commerce is characterized by a multiplicity of exciting new technologies, applications, and services. Among the most promising ones will be the ability to identify the exact geographical location of a mobile user at any time. This ability opens the door to a new world of innovative services, which are commonly referred to as Mobile Location Services (MLS). *This chapter aims at exploring the fascinating world of MLS, identifying the most pertinent issues that will determine its future potential, and laying down the foundation of a new field of research and practice. The contribution of our analysis is encapsulated into a novel classification of mobile location services that can serve both as an analytical toolkit and an actionable framework that systemizes our understanding of MLS applications, underlying technologies, business models, and pricing schemes.*

INTRODUCTION

The term 'mobile era' as a characterization of the 21st century can hardly be considered as an exaggeration (Kalakota and Robinson, 2001). In times where mobile phone penetration is well above the 50% mark in some countries, and has even surpassed fixed line penetration in a few cases (Nokia, 2001), it is not surprising that *wireless applications* are claiming much of the industrial, academic, and even popular media attention.

Mobile (or wireless) applications, despite potentially being very different in nature from each other, all share a common characteristic that distinguishes them from their wireline counterparts: they allow their users to move around while remaining capable of accessing the network and its services. With the ability of mobility, *location identification* has naturally become a critical attribute, as it opens the door to a world of applications and services that were unthinkable only a few years ago (May, 2001). The term *Mobile Location Services (MLS)* (or *Location Based Services – LBS*, as they are commonly referred to) has been coined to group together all those applications and services that utilize information related to the geographical position of their users in order to provide value-adding services to them.

MLS is perhaps the latest entrant to the world of mobile applications, and hence limited work to date has addressed its real potential and implications. For the most part, perhaps with the exception of a few in-car services, most MLS applications are still at a trial stage with service definitions, revenue models, pricing, and business relationships largely undefined (UMTS, 2001c). However, the market promises to be lucrative. According to the UMTS Forum, the worldwide size of the mobile location services market is expected to increase from US$0.7 billion in 2003 to US$9.9 billion in 2010 (UMTS, 2001b).

One of the main enablers of MLS proliferation in late years was the 1999 mandate of the US Federal Communications Commission (FCC) that, by October 2001, emergency services should be able to automatically position any citizen dialing 911 to within 125 meters in two-thirds of cases. The reasoning behind this mandate is that people who are injured or in some other need do not necessarily know exactly where they are, and hence the emergency services should be able to locate them in an automatic way so that help can be sent out to them. This has placed a legal obligation on mobile networks to support location identification provision in their service portfolio. Given this legal obligation, many network providers have seized the opportunity to design and implement further mobile location services that will commercially exploit the ability to know the exact geographical location of a mobile user.

From the consumer/citizen point of view, the peace of mind provided by a cellular phone that is capable of revealing their exact position in case of an emergency is usually welcomed. However, this is not necessarily the case when considering additional mobile location services, such as personalized mobile advertisements. The emergence of MLS has paved the way for innovative marketing strategies that, unless carefully monitored and exercised, can become extremely intrusive and even jeopardize the privacy of a mobile user's personal data. Hence, besides their potential benefits, mobile location services open up a number of additional issues (privacy protection being perhaps the most critical of them) that need to be examined and managed.

Moreover, the complexity of studying MLS is further exacerbated by the fact that most marketable services do not come as a direct consequence of our ability to identify someone's location through a mobile device, but rather through combining location identification with additional data to provide added value to the user (Dix et al., 2000). For example, just knowing a mobile user's location may not be sufficient in order to assist them in locating a suitable nearby restaurant. Such service provision requires the user's location information to be combined with knowledge of the exact locations of local restaurants, navigation routes between points, and, perhaps more importantly, knowledge of the user's dietary preferences. It is evident that location identification is only one of several elements (technological, contextual, and others) that need to be combined to provide innovative and compelling services to the user (Long et al., 1996). The study of these elements is the main scope of this chapter.

The chapter is organized in to a number of sections. The following section is primarily concerned with establishing a baseline understanding of potential marketable services and applications for mobile location services. The section that follows presents an overview of the most pertinent technological developments that have made the emergence of MLS possible. Next, alternative business models for the provision of MLS are discussed, together with an analysis of alternative pricing schemes for the services. Technological, service, and business model findings are then synthesized into a classification framework for MLS. Finally, a number of critical issues, such as consumer privacy and standardization, which are not covered within the framework, are presented before discussing some final concluding thoughts regarding future research on MLS.

MLS APPLICATIONS AND SERVICES

Emergency Services

Perhaps the clearest market application of MLS, as already discussed in the previous section, is the ability to locate an individual who is either unaware of his/her exact location or is not able to reveal it because of an *emergency situation* (injury, criminal attack, and so on). Mobile location services are even applicable as a means of overcoming one of the most common problems of motorists, namely the fact that most often than not they are unaware of their exact location when their vehicle breaks down. The ability of a mobile user to call for assistance and at the same time automatically reveal their exact location to the *automotive assistance* agency is considered one of the prime motivators for signing up subscribers to mobile location services (Hargrave, 2000).

Navigation Services

Navigation services are based on mobile users' needs for directions within their current geographical location. The ability of a mobile network to locate the exact position of a mobile user can be manifested in a series of navigation-based services:

a)   By positioning a mobile phone, an operator can let the user know exactly where they are as well as give him/her detailed *directions* about how to get to a desirable destination.

b)   Coupled with the ability of a network to monitor traffic conditions, navigation services can be extended to include destination directions that take account of current *traffic conditions* (for example, traffic congestion or a road-blocking accident) and suggest alternative routes to mobile users.

c)   The possibility to provide detailed directions to mobile users can be extended to support *indoor routing* as well. For example, users can be assisted in their navigation in hypermarkets, warehouses, exhibitions, and other information-rich environments to locate products, exhibition stands, and so on.

d)   Similarly, *group management* applications can be provided to allow mobile users to locate friends, family, coworkers, or other members of a particular group that are within close range and thus, create *virtual communities* of people with similar interests.

Information Services

Location-sensitive information services mostly refer to the digital distribution of content to mobile terminal devices based on their location, time specificity, and user behavior. The following types of services can be identified within this category:

a)   *Travel services* such as guided tours (either automated or operator-assisted), notification about nearby places of interest (monuments, etc.), transportation services, and other services that can be provided to tourists moving around in a foreign city.

b)   The application of *mobile yellow pages* that provide a mobile user, upon request, with knowledge regarding nearby facilities is another example of information services.

c)   *Infotainment services* such as information about local events, location-specific multimedia content, and so on, can also be provided to interested users.

Advertising Services

Mobile advertising is among the first trial applications of MLS, due to its promising revenue potential as well as its direct links to mobile commerce activities. Furthermore, mobile advertising has gained significant attention because of the unique attributes, such as *personalization* (Kalakota and Robinson, 2001), that offer new opportunities to advertisers to place effective and efficient promotions on mobile environments. There are various mechanisms for implementing mobile advertising coupled with MLS. Examples of mobile advertising forms include *mobile banners*, *alerts* (usually dispatched as SMS messages), and *proximity-triggered advertisements*.

Due to the potentially intrusive nature of mobile advertising services, it is generally acknowledged that users will have to explicitly 'opt in' or register to receive such services (UMTS, 2001c), perhaps in exchange for other benefits (for example, reduced call rates or special offers).

Tracking Services

Tracking services can be equally applicable both to the consumer and the corporate markets. As far as consumers are concerned, tracking services can be utilized to monitor the exact whereabouts of, for example, *children and elderly people*. Similarly, tracking services can be effectively applied in corporate situations as well. One popular example refers to tracking *vehicles* so that companies know where their goods are at any time. Vehicle tracking can also be applied to locating and dispatching an ambulance that is nearest to a given call. A similar application allows companies to locate their field *personnel* (for example, sales-people and repair engineers) so that they are able, for example, to dispatch the nearest engineer and provide their customers with accurate personnel arrival times. Finally, the newfound opportunity to provide accurate *product tracking* within the supply chain offers new possibilities to mobile supply chain management (m-SCM) applications (Kalakota and Robinson, 2001).

Billing Services

Location-sensitive billing refers to the ability of a mobile location service provider to dynamically charge users of a particular service depending on their location when using or accessing the service. For example, mobile network operators may price calls based on the knowledge of the location of the mobile phone when a call is made. *Location-sensitive billing* includes the ability to offer reduced call rates to subscribers that use their mobile phone when at their home, thereby allowing mobile operators to compete more effectively with their fixed telephony counterparts.

Table 1: A taxonomy of mobile location applications and services

SERVICES	EXAMPLES	ACCURACY NEEDS	APPLICATION ENVIRONMENT
EMERGENCY	Emergency calls	Medium to High	Indoor/Outdoor
	Automotive Assistance	Medium	Outdoor
NAVIGATION	Directions	High	Outdoor
	Traffic Management	Medium	Outdoor
	Indoor Routing	High	Indoor
	Group Management	Low to Medium	Outdoor
INFORMATION	Travel Services	Medium to High	Outdoor
	Mobile Yellow Pages	Medium	Outdoor
	Infotainment Services	Medium to High	Outdoor
ADVERTISING	Banners, Alerts, Advertisements	Medium to High	Outdoor
TRACKING	People Tracking	High	Indoor/Outdoor
	Vehicle Tracking	Low	Outdoor
	Personnel Tracking	Medium	Outdoor
	Product Tracking	High	Indoor
BILLING	Location-Sensitive Billing	Low to Medium	Indoor/Outdoor

ENABLING TECHNOLOGIES OF MLS

The applications and services that were discussed in the previous section are based on underlying technological capabilities that enable the identification of the location of a mobile device, thereby making the provision of MLS possible. Location technologies can be divided into two basic categories: *enabling* and *facilitating*. The former refer to the basic technologies that allow for obtaining location information from a mobile user, while the latter refer to complementary technologies that provide the contextual and/or infrastructural environment within which MLS can be implemented in a value added fashion (Johnson, 1998). Only the former category (enabling technologies) is dealt with in this section, while facilitating technologies are discussed later in the chapter.

A number of different enabling technologies exist, each with its own advantages and disadvantages. The basic technology assessment criteria refer to *coverage range, accuracy support,* and *application environment*. The enabling technologies can be further divided into two sub-categories. *Mobile network-dependent* technologies depend on the ability of a mobile device to receive signal from a mobile network covering its area of presence. Such technologies can naturally perform better in densely populated environments where network base stations are closer to each other. Conversely, *mobile network-independent* technologies can provide location identification information even in the absence of mobile network coverage. These technologies can be further divided into *long-range* and *short-range* ones, depending on their range of coverage, as discussed later in this section.

The most popular enabling technologies for mobile location services are discussed in this section.

Mobile Network-Dependent Technologies

Cell Identification (Cell-ID): The Cell-ID (or *Cell of Origin, COO*) method is the most basic manifestation of the ability to provide location services. The method relies on the fact that mobile networks can identify the approximate position of a mobile handset by knowing which cell site the device is using at a given time. The main benefit of the technology is of course that it is already in use today and can be supported by all mobile handsets. However, the accuracy of the method is generally low (in the range of 200 meters and even up to several tens of km), depending on cell size. Generally speaking, the accuracy is higher in densely covered areas (for example, cities) and much lower in rural environments. It is, however, expected that the advent of new generations of mobile networks (such as GPRS or UMTS—see below) will support smaller cell sizes and thus contribute to better accuracy than existing networks.

Time of Arrival (TOA): The TOA technology locates a mobile device by measuring the time it takes its signal to reach three different cell sites. This method is known by the term 'triangulation' to denote the need to have at least three different measurements from different cells in order to locate the Cartesian coordinates of a mobile device. Although the TOA method is fairly accurate (in the range of 10 to 100 meters), its major disadvantage lies on the investments that need to be employed by the network operator. More specifically, all the cell sites of the network have to be equipped with Location Measurement Units (LMUs), which additionally need to have their internal clocks synchronized by GPS (Global Positioning System) in order to effectively triangulate a user's location. Both the equipment and the synchronization process can be quite costly.

Observed Time Difference (OTD): The method is similar to TOA, although the OTD method places additional emphasis on the mobile device itself in order to reduce the necessary investment on the network side. The method is again based on the principle of triangulation; however, in this case it is the mobile device that measures the time it takes for a signal from three cell sites fitted with LMUs to reach it. The benefit of the method is that only a limited number of network cell sites have to be fitted with LMUs, thereby reducing the implementation cost significantly. The main drawback of the method is of course the need to have mobile devices that are capable of performing the necessary calculations, which would require significant investment on behalf of the end-users.

Long-Range Mobile Network-Independent Technologies

GPS and Assisted GPS (A-GPS): The Global Positioning System (GPS) was developed by the US military as a satellite-based mechanism for locating the exact Cartesian coordinates of a given target. The system is now also open for commercial use and is heavily employed by a number of communities (the maritime community being perhaps the most well-known example) that need to know their location for security and/or navigation services. It is not surprising that many mobile phone operators are considering the possibility of incorporating GPS receivers into their devices so that they can utilize the GPS network of satellites to compute the mobile phone's position. The main advantage of the method is of course that the GPS system is already in use for many years. On the other hand, the main disadvantage is that the existing devices will not be sufficient and users have to buy a new generation of mobile phones with built-in GPS support. A second drawback of the approach is that the handset must be 'visible' to at least three satellites at all times. In many cases (for example, in indoor environments, in cities with tall buildings, when inside tunnels, or even in very cloudy weather) the functionality of the system may be hampered. In the Assisted-GPS (A-GPS) method, the mobile network or a third-party service provider can assist the handset by directing it to look for specific satellites and also by collecting data from the handset to perform location identification calculations that the handset itself may be unable to perform due to limited processing power. The A-GPS method can be extremely accurate, ranging from one to 10 meters.

Short-Range Network-Independent Technologies

The technologies discussed above, either dependent on the mobile network or not, all share a common characteristic: their range of coverage is relatively long, meaning that they can be useful in identifying the location of a target within a region, city, neighborhood, or other wide spatial area. However, some of the mobile location services discussed in the previous section (such as indoor routing and product tracking) have a more 'microscopic' view of the world and require that the location of a target is identified within the boundaries of a limited coverage range, such as a building or a large room. A number of so-called *short-range* location identification technologies can be applicable in these cases. These technologies are by definition mobile network-independent since they do not refer to locating a mobile phone handset but rather a different sort of target entity (for example, a product).

In the short-range technologies, location identification is a function of the distance between the moving target object and a fixed reference point. As a result, the accuracy of these technologies is always very high when compared to the long-range methods discussed earlier (ranging from a few cm to some tens of meters in all cases). However, since their coverage range ability is restricted, the accuracy of these methods is usually measured in relative rather than in absolute terms.

The most important short-range technologies for mobile location services are discussed in this section.

Bluetooth: Bluetooth is a Radio Frequency (RF) specification for short-range, point-to-multipoint data transfer. Its nominal link ranges from 10cm to 10m, but can be extended up to 100m by increasing the transmit power. It is based on a low-cost, short-range radio link, and facilitates ad hoc connections for stationary and mobile communication environments. Although Bluetooth was originally designed to connect different devices wirelessly, the potential uses of the technology are countless. For example, printers, desktop computers, fax machines, keyboards, joysticks, even home and office alarm systems, all could eventually be Bluetooth-compliant. The location capability is based on a schema similar to the Cell-ID method, however the location accuracy is high due to the very narrow coverage range.

Wireless Local Area Networks (WLANs): WLANs are substituting cable-based Local Area Networks. The WLAN infrastructure is similar to cellular systems where the terminal communicates with the base station over an air interface at a certain frequency band. WLAN advantages compared to Bluetooth can be summarized to higher bandwidth capabilities and communication range (more than 100m). Nevertheless, WLANs do not provide the proximity accuracy of Bluetooth, therefore their use on MLS may ultimately prove to be limited. However, the small implementation cost can render the technology applicable for services that do not require high levels of accuracy (such as product tracking).

Radio Frequency Identification (RFID): RFID is a relatively new automatic identification and data capture technology, first appearing in tracking and access applications during the 1980s. These wireless systems allow for non-contact reading of RF-enabled tags and are therefore effective in environments where other identification mechanisms (such as barcode labels) may not be sufficient. RFID has established itself in a wide range of markets, including livestock identification and automated vehicle identification systems, because of its ability to track moving objects. The technology is also expected to play a primary role in mobile supply chain management (m-SCM) applications since it enables automated data collection and identification of products as they move through the value chain. The accuracy of the method can be extremely high (from a few cm to one meter) due to the very short operating range.

Table 2 summarizes the above discussion into a taxonomy of enabling technologies of mobile location services according to the categories and assessment criteria discussed earlier in the section.

Table 2: A Taxonomy of Enabling MLS Technologies

TECHNOLOGY CATEGORY	TECHNOLOGY	COVERAGE RANGE	ACCURACY SUPPORT	APPLICATION ENVIRONMENT
MOBILE NETWORK-DEPENDENT TECHNOLOGIES	CELL-ID	Long	Low	Indoor/Outdoor
	TOA	Long	Medium	Indoor/Outdoor
	OTD	Long	Medium	Indoor/Outdoor
MOBILE NETWORK-INDEPENDENT TECHNOLOGIES	GPS / A-GPS	Long	High	Outdoor
	BLUETOOTH	Short	High	Indoor
	WLANs	Short	Low to Medium	Indoor
	RFID	Short	High	Indoor

FACILITATING TECHNOLOGIES

In addition to the enabling technologies of location identification that were analyzed in the previous section, a number of complementary technological capabilities may play a particularly important role in our ability to provide added value to MLS. These technologies facilitate the provision of contextual information, which (as already discussed earlier in the chapter) is of paramount importance for innovative MLS service provision, while at the same time may also enhance the location accuracy in terms of proximity. These facilitating technologies are briefly discussed in this section.

Wireless Application Protocol (WAP): The WAP protocol allows wireless networks and mobile devices to access information from the Internet taking into account the inherent limitations of processing power and display capabilities that characterize mobile handsets. Coupling location identification technologies with WAP allows for a direct semantic link between location and content, thus opening a wide spectrum of opportunities for innovative service provision to mobile users (for example, tourists on the move).

General Packet Radio Service (GPRS): Also known as 2.5-generation mobile telephony (2.5G), GPRS is the successor to today's prevailing mobile telephony technologies (for example, the Global System for Mobile Communication or GSM). GPRS allows for much faster data communication speeds compared to the traditional mobile technologies, thereby opening up the possibility for developing many added value services to complement mobile location services. Still, the most important difference between GSM and GPRS is perhaps the capability of the latter to support *packet-based* connections, as opposed to the existing *circuit-based* connections of second-generation networks. The packet-based connection minimizes time-to-connect and allows for charging based on data volume exchanged instead of airtime used.

Universal Mobile Telecommunications System (UMTS): Also known as third generation (3G) mobile telephony, UMTS is also utilizing the packet-based connection schema and is promising to become a very high speed universal standard that will enable the full potential of mobile multimedia services to be realized. According to the UMTS Forum, 3G services represent a cumulative revenue potential of one trillion US$ for mobile service providers between 2001 and 2010 (UMTS, 2001a). Although the proliferation of UMTS networks will not enhance the location identification mechanism *per se*, it will allow for augmenting the content delivered to a mobile user, for example through displaying high-resolution pictures of the surrounding road network or playing videos of the facilities provided by a nearby hotel.

Geographic Information Systems (GISs): GIS represent computer systems capable of storing, manipulating, and presenting geographically related information.

Such information is described either explicitly in terms of geographic coordinates (latitude and longitude or some other grid coordinates) or implicitly in terms of a street address, postal code, or other landmarks. GISs are crucial to mobile location services because they are used to associate geographic coordinates with their respective information context in the physical environment.

BUSINESS MODELS AND PRICING SCHEMES

The technological developments and market potential of mobile location services have created a complex web of different stakeholders that claim a role in the provision of the services and the distribution of the market revenues. We envisage that the mobile location services market will result in a transformation of the 'traditional' value chain of mobile telephony and will result in the emergence of a number of third-party facilitators of business and service provision. Strategic partnerships, revenue sharing agreements, and license fee-based deals will probably define the relationships between market participants, at least for some MLS (Vos and de Klein, 2002). To start with, we can identify the following major roles in the mobile location services market:

a) *Positioning Technology Developers* that design and implement the underlying technologies of mobile location services. These actors can be further divided into *positioning technology providers* and *positioning infrastructure providers*. The former provide the core technology that locates the position of a dedicated mobile handset, while the latter provide the client software that must be incorporated in the mobile terminal devices.

b) *Mobile Network Operators* that provide basic and enhanced access to wireless networks.

c) *Mobile Device Manufacturers* that manufacture the handsets with which mobile users access the services provided by wireless networks. They collaborate with the *positioning infrastructure providers* in order to manufacture active location-sensitive devices.

d) *Third Party Service Providers* that provide additional services that cannot be economically internalized by mobile network operators (for example, geographic information providers, advertising agencies, Wireless Application Service Providers – WASPs, content providers, and so on).

e) *Customers (corporate and individual)* that constitute the ultimate recipients of mobile location applications and services.

It must be noted that the above roles are neither mutually exclusive (for example, a network operator can assume some of the tasks of a third party provider) nor comprehensive in coverage. The provision of different services will undoubtedly require different roles and relationships between the above stakeholders, thereby influencing the structure of MLS markets. Similarly, the revenues to be generated from each service will also be dependent not only on the nature of the service itself, but also on the business model under which it will be provided. Network providers will benefit by direct revenues from subscription and usage fees for nearly all MLS scenarios, but they may need to share these revenues with other parties in many instances. For example, network operators will not typically own the contextual information needed to provide added value navigation services, neither will they be the producers of content in mobile advertisements.

Two major charging schemes for mobile location services can be identified:

a) *User-Charged Services*. Under this model, users are charged for accessing and using the mobile location services, most likely through a service subscription fee. For example, most navigation and information services are likely to be provided under some form of user charging mechanism. The revenues resulting from user charging may benefit the mobile network operators only, but in many cases will have to be shared between network operators and third parties that provide contextual information or other support needed for location-sensitive service provision.

b) *Free-of-Charge Services*. A number of services may be provided without charge to the end user because the service will be paid for by a third party. For example, advertising agencies (or their clients) will normally have to pay for the ability to send mobile advertisements to consenting end users. In other cases, mobile network operators may choose to bear the full cost of implementing mobile location services, expecting to benefit from service differentiation and/or customer churn reduction, as for example in the case of location-sensitive billing. In other cases, some mobile location services will be provided free-of-charge because of regulatory obligations enforced on the mobile network providers (for example, emergency services).

Needless to say that different revenue models and pricing mechanisms may be mostly applicable to different mobile location services, therefore stakeholders need to make careful choices when implementing such services. The next section synthesizes the aforementioned analysis of mobile location services and technologies with their most likely corresponding business models and pricing schemes.

SYNTHESIS: A CLASSIFICATION FRAMEWORK FOR MLS

Although the technologies that enable the provision of mobile location services are already in place, the commercial success of such services is primarily dependent

Table 3: A Classification Framework for Mobile Location Services

SERVICES		CORRESPONDING ENABLING TECHNOLOGIES	CORRESPONDING FACILITATING TECHNOLOGIES	LIKELY PRICING SCHEME
EMERGENCY	Emergency calls	TOA / OTD / A-GPS	---	Free-of-charge
	Automotive Assistance	TOA / OTD / A-GPS	---	User-charged
NAVIGATION	Directions	A-GPS	GIS	User-charged
	Traffic Management	TOA / OTD / A-GPS	WAP / GIS	User-charged
	Indoor Routing	BLUETOOTH / WLANs / RFID	WAP / GIS	Free-of-charge
	Group Management	CELL-ID / TOA / OTD / A-GPS	---	User-charged
	Travel Services	TOA / OTD / A-GPS	WAP / GPRS / UMTS / GIS	User-charged
INFORMATION	Mobile Yellow Pages	TOA / OTD / A-GPS	WAP / GIS	User-charged
	Infotainment Services	TOA / OTD / A-GPS	WAP / GPRS / UMTS	User-charged
ADVERTISING	Banners, Alerts, Advertisements	TOA / OTD / A-GPS	WAP / GPRS / UMTS / GIS	Free-of-charge
TRACKING	People Tracking	OTD / A-GPS	---	User-charged
	Vehicle Tracking	CELL-ID	GIS	Corporate User-charged
	Personnel Tracking	TOA / OTD / A-GPS	GIS	Corporate User-charged
	Product Tracking	BLUETOOTH / RFID	---	Corporate User-charged
BILLING	Location-sensitive billing	CELL-ID / TOA / OTD	---	Free-of-charge

on the correct matching between technological capabilities and service offerings. Different technologies may be applicable to different services, and different services may require different business models in order to be commercially viable.

To assist the process of matching mobile location services and applications with their corresponding underlying technologies and alternative business models, this section introduces a novel classification framework of MLS that takes into account the previous discussion and findings. This framework is illustrated in Table 3.

The framework is service-oriented and its major objective is to encapsulate in a summarized form the findings of the discussion of the preceding sections. This is achieved through a mapping of each mobile location service identified earlier in terms of:

a)   *Corresponding enabling technologies* (i.e., those technologies that effectively cover the service's needs for location accuracy and application environment).

b)   *Corresponding facilitating technologies* (based on each service's corresponding contextual needs).

c)   The most likely *user-charging scheme* for each mobile location service.

Apart from its theoretical value as a comprehensive classification source for MLS, the framework can also assist potential providers of mobile location services at placing themselves and their product or service offerings within a wider frame that will help them design better applications, match them with their technological capabilities, and launch them within an appropriate business model and user charging scheme. For example, a mobile network operator can utilize the framework in order to:

a)   Decide (based on its existing business strategy and perceived window of opportunity) on which mobile location services to develop and launch to the market.

b)   Identify the technological requirements associated with the chosen portfolio of services and perhaps redefine its market development plan so as to minimize the total investment required for service provision. For example, if a common corresponding enabling technology exists that can effectively support all planned services, the network will have a strong financial incentive to invest on the specific technology to achieve economies of efficiency and to avoid the threat of fragmentation.

c)   In a similar vein, by investigating alternative business models and pricing schemes associated with its chosen service portfolio, the network operator will be better positioned to proactively seek alliances and develop realistic business plans.

Although the framework is a valuable cross-reference analytical tool for designing and/or assessing mobile location services, a number of other critical issues related to MLS cannot be included within the framework in a meaningful fashion. For the sake of presentation completeness, the most important of these issues are summarized in the following section.

OTHER CRITICAL ISSUES RELATED TO MLS

Privacy Protection

According to Nokia (2001), "*Of all the challenges facing mobile location service providers, privacy is undoubtedly the biggest single potential barrier to market take-up.*" For example, mobile advertising based on a user's location is a sensitive issue and has to be provided, as discussed earlier, only with the explicit consent of the user.

However, even in such a case, the likely exchange of information between third parties (for example, network operators and advertising agencies) may hamper the privacy of the user's personal data. To ensure commercial success of mobile location services, user trust must be ensured. A clear prerequisite of the trust-building mechanism is that the control over the use of location information is always on the hands of the user, not of the network operator or the service provider.

Regulation

The role of regulatory and policy-making bodies is substantially enhanced in the case of mobile location services, as opposed to mobile commerce applications in general. It is not surprising that the initial boost to the market has come from such bodies (the US FCC mandate for emergency services) and that the European Commission has had a very active role in the development of the market on the other side of the Atlantic. However, the issue of analyzing the enabling (as well as the potentially disabling) role of regulation in the proliferation of the mobile location services market goes beyond the scope of this chapter.

Standardization

Standardization can also be a serious success or failure factor for any new technology, and mobile location services are not an exception to this rule. This is especially true given that, as discussed earlier, MLSs are dependent not only on a number of direct enabling technologies, but also on a number of indirect facilitating technologies of added value services; the Babel tower of existing standards can potentially hamper the market potential of MLS. A number of bodies worldwide are working towards defining commonly accepted standards for the mobile industry, but prior experience has shown that standardization efforts may have a regional, rather than a global, scope. For example, the presence of incompatible standards for second-generation mobile telephony in Europe and the Americas has created considerable problems to users and the industry alike. Worldwide efforts to define universal standards, such as UMTS, provide an optimistic view of the future; however, the danger of incompatibility and technological 'islands' remain. To this end, a number of standardization initiatives are under way, sometimes initiated by the industry itself. For example, Ericsson, Motorola, and Nokia, have joined forces to establish the Location Interoperability Forum (LIF) with the purpose of "*developing and promoting common and ubiquitous solutions for mobile location services.*"

The importance of standardization becomes even more evident when we think of what can be termed as *the paradox of mobile location services*. Although these services are by definition local, any given user will most probably need them when in a non-local environment. We can envisage tourists outside the familiar surroundings of their local residence relying on MLS to obtain assistance and directions, and companies also utilizing MLS to track their goods in distant lands. To be useful to their users, mobile location services must therefore be provided in a location-independent and user-transparent fashion. From a standardization and technological point of view, this requirement poses a difficult problem: service portability and roaming issues have to be resolved in order for MLS to be compelling to users (UMTS, 2001c).

CONCLUSIONS AND FUTURE WORK

Although the technology of mobile location services has been proven in a number of trials (Hargrave, 2000), it still remains unclear whether a market will be created that will take advantage of the technological capabilities. Achieving mass-market acceptance for MLS is dependent on a complex web of relationships between the various market stakeholders. A number of 'basic' applications can be envisaged (roadside assistance, emergency calls, navigation services), but, as with electronic commerce, the real push to the market will happen if and when innovative service provision is matched with real market demand.

The classification framework presented in this chapter is based on a series of taxonomies for MLS services, technologies, market stakeholders, and pricing schemes. The framework aims at assisting stakeholders at placing themselves and their service offerings within a wider frame that will help them design better applications, match them with the most suitable underlying technology, and direct them to the most receptive target market base. Furthermore, the framework is also aimed at assisting the process of understanding the dynamics of the emergent phenomenon of mobile location services, with a view to realizing its true added value. To this end, the framework is a valuable tool for theorists and practitioners alike, as it can be used both as an extensible analytical instrument and as a deductive actionable toolkit.

The MLS field, as with all new technologies in the mobile world, is progressing at an extremely fast pace; therefore, the static picture of the framework may be of limited value in a short time. Future work is continuously needed to place new technological and service developments within the framework dimensions so that its sustained usefulness and validity remains. Furthermore, further research work will need to identify and explore further issues that were beyond the scope of this chapter. For example, the issue of billing for mobile location services, the issues of standardization and interoperability, and the exploration of mechanisms to ensure privacy protection, were deliberately left out of this analysis. However, a detailed analysis of MLS will inevitably lead to such questions, which can be the starting point for future investigations.

REFERENCES

Dix, A., Rodden, T., Davies, N., Trevor, J., Friday, A. and Palfreyman, K. (2000). Exploiting space and location as a design framework for interactive mobile systems. *ACM Transactions on Computer-Human Interaction, 7*(3), 285–321.

Hargrave, S. (2000). *Mobile Location Services: A Report into the State of the Market*. White Paper, Cambridge Positioning Systems.

Johnson, C. (Ed.). (1998). *Proceedings of the 1st Workshop on Human Computer Interaction with Mobile Devices*, May, 21-23. GIST Technical Report G98-1, University of Glasgow.

Kalakota, R. and Robinson, M. (2001). *M-Business: The Race to Mobility*, New York: McGraw-Hill.

Long, S., Kooper, R., Abowd, G.D. and Atkeson, C.G. (1996). Rapid prototyping of mobile context—Aware applications: The cyberguide case study. *Proceedings of the 2nd Annual International Conference on Mobile Computing and Networking*, New York: CM Press, 97–107.

May, P. (2001). *Mobile Commerce: Opportunities, Applications, and Technologies of Wireless Business*, UK: Cambridge University Press.

Nokia Corporation. (2001). *Mobile Location Services*. White Paper of Nokia Corporation [available online at http://www.nokia.com].

UMTS Forum. (2001a). *3G: How to Exploit a Trillion Dollar Revenue Opportunity*. UMTS Forum Position Paper #1, August [available online at http://www.umts-forum.org].

UMTS Forum. (2001b). *Ranking of Top 3G Services*. UMTS Forum Position Paper #2, August [available online at http://www.umts-forum.org].

UMTS Forum. (2001c). *The UMTS Third Generation Market–Phase II*, UMTS Forum Report #13, April [available online at http://www.umts-forum.org].

Vos, I. and de Klein, P. (2002). *The Essential Guide to Mobile Business*, New York: Prentice Hall.

Chapter V

Wireless Personal and Local Area Networks

Thomas G. Zimmerman
IBM Almaden Research Center, USA

ABSTRACT

Wireless communication is a technical and business revolution. Mobile phones are a common site in most cities around the world. Wireless personal and local area networks provide digital connectivity among mobile computing devices, including desktop, laptops and personal digital assistants (PDA). This chapter focuses on the competing standards that are vying for dominance in the booming wireless market. To prepare the reader, a broad review of wireless technology is provided. The various organizations that support the competition standards are outlined. The chapter concludes with some predictions, anticipating the outcomes of a very volatile marketplace.

INTRODUCTION

Wireless Personal Area Network (WPAN) refers to short-range wireless communication, typically less than 10 meters. This small area corresponds to a person's immediate environment, their personal operating space. Any device that physically interacts with a person will typically be within this area. This includes devices worn, carried by, or in close proximity to a person.

Wireless Local Area Network (WLAN) refers to modest-range wireless communication, typically less than 300 meters. This area corresponds to an office, home, small building, or factory floor. The devices can be stationary or mobile, for information processing, monitoring, or entertainment.

In this chapter we will first familiarize ourselves with the terms and concepts of wireless technology. Since networks imply inter-device communication, we will spend most of the chapter discussing wireless standards and interoperability, the ability of devices to talk to each other. Device interoperability is a daunting yet necessary task to make devices and applications useful, convenient, and ubiquitous. We will survey some of the industry associations and standard groups that do this difficult work. Finally we will make some predictions of which standards will dominate the marketplace.

BACKGROUND

Information is transmitted on a medium. The properties of various media for wireless communication are summarized in Table 1.

We shall briefly review the development of wireless networks on these media.

Zimmerman (1996) and Gershenfeld use the human body as a "wet wire" to send data through the body, creating a WPAN with a 2 meter range. A modulating electric field induces a small current in the human body, representing data. The body, however, also acts as a shield. Placing a hand on the transmitter blocks any electric field from leaving the device.

Richley and Butcher (1995) demonstrate a WPAN using magnetic fields. A magnetic field is unimpeded by the human body and can achieve a 6 meters range. However, the antennas to achieve this range are large in size.

Table 1. Wireless Communication Media

Medium	Range (m)	Data (kbps)	Advantages	Problems	Application
Electric	2	20	Low cost International	Blocked by conductors	ID badges
Magnetic	6	200	Low cost International	Antenna size	Control Messaging
Optical InfraRed	30	1,000	Very low cost International	Directional Power	Control IRDA
RF UHF	30	100	Low cost Good range	Regulations	Control Messaging
RF ISM	400	11,000+	International High data rate	Cost Power	Networking

Infrared (IR) is widely used in television remote controls as an inexpensive method for low data rate communication. Laptop computers use faster IR transceivers (transmitter + receiver) to provide data exchanges in excess of 1 Mbps (million bits per second) within 1 meter. The IR light is blocked by opaque objects and must be line-of-site or use powerful transmitters to bounce off walls.

Radio frequency (RF) relies on the propagation of high frequency energy and is well suited for WPAN and WLAN. The greatest limitation is the lack of uniform international regulations. Each country imposes RF transmission regulations. Radios operating in the UHF (Ultra High Frequency) band (300-450 MHz) are inexpensive and widely used for remote car access and wireless alarm systems. However there is no single UHF frequency available for worldwide use.

In the late 1970s, Hewlett-Packard (HP) began experimenting with direct sequence spread spectrum (DSSS) transmission for wireless inter-terminal networking. Spread spectrum is a method of modulating a radio signal so it occupies a wide band of radio frequencies. This method makes the signal more immune to interference since it does not rely on a single frequency. HP successfully petitioned the Federal Communications Commission (FCC) to release the ISM (Industrial Scientific Medical) bands that are available unlicensed worldwide and offer large data rate capability, ideal for WLAN. All of the radios we will now discuss operate at the 2.4 GHz or 5 GHz ISM bands.

RADIO STANDARDS

Wireless data networks require radios that take in digital data, zeros--and ones, modulate and transmit the data as radio waves, receive the radio waves, demodulate the signal, and convert them back to zeros and ones. Coexistence is the ability to have many radios operating without interfering with each other. Interoperation is the ability of radios to share data. These capabilities must be explicitly defined in a specification, confirmed by standard test procedures and adhered to by manufacturers.

Defining the specification is the job of industry and standards groups. We shall examine four contenders for the 2.4 GHz band and three for the 5 GHz band. To understand why there is not one standard, we must consider wireless network applications and geographies summarized in Table 2.

The cost and power consumption of radio hardware increases with frequency and data rate, so the items listed in the table generally reflect ascending power consumption and cost as one moves down the list. Bluetooth is designed primarily for mobile phones where power, cost, and radio component size are at a premium. OpenAir and HomeRF are designed to deliver wireless Internet, digital video, and

Table 2. Wireless Applications and Geographies

Technology	Freq (GHz)	Data (Mbps)	Target Application	Target Country
Bluetooth	2.4	0.8	Mobile Phone	World
OpenAir	2.4	1.6	Home	World
HomeRF	2.4	10	Home	World
802.11b	2.4	11	Office	N. America
802.11a	5	54	Office	N. America
HiperLAN1	5	18	Office	Europe
HiperLAN2	5	54	Office	Europe

audio throughout the home where cost and ease of installation are the determining factors.

The office environment demands high data rates for many people. The 802.11 standards are optimized to deliver data packets (e.g., Internet traffic), while the HiperLAN standards are further optimized to deliver audio and video. The standards are also distinguished geographically; the 802.11 standards were developed in North America while the HiperLAN standards were developed in Europe.

With so much money at stake, particularly in the home, there is fierce rivalry among supporters of competing standards. The wireless market is more volatile, complex, and unpredictable than a technical evaluation would suggest. As we shall see in our discussion of the various radio standards, the time-to-market and the issues associated with attracting a large market share are as important as data rates and power consumption.

Bluetooth

Bluetooth is a slow speed (0.8 Mbps) short-range (10 meters) WPAN specification operating in the 2.4 GHz band created by the Bluetooth Special Interest Group (SIG). Bluetooth takes its name from King Harald Bluetooth who unified Denmark in the 10th Century. The name implies Bluetooth will unite all the small devices of the world. Bluetooth has its origin as a low-cost cable replacement for mobile phone headsets. Notebook manufacturers were early supporters, preferring a universal low cost WPAN to a regional and carrier-specific mobile phone for wireless connectivity. Once a low-cost WPAN was announced with a suggested cost of $5, over two thousand vendors provided their support for this technology.

HomeRF

HomeRF is a WLAN specification operating in the 2.4 GHz band, designed for the home to integrate voice, data, and video over inexpensive hardware. The specification provides six simultaneous voice connections and enough data capacity to satisfy the phone and Internet needs of a typical household.

Success in the home is determined by price, ease of installation, and reliable delivery of services. The consumer market requires an "unwrap and play" experience. A complex full-featured WLAN designed for the office cannot survive in this environment.

In August 2000 an FCC ruling allowed the HomeRF data rate to jump from 1.6 Mbps to 10 Mbps, assuring sufficient data capacity for most home applications. The home wireless market is estimated to grow to over $6 billion in 2004 (Anonymous, 2001a; Wong, 2000). In March 2001, the competitors for this lucrative space began a legal war over intellectual property, complicating deployment plans (Anonymous, 2001a, 2001b).

OpenAir

OpenAir is a WLAN specification operating in the 2.4 GHz band, designed for low cost office and vertical networking solutions. The specification began as a proprietary protocol from Proxim and later was adapted by the Wireless LAN Interoperability (WLI) Forum and placed in the public domain.

High Performance Radio Local Area Networks (HiperLAN1)

HiperLAN1 is a high speed (20 Mbps) WLAN operating in the 5 GHz band under development by the European Telecommunications Standards Institute (ETSI). The functional specification was released in 1999 after eight years of development. The 5 GHz band is relatively interference-free compared to the 2.4 GHz band which is congested by Bluetooth, 802.11b, microwave ovens, cordless phones, security cameras, and baby monitors. The United States, Europe, and Japan are making available a large section of the 5 GHz band (about 300MHz) with more lenient operation rules to enable data rates in excess of 100 Mbps. Currently there are no HiperLAN1 products, and it appears the European wireless community is bypassing HiperLAN1, focusing support for the much faster HiperLAN2 standard.

HiperLAN2

HiperLAN2 is a high speed (54 Mbps) WLAN operating in the 5 GHz band, designed to carry Internet traffic, video (Firewire-IEEE 1394), and digital voice

(3G, third-generation mobile wireless technology). HiperLanII includes quality of service (QoS) important for real-time audio and video. When you download a file such as a document, you generally don't notice if the data arriving pauses for a fraction of a second. However, if you are watching a video or listening to a song, you do mind if the picture sporadically freezes or the music occasionally skips. Quality of service guarantees data will get there within an acceptable time delay.

The HiperLAN2 standard enjoys support from many European companies including Ericsson, Nokia, and Philips. Anticipated deployment locations include offices, classrooms, homes, factories, and public areas.

802.11b

In 1991 the Institute of Electrical and Electronics Engineers (IEEE) formed a committee to develop a specification for a WLAN. Eight years later the 802.11b WLAN specification was approved providing 11 Mbps in the 2.4 GHz band. However, the 802.11b standard did not specify nor establish a procedure to test interoperation of 802.11b products, and as a result radios from different vendors were not compatible. This meant that an information technology (IT) manager had to choose one vendor and stay with that vendor for all their WLAN equipment. One of the foundations of the Industrial Revolution was interchangeable parts. The same principle of interoperation applies to the wireless revolution. To remedy this problem, the Wireless Ethernet Compatibility Alliance (WECA), an industrial association, developed requirements and testing programs to certify the interoperability of 802.11b products. Within two years, the Wi-Fi (Wireless Fidelity) certification program awarded over 200 product certifications to over 60 companies. The resulting interoperation and competition led to a dramatic drop in 802.11b prices, fueling wide deployment (Anonymous, 2001c).

In July 2001 a security flaw in the Wired Equivalent Privacy (WEP) encryption component of 802.11b was uncovered and published. WEP was intended to be as secure as wired Ethernet. The problem is a wireless network is less secure than a wired network since the latter requires gaining physical access to wires. With a proper antenna, wireless network traffic may be monitored from outside a building since it is the nature of radio waves to propagate. By monitoring enough network traffic (e.g., 5 hours of heavy traffic), an intruder is able to recover text messages. One solution is to change the key faster than a hacker can figure it out, for example every five minutes. A better solution is to assume intruders and run security on top of the wireless network such as virtual private network (VPN).

Bluetooth and 802.11b both operate on the unlicensed 2.4 GHz band and unfortunately interfere with each other. The FCC will not resolve the interference problem since the band is unlicensed. Interference can be resolved by co-locating the Bluetooth and 802.11b radios on the same device so they take turns sharing the airwaves. A longer-term solution will be provided by the 802.15 WPAN Working Group, which is developing a Bluetooth standard to assure compatibility and co-existence among 802.11 WPAN devices.

802.11a

The 802.11a standard specifies a high speed (54 Mbps) WLAN operating in the 5 GHz band, designed for efficient distribution of Internet protocol packets. In August 2001 the hardware manufacturer Atheros demonstrated the AR5000 chip set with a data rate of 54 Mbps, with a "turbo" mode of 108 Mbps. According to laboratory data from Atheros (Stevenson, 2001), the 802.11a WLAN provides 54 Mbps at 20 feet while 802.11b provides 11 Mbps at 65 feet. Based on this data and the fact that the coverage area is the square of the distance, an information technology manager must deploy about 10 times the number of 802.11a access points to achieve about 5 times the data rate. Assuming the access points are the same cost, 802.11a delivered bits at twice the cost of 802.11b. Therefore 802.11a only makes sense if you really need high speed and are close to the access point.

HiperLanII and 802.11a operate in the same 5 GHz band and have similar data rates, but they are incompatible. A 5 GHz Wireless LAN Industry Advisory Group, headed by Compaq, Intel, and Microsoft, is working to encourage some interoperation between the two standards. If this effort is unsuccessful and the marketplace adopts both standards, hardware vendors will probably produce a radio with configurable protocol, programmable for either standard.

Europe and North America are headed towards incompatible standards for next generation mobile phones. Europe is promoting third-generation (3G) for voice and data, while North America relies on a mix of standards (CMDA, TDMA, GSM). For WLAN, Europe will likely use HiperLan2 as North American widely deploys 802.11b, migrating to the faster 802.11a standard. Europe and North America will probably converge on the Bluetooth-based 802.15 WPAN standard once coexistence is solved, since both were involved in the development and promotion of Bluetooth.

802.11e

802.11e is a modification to provide quality of service (QoS) to all 802.11 WLANs. Specifically, it modifies the radio component to better handle time-sensitive traffic. The draft specification is based on work from ShareWave, Lucent, and AT&T (Wuelfing, 2001).

If successful, the standard will enable 802.11e WLAN to deliver multimedia (audio and video) into the home, capitalizing on low price 802.11 hardware. The remaining factor necessary would be "unwrap and play" usability. If both these goals are achieved, 802.11 will probably succeed in North American homes.

Europe may succumb to this solution if HiperLAN2 does not deliver a competitive service.

802.11g

802.11g is a high speed (20 Mbps) WLAN operating in the 2.4 GHz band. The WLAN is proposed as a transition step from 802.11b to 802.11a, offering information technology managers an opportunity to introduce higher speed WLAN while maintaining backward compatibility with slower 802.11b WLAN. However, the arrival of faster 802.11a products operating in the less congested 5 GHz band will eclipse the need for 802.11g.

802.15

IEEE P802.15 is a working group of the IEEE 802 Standards Committee, developing a standard for short-distance wireless personal area networks (WPAN). In 1999 the group established the needs and requirements for a WPAN citing the following six criteria;

- Broad market potential for many wearable and hand-held devices
- Compatibility with the 802 family of wireless networks
- Low power, inexpensive, and simple
- Technical feasibility using proven technology for a reliable solution
- Economic feasibility providing a high volume, low cost product
- Coexistence with other WLANs

Task Group 1 is developing a WPAN standard based on Bluetooth. The resulting specification should boost the acceptance and deployment of WPAN solutions.

Task Group 2 is charged with facilitating the coexistence of WPAN and WLAN, by quantifying the interference and developing a set of coexistence mechanisms, essential for the survival of Bluetooth. Otherwise information technology managers may ban Bluetooth radios from their enterprise to avoid interference with their 802.11b-based WLAN.

Task Group 3 is chartered to draft and publish a new standard for high-rate (>20 Mbps) WPANs based on the Bluetooth specification, preferably backward compatible with the Task Group 1 specification.

Task Group 4 is investigating extremely low power, (months to years on battery power) low data rate (<200 kbps) WPANs intended for sensors, toys, remote controls, and home automation. This is a very exciting area for innovation, enabling for example networked toys that learn behavior from each other. The challenge for this market is to make a radio for less than one dollar.

ORGANIZATIONS

It has long been recognized that wireless devices must comply with standards recognized by multiple vendors. In 1991 the IEEE held a workshop to establish wireless networking technology for the information technology industry. Since then new specifications have been introduced by industry alliances. A common scenario for the creation of a wireless standard is as follows. An industry alliance defines and promotes a specification. Companies manufacture and sell products based on the specification. The market and industry demonstrate a strong need for the product. A standards group examines the needs, technical problems, and solutions provided by the competing methods. The standards group formulates a detailed specification and places it in the public domain, providing a robust, interoperable solution for industry and customers. The two most important standards groups in the wireless field are the IEEE and European Telecommunications Standards Institute (ETSI).

The various organizations involved in WPAN and WLAN specifications and standards are summarized in Table 3 (Anonymous, 2002).

Wireless Ethernet Compatibility Alliance (WECA)

WECA is an industry alliance of over 200 wireless networking and software companies. WECA's mission is to certify the interoperability of 802.11b WLAN products, under a trademarked logo "WiFi" (Wireless Fidelity). Products that bear this label must pass a series of standardized interoperability tests administered by an independent lab. The endorsement assures customers that the wireless product will interoperate with all other Wi-Fi certified products.

WECA has had a tremendous effect on the wireless market, turning a specialty technology into a commodity product. While IT managers are waiting for 802.11a, end users have been buying Wi-Fi WLANs. However Wi-Fi is not a consumer

Table 3. Specification and standards

Name	Type	Purpose	Technology
Wireless Ethernet Compatibility Alliance	Industry Alliance	Interoperability	802.11
Bluetooth Special Interest Group	Industry Alliance	Promote, Interoperability	Bluetooth
HomeRF Working Group	Industry Alliance	Specification	HomeRF
HiperLAN Alliance	Industry Alliance	Promote	HiperLAN1
HiperLAN2 Global Forum	Industry Alliance	Promote	HiperLAN2
IEEE 802	Standards Body	Specification	802.11, 801.15
ETSI HiperLAN	Standards Body	Specification	HiperLAN1
ETSI BRAN	Standards Body	Standardize	HiperLAN2

device suitable for general home use. It requires networking knowledge, including familiarity with terms like IP address, DCHP, domain suffix, port, and protocol. These words are not in the vocabulary of an average homeowner. Further, the lack of guaranteed quality of service (QoS) limits the audio and video streaming performance of 802.11. The QoS should improve with the release and adoption of the 802.11e specification.

Bluetooth Special Interest Group (SIG)

Ericsson, IBM, Intel, Nokia, and Toshiba formed the Bluetooth SIG in 1998 that has grown to over 2,000 members. The SIG released a specification in 1999. A year later Ericsson released the first Bluetooth product, a wireless headset for mobile phones.

HomeRF Working Group (HRFWG)

HRFWG is an industry alliance primarily of radio manufacturers formed in 1998 to promote a wireless standard for the home for computer networking, cordless phones, multi-player games, and toys.

HiperLan Alliance

The HiperLan Alliance is an industry organization that promotes the ETSI HiperLan standard. Key members include Apple, Hewlett Packard, Harris Semiconductor, IBM, Nokia, Proxima, Intermec, and STMicroelectronics.

HiperLan2 Global Forum

Formed in 1999, the HiperLan2 Global Forum is an industrial alliance of communication and information technology companies organized to ensure the completion of the HiperLan2 standard and promotion as a worldwide standard for corporate, public, and home environments.

Institute of Electrical and Electronics Engineers (IEEE)

The IEEE is an international non-profit technical association with more than 375,000 members in 150 countries that has produced over 800 standards. IEEE 802.11 is collection of over sixteen WLAN standard groups, producing standards like 802.11b and 802.11c from Task Group B and C, respectively.

European Telecommunications Standards Institute (ETSI)

ETSI is a European standards group based in France composed of manufactures, service providers, and network operators. The group develops standards for the telecommunication, broadcasting and information technology fields. The ETSI

Broadband Radio Access Network (BRAN) develops test specifications to insure interoperation of HiperLan2 products.

CONCLUSION

The field of wireless local area networking has blossomed in the past few years due to the availability of inexpensive hardware and inter-operational devices. In the office, 802.11b has established itself as the WLAN solution. There is a great sense of liberation to be able to roam around an office with a WLAN-enabled notebook computer. Engineers have access to data books, email, databases, and other computers from their lab bench. Managers, designers, and planners have access to emails, databases, and web sites during meetings, providing quick factual answers. A casual meeting in a coffee house can turn into an impromptu sales pitch when presentations are available on-line.

802.11a will eventually replace 802.11b, as multimedia conferencing, messaging, and video cameras become standard applications and equipment on notebook computers, and next generation PDA devices. The 802.11e modifications will provide 802.11a the quality of service necessary to deliver streaming audio and video in the office and home. The battle between 802.11a and HiperLAN2 is a marketplace race; the first to deliver good performance at commodity prices will win. If the market does split on geographic boundaries, radio manufacturers will produce programmable devices capable of operating either standard.

Bluetooth is optimized for networking mobile phones to personal digital assistants (PDAs) and notebook computers. Notebook computers can tolerate the size, power, and cost of an 802.11b solution. The deployment of 802.11b networks in public spaces, so called "hot spots" such as airports, hotels, and coffee houses, threatens the model of the mobile phone as the wireless access device for notebook computers. As mobile phones and two-way pagers incorporate personal information management (PIM) functions and better user interfaces (screens and keypad entry), the need to network PDAs to mobile phones disappears. On the technical front, as Bluetooth increases range and capability, so will the cost, power, and size, threatening the attributes that won tremendous support for Bluetooth.

To understand the battle for the home, one must look closely at the use scenarios and value propositions. What are the real needs being solved by wireless home networking and will consumers pay for it? It is hard to justify adding $30 of WLAN hardware to a VCR or DVD player that retails for $50 and $100 respectively. Another dimension to the problem is complexity. We are all familiar with the classic problem of setting a clock on a VCR. Networking a cordless phone to a computer seems to complicate what is currently a workable inexpensive solution. Early adopters of home networking will be technophiles who use a WLAN at work and want to read email and surf the Internet with their notebook computer at their dinner table or couch at home.

In the uncertainty of the market for specific wireless standards and solutions, there are some fundamentals we can count on. Wireless devices follow the classic trend of electronics, dropping prices and increasing capabilities. Humans are mobile creatures with a voracious appetite for information and entertainment, with an impatience that only increases. These factors will assure a bright and expanding future for wireless networking and applications in the office, home, and environment.

REFERENCES

Anonymous. (2001a). *Analyst: Intel Defection Hurts HomeRF.* http://www.internetnews.com/bus-news/article/0,,9_720021,00.html (accessed January 2002).

Anonymous. (2001b). *Proxim Downplays Intel Move, Stock Shock.* http://www.10meters.com/intel_homerf.html, (accessed January 2002).

Anonymous (2001c). Wireless LAN standards, *Frontline Solutions,* http://www.frontlinemagazine.com/art_th/FebMWguide.htx (accessed January 2002).

Anonymous. (2002). Matrix of wireless organizations. *The Wireless LAN Association,* http://www.wlana.org/direct/matrix.htm (accessed January 2002).

Richley, E. & Butcher, L. (1995). *Wireless Communications Using Near Field Coupling.* US. Patent No. 5,437,057 (issued July 25).

Stevenson, T. (2001). 802.11a: First glimpse - Part 1. *802.11 Insights,* http://www.80211-planet.com/columns/article/0,4000,1781_873181,00.html (accessed January 2002).

Wong, W. (2000). Future of home networking rests on FCC. *CNet News,* http://news.cnet.com/news/0-1004-200-1923014.html. (accessed January 2002).

Wuelfing, J. (2001). 802.11 standards and specifications, *Home Toys Article,* http://www.hometoys.com/htinews/aug01/articles/wifi/wifi.htm (accessed January 2002).

Zimmerman, T.G. (1999). Wireless networked digital devices: A new paradigm for computing and communication. *IBM Systems Journal,* 38(4).

Zimmerman, T.G. (1996). Personal area networks (PAN): Near-field intra-body communications. *IBM Systems Journal,* (35)3&4. 609-617.

OTHER SOURCES OF INFORMATION
Technical details of 802.11:
Geier, J. (1999). *Overview of the IEEE 802.11 Standard,* http://www.wireless-nets.com/whitepaper_overview_80211.htm (accessed January 2002).

History of 802.11 developments:
Champness, A. (1999). IEEE 802.11 Is the Path to High-Speed Wireless Data Networking. http://www.parksassociates.com/events/forum99/F99papers/ieee802.11.htm (accessed January 2002).

Summary of the 802.11 working groups:
http://grouper.ieee.org/groups/802/11/

Chapter VI

The Impact of Technology Advances on Strategy Formulation in Mobile Communications Networks[1]

Ioanna D. Constantiou
ELTRUN (The eBusiness Center), Athens University of Economics
and Business, Greece

George C. Polyzos
Mobile Multimedia Lab, Athens University of Economics and
Business, Greece

ABSTRACT

Over the last couple of years, we have been witnessing the process of convergence between wireless and wired networks, under intense technological innovation and rapid market evolution. Mobile operators are trying to maintain a leading role in the market as it evolves towards integration with the Internet. However, the development of a multitude of new services will require the growth of a whole new market for network connectivity, applications and content. In order to understand the business opportunities arising and the ensuing competitive dynamics, this chapter explores the evolution of key players' business relationships and strategies along with a range of critical issues that will be faced by companies and regulators alike.

INTRODUCTION

The rapid development of new mobile and wireless technologies and integration with the Internet are major challenges for mobile communications and Internet businesses. Forthcoming generations of mobile technologies are expected to change mobile communications infrastructure both in access and core networks, providing both higher data rates and packet access with Quality of Service (QoS). The ultimate goal is the development of technological infrastructures for universal communication networks, which will be based on the convergence of existing networks (fixed and wireless). Such ubiquitous, but heterogeneous networking is expected to be fully operational in the Fourth Generation (4G) of mobile networking. In the new environment many business opportunities will emerge, as technological innovation is expected to enable the development of new value added services (UMTS Forum, 2000).

In this context, two main questions arise: Who will the key market players be? How will business models evolve due to the introduction of new technologies? In this early stage, it is not useful to predict potential growth rates and size of the new market. Several attempts to do this have failed to predict the potential of highly innovative markets. For example, the popularity of the Short Message Service (SMS) had been underestimated whilst Wireless Applications Protocol (WAP) services over GSM have been overestimated.

In the following sections we attempt a systematic account of key industry players and their likely roles in future generations of wireless technologies, leading up to 4G. In doing so, we sketch likely scenarios of market dynamics based on mobile networks' technological evolution. In particular the following section introduces key mobile communications players (e.g., mobile operators, Internet service providers). Then, starting from the Second Generation to the Fourth Generation, we present technological innovation in mobile networks and its impact on the key players. We mainly focus on mobile operators' strategies and their business relationships with other market players.

STAKEHOLDERS AND BUSINESS RELATIONSHIPS IN MOBILE AND DATA NETWORKS

For the purposes of this discussion, we define stakeholders in the narrow sense as economic entities that could take on one or more roles in the market and develop business activities to exploit them. Stakeholders in mobile and wireless data networks can be classified according to their position in a business transaction. It is worth mentioning here that stakeholders' groups are presented at a high level of abstraction, seeking to emphasize the differences between network services and information services.

In the services provision value chain of both mobile communications and the Internet, the key legacy players that are expected to sustain or even improve their market position in the dynamic environment of technological innovation are the following:

- *Mobile operators* are connectivity providers that own wireless network infrastructure and have large customer bases for mobile communication services. These players have been the leaders of the development and success of mobile communication markets in Europe, by providing personal communication services and information services through wireless (e.g., GSM) networks. They created the critical mass of mobile customers. They have been investing large amounts on upgrading, maintaining, and expanding their networks, while developing competence in managing customer relations and pricing mechanisms. The mobile communications market presents intense competition, which in combination with continuous technological innovation, is shrinking profit margins as communication services are becoming commodities (Laffont & Tirole, 2000). Communication services remain the primary revenue source for mobile operators. However, they already face the challenge of developing new strategies towards providing value-added services, content and applications, in order to sustain their profit levels.

- *Internet Service Providers (ISPs)* are connectivity and, sometimes, information providers that have IP network infrastructure, Internet know-how, and high speed backbones, which may be integrated as the core network of 3G and 4G technologies. They have a large customer base of fixed Internet subscribers. They are currently managing their own networks locally (traditional ISPs) or internationally (backbone providers). They provide connectivity services (local and global) Internet access, through packet switched networks. Many ISPs also provide information services to both individuals and corporate customers. They have been developing, in collaboration with technology integrators, various types of value added services covering security, QoS and a wide range of commercial needs. National markets have numerous ISPs that either provide competitive network services or focus on niche markets. However, intense competition is driving the market to consolidation.

- *Content providers* are information providers that are currently active mostly in the fixed Internet. They provide services at the applications level, typically on top of a TCP/IP communication infrastructure. Their role will be empowered in the new market, as 3G and 4G network technologies will provide the necessary infrastructure for wider use of multimedia content and applications, as well as for provision of high value added personalized services. Their services are usually bundles of information or content, customized to individual needs. Some content providers have developed mediating roles that facilitate customer exploitation of the Internet by providing suitable online facilities for communication and/or business transactions (e.g., electronic marketplaces and virtual communities).

There are two additional players having a significant impact on technology deployment and, hence, market evolution, namely:

- the *mobile device and infrastructure vendors* (e.g., Nokia, Ericsson, Palm etc.) and
- the *software vendors and communities* (e.g., Microsoft, Symbian, Linux, etc.).

As mobile and wireless IP network technologies are not fully deployed and no specific standards are widely adopted yet, the role of these market players will be critical, through the selection and timing of the technology that will be implemented on end user devices. Even though they do not participate directly in our models of future business relationships, they are expected to influence market evolution by supporting or constraining the deployment of services to end-users. For example, if a vendor restricts the usage of a device by making it specific to a certain mobile operator, it will reduce competition and will restrict innovation.

The market position of each key player is determined by their respective market power or, in other words, their ability to command superior revenues and profits. The main revenue source is the customer base. In many cases, the key player who owns the customer base has the market power to decide on new business relationships with other players. Mobile operators have large customer bases already created from the provision of mobile telephony. ISPs and information providers also manage large customer bases on the Internet. Customers currently subscribe to mobile operators, for mobile communication services, to ISPs for Internet access services and to information providers for content or application services over the Internet.

The convergence of the mobile and Internet markets will facilitate provisioning of the above-mentioned services through a common network infrastructure. Thus, customers will not need to subscribe to more than one operator for access and network services. Instead, they will buy network access and probably end-to-end services with QoS via a single contract. In this context, the strategic challenges are reframed as follows. Who is going to provide network access and services to the customer, and what are the other key players going to do? Will they adapt their strategy by developing different value added services (i.e., information and application services)? Are ISPs going to become mobile operators? Are mobile operators going to become ISPs and squeeze the market share of existing ISPs, as is already happening in some European markets? Since connectivity will become a commodity, will mobile operators and ISPs alike succeed in moving up the value chain and, if so, how will they accomplish this move? Will mobile operators restrict customer choices?

BUSINESS RELATIONSHIP MODELS IN 2G AND 2.5G TECHNOLOGIES

Communication services in 2G are based on circuit switched radio channels (and in Europe the GSM standard). Kano (2000) identifies the driving force of GSM success as standardization vis-à-vis the proliferation of platforms that defined First Generation analogue systems (*inter alia*, NMT, TACS and AMPS). 2G standardization in Europe occurred for three substantive reasons. First, over the European continent, individuals' mobility makes the case for regional "roaming" compelling. National boundaries in Europe in terms of mobile communications are less meaningful than in the United States where, for example, international mobility in the First Generation was negligible. Second, mobile telephony was viewed in Europe as a technological opportunity to be grasped in order to elevate the continent's high technology developers and vendors into global businesses able to compete with the United States and Japan. Third, Europe has a predilection towards central standards setting and adoption in contrast to the United States where the market is a significant determinant of outcomes.

Although GSM is the outcome of a pan European standards effort (in contrast to the market-driven standards in the United States), and initially spearheaded by the continent's state telecom monopolies–the outcome was competition on a grand scale in terms of equipment and service provision. It also paved the way for liberalization of the wired telecom world across Europe. Competition for subscribers to mobile networks drove down tariffs creating in turn a mass market. It is also worth mentioning that uncertainty pervades much of GSM history. In particular the dependence of operators on the technology vendors exposes them to considerable risk. The example of GSM rollout was dominated by the failure of vendors to adequately test their phones.

In the GSM example, the drivers were two-fold. First, there was a clear technology push–this was arguably feasible because of the modest market expectations for the technology. Second, there were no predictions for the generation of a mass market; moreover, the existence of the mass market in Europe cannot be explained by consumer demand nor is the explanation rooted in technology. The answer lies somewhere in-between. For example SMS, the GSM short messaging service, was not demanded by consumers, but once available became a cultural phenomenon, especially in the youth market. Mobile operators have generated high profits via SMS, which is considered to be a "killer" application in 2G.

Wireless Application Protocol

Recently, mobile operators invested in Wireless Application Protocol (WAP) technologies that permit an embryonic (i.e., limited) interconnection with the Internet. They expected to generate high profits by providing information services that are enabled through WAP. WAP is an open, global specification that empowers mobile users with wireless devices to access and interact with information services. WAP can be used on top of various communication systems. WAP was designed for the current generation of wireless devices. It adds a relatively small additional memory requirement to non-WAP mass-market products. Therefore, in volume production, WAP devices were able to reach mass market prices. Equipment manufacturers and software developers were thus encouraged to develop, deploy, and support applications for users of wireless devices by extending their existing tools. Content providers saw an opportunity to extend their business model to include an untapped market of mobile customers. As mobile operators own the access network, ISPs and content providers often enter into agreements with them to effectively exploit their investments in WAP.

Regarding security, WAP includes a specification called WTLS, which implements options for authentication and encryption that are optimized for use in the mobile environment. WAP is using existing Internet standards such as XML, User Datagram Protocol (UDP) and IP. The WAP architecture was designed to enable standard off-the-shelf Internet servers to provide services to wireless devices. WAP is based on Internet standards such as HTTP and TLS, but has been optimized for the unique constraints of the wireless environment. The Wireless Markup Language (WML) used for WAP content makes optimum use of small screens and allows easy navigation. However, available bandwidth is very limited, circuit switching is not efficient for data transfers, and mobile devices have limited capabilities and are unable to support a wide range of more complex and demanding WAP applications.

Having said all that, mobile operators may act in collaboration with device manufacturers in order to limit the accessibility of end-users to online content, thus safeguarding their privileged market position. With respect to end-to-end services, mobile operators may enter into private agreements with specific ISPs and Internet Backbone Providers (IBPs) in order to provide dedicated networks for "specific servers." Such initiatives may stifle growth and market penetration of new services. Although strengthening the long-term growth of the market is in their interest, mobile operators may discount future returns too heavily and focus on short-term profitability by making exclusive arrangements with device manufacturers and content providers, in a way similar to current practices. This is a contestable issue that may call for regulatory intervention. Such intervention at the national or EU level should attempt to balance investment and innovation incentives for operators while promoting long-term market growth and consumer surplus (i.e., the EU Court has already decided against France Telecom for trying to force their users to go through its WAP gateway).

"2.5" Generation Technologies

With the introduction of 2.5G technologies, GPRS (General Packet Radio Service) and EDGE (Enhanced Data Rates for Global Evolution), the technical limitations of 2G will be partially removed. 2G and 2.5G technologies are empowering mobile operators by enabling them to expand their business scope towards information services, notably information portals. GPRS deployment will enable customers to have efficient (because of packet switching) and effective (because of higher data rates) Internet access. According to Bettstetter et al. (1999), GPRS "is an important step in the evolution toward Third Generation mobile networks." The challenge in the development and implementation of GPRS has been in integrating the circuit switching mode technology of GSM and the packet switching mode of GPRS. Bringing GPRS online involves operators overlaying a packet-based infrastructure over the GSM circuit switched network infrastructure. This is both a major upgrade and a step towards UMTS rollout.

GPRS does not require that mobile devices dial-up to access the Internet because users are always connected. The service enhancement is illustrated by GPRS's ability to allow users to capitalize on services currently available on the fixed Internet, namely, file transfer, Web browsing, chat, email, telnet, corporate LAN access; location-dependent information services; and WAP. Another key issue is the expectation of operators on equipment vendors, particularly device manufacturers, with respect to the provision of mobile devices that can exploit the full range of services they expect to offer. The recent experience of GPRS phones suggests that operators are vulnerable to the withdrawal of vendors' support for new services should they choose not to market the application-supporting devices.

It is also worth mentioning that 2G operators without 3G licenses will attempt to compete with 3G operators on a range of non-voice services (e.g., information services). GPRS could be the top of the value chain for many consumers who will not see the need to migrate to 3G.

In Figure 1 we present the 2G-business relationship model, where players retain their traditional business activities. Mobile operators provide network and information services to the customer. Any player wishing to offer services to the mobile customer will need to interconnect to the customer's mobile operator. On the other hand, if a more powerful user device is used, and standard TCP/IP protocols and applications are used, the business model could change drastically and start resembling what we expect for the early 3G phase (this is described further in section 5). With the introduction of GPRS, the business relationship model could remain basically the same as in 2G. However, even in this case the service mix would probably change with information services having a higher share than in 2G, just because of economic considerations and the (slightly at the beginning) higher data-rate.

Figure 1: Current Business Relationships' Model in 2G and 2.5G

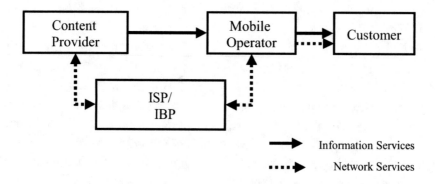

"2.5G" in Japan: The Case of DoCoMo and i-mode Deployment

"i-mode" is a packet switched service developed by DoCoMo to provide mobile Internet access to its customers. DoCoMo has developed a very successful business model. Its rapid success provides useful insights for mobile operators, who are watching DoCoMo preparing its entry into many markets, beyond Japan. i-mode uses compact HTML for content delivery, and packet switching at a data rate of 9.6 kbps. Customers are charged based on the data volume they transfer, plus a low monthly subscription fee. DoCoMo charges only for network access. Independent content providers may levy additional charges to those I-mode customers who subscribe to their content.

As of December 2001, DoCoMo's customer base includes 30 million subscribers. This represents an average of 50,000 new subscribers per day over the two years since i-mode was launched, an impressive growth rate. DoCoMo has developed its own mobile Internet portal, through which subscribers get access to a broad variety of content. Monthly subscription to i-mode services is lower than mobile telephony services (around $3).

Content and application service providers link their i-mode Web pages with DoCoMo's portal after entering a collaboration agreement. Several content providers and individuals have independently launched i-mode compatible content, and many application providers and ecommerce companies sell products and services to i-mode customers. DoCoMo revenues involve both mobile communication and data transfer services. Alliance Partners pay advertisement fees and commission (9%) on every commercial transaction that takes place through i-mode.

i-mode connects customers to a wide range of handy online services, many of which are interactive, including mobile banking, news and stock updates, telephone directory service, restaurant guide, ticket reservations and much more. All services linked directly to the i-mode portal website can be accessed virtually instantly by simply pushing the mobile phone's dedicated i-mode button. Customers can also access hundreds of other unlinked i-mode sites via URLs (DoCoMo Report, 1999).

NTT DoCoMo is at present entering into a number or partnerships in the United States, the Americas, Europe, Asia and Australia such that DoCoMo will seek to introduce i-mode based services, or services similar to i-mode, to other countries. Several components of the i-mode business model could seemingly be transferred to other countries, but some specific usage patterns and business models may be uniquely applicable to Japan's circumstances. For example, in Japan commuters usually spend a long time on trains going to work or school, in Europe and the United States, a much higher proportion of workers take their car to work and cannot use their mobile phone for data services while driving the car.

There is not one single reason that explains i-mode's phenomenal success. NTT-DoCoMo enables easy development of i-mode sites and content by using cHTML. This has resulted in an explosion of available content. A menu gives users access to a list of selected content on partner sites that are included in the micro-billing system and can sell content and services. The micro-billing system enables the subscribers to pay for value-added and premium sites through their telephone bill and is attractive for site owners to sell information to users. In addition, market characteristics of fixed telecommunication and information technology in Japan facilitated i-mode rapid market success. There is relatively low PC penetration in Japanese households and high mobile phone penetration (60 million). Local access charges are high in Japan, so that Japanese people do not use PCs for Internet access as much as in the US or Europe. Finally, the street price of i-mode-enabled handsets at points of purchase is relatively low.

Comparison of WAP and i-mode
in Technical and Business Terms

Comparing i-mode and WAP is not straightforward. In one sense, i-mode and WAP based services are in competition worldwide. Both i-mode and WAP are complex systems, but we will attempt to compare present deployments of i-mode and WAP, as well as their business models, pricing and marketing.

From a technical perspective i-mode is deployed with a packet switched system, which is in principle "always on," while WAP systems over 2G are circuit-switched and involve dial-up. DoCoMo already had a fully functional packet switched network installed before introducing i-mode. It is important to keep in mind that WAP is a protocol, while i-mode is a complete wireless Internet service covering almost all of Japan. It is pertinent to compare WML with cHTML, or to compare a particular WAP implementation (for example, Japan's EZnet or T-Mobil's WAP service in Germany) with i-mode. One important difference from the user and site developer perspective of wireless services is that sites for i-mode are very similar to ordinary HTML-based Internet websites. i-mode sites can also be viewed with ordinary Internet web browsers. Websites for WAP-based services on the other hand need to be written in WML.

From a business perspective, in the case of WAP as implemented in Europe, in principle anyone with an Internet connection could operate a WAP portal; there is also the possibility that multiple WAP portals could be accessed. In Japan, NTT-DoCoMo operates the "official menu" and "i-mode center(s)." Anyone can operate an i-mode site, but in order to do so, one has to enter into a partnership with DoCoMo if the site is to appear on the "official" i-mode menu. With respect to customer base, WAP is centered on business users whilst i-mode is mainly directed at consumers. More specifically, marketing of WAP based-services presently focuses on business customers applications (e.g., banking, stock portfolio, business news, flight booking), while marketing of i-mode in Japan focuses on fun and lifestyle: restaurant guides, games, images, ring tones.

With respect to pricing, an i-mode user is charged for the amount of information downloaded plus various premium service charges (if used), while WAP services are currently charged by the connection time. WAP has no billing elements attached to it which means that it is somewhat inflexible and relatively expensive, as charges are levied according to connection time rather than data volumes.

NTT DoCoMo's is an interesting paradigm for European operators. First, it points to cultural differences between Japan and Europe. The Japanese have embraced i-mode for its gadgetry, ease of content composition and display, and its 'fun' value. Europe is not a homogenous culture. The proliferation of mobile communications across the continent is not uniform; service providers clearly have to understand cultural and national differences in the absence of demand for any particular service or a "killer application."

BUSINESS RELATIONSHIP MODELS IN 3RD GENERATION TECHNOLOGIES

Third Generation wireless technology will enable high-speed mobile connections to the Internet while offering customers full access to rich content, applications and value added services. An important issue to consider is Fransman's observation that mobile communication technologies have in the past developed well in advance of consumer demand (Fransman, forthcoming). There is little evidence that there exists a demand for 3G and its supporting hardware. Operators, however, need to maximize revenue through services and applications in order to recoup the often huge investments ploughed into network development and licenses.

Recent stock market declines have been fuelled by telecom stocks, in turn informed by the liabilities that some operators have taken on as a result of license auctions, notably those in Germany and the UK. The shocks in the financial markets are rooted in uncertainty about business and consumer interest in the technology and its capabilities especially in the context of modest but adequate competition from innovations that significantly enhance the GSM platform, notably GPRS, which fully captures the essence of incremental innovation in its building on existing technology whilst increasing capacity.

The perceived problem with 3G is the absence of a "killer application" sufficient to secure a migration from 2G to 3G, in particular from GSM (and GPRS) to Universal Mobile Telecommunications System (UMTS). This illustrates an interesting phenomenon, namely the interdependence of actors in the chain of service delivery. Generation of services, colonization of the value chain by operators through acquisition and joint ventures, represents a departure from the 2G systems where, at least initially, content was simply absent since services were predominantly voice based. The success or failure of these strategies remains to be seen, but operators are clearly under siege. On the one hand as license holders they have exclusive rights to the radio spectrum. On the other hand, those rights are not in themselves revenue generating. Operators' core competence is in infrastructure and service provision based on voice transfer.

3G is not about voice (to the extent that it may be provided free of charge in future tariff packages), it is about data communications and multimedia based packet switched, volume charged models. To deliver on this a plurality of technical and content skills have to be deployed in some form of collaborative effort. It is not yet clear what the terms of such collaborative efforts will be. What is clear is that operators are in danger of entering parts of the value chain in which they have no competence simply to control the arena and avoid having to relinquish control over their customer information–probably the most valuable asset in the whole scheme.

The examination of key player's strategies points to the importance placed by operators and equipment vendors on applications rather than in the inherent value to consumers of the technology itself. In addition, business opportunities are located outside of the mobile arena, often in the computing rather than the mobile world. Furthermore, development time is short and external intellectual capital has to be deployed alongside that held by corporations associated with infrastructure, handsets and operational business models.

Strategies employed by vendors and operators for the rollout of 3G services cover both ends of the value chain: infrastructure sharing at one extreme, and application writing at the other. At the heart of the debate is the cost of establishing 3G networks both in terms of direct investment costs and license fees, and the realization amongst operators and investors that there is considerable uncertainty about market demand for high value multimedia services. Applications are important, though no one expects there to be a "killer application," hence there is a need to spread risk. Having said that, the mobile Internet is the focus of much additional "venture" funding from both operators and equipment vendors. A migration of customers from existing 2G services to 3G services is by no means a given.

Another important strategic issue for operators is how they manage access to their networks in the Internet world. As gatekeepers by virtue of their control over networks and subscriber and customer information, analysts have argued that they may well be tempted to try to control and/or own the whole value chain. However, content is very important if operators are to attract additional users. Content needs to be relevant, interesting, useful and easy to access. The operator's goal is to own as much of the value chain as possible, or at least a considerable stake in it. The majority of users are prepared to accept the trade-off between universal access and ease when presented with end-to-end solutions by operators. Mobile operators may need the support of Wireless Application Service Providers (WASPs) in order to provide the applications and services, meet time-to-market objectives, lower predictable cost, maintain a focus on their core business and hedge their bets against the failures of particular service ventures.

Circumventing the power of the operators may also feature in the future scenarios. If operators try to control the content access by users–for example, by directing customers through particular portals rather than simple access to the Internet–handset manufacturers may well exploit this constraint and differentiate their products by making it easy for users to reconfigure their devices to suit their own needs rather than those of the operators. The trade-off, however, is whether users have the inclination and motivation to do this.

In October 2001, NTT DoCoMo launched its 3G wireless communications service "FOMA"(Freedom on Mobile Multimedia Access). Currently FOMA offers high quality voice communication, videophone (64Kbps for real-time video), data communications (packet switched, 384 Kbps downlink, maximum 64Kbps uplink, high-speed connections and circuit switched connections 64Kbps-uplink and downlink speed), short messaging and multi-access through simultaneous voice and packet communications. In addition, FOMA offers *i-mode* services along with, *mopera* that connects through the Internet (at 64Kbps) a PC to a FOMA handset and offers mail service. FOMA also offers two types of leased line services.

With respect to pricing, the FOMA tariff structure is designed to be suitable for the age of mobile multimedia and is based on a comprehensive subscription package allowing the use of all communication modes and on a cheaper packet communication package with charges suitable for high-speed and large volume transmission. The FOMA pricing strategy aims at facilitating a smooth migration from existing phones, through similar charges for voice communications to those for current cellular phone service and provision of incentives when migrating from the current phones. In order to promote FOMA service for the next six months (until March 2002), contract-handling charges were waived for all new customers. Additionally to encourage customers to use FOMA's special features, such as data communications, videophone and multi-access, customers received an extra ¥1,000 (7.50 USD) worth of bundled free service with their new contracts.

The Case of Network Services Provision in Universal Mobile Telecommunications System (UMTS)

The UMTS will enable the provision of high quality communication and multimedia services. The development of the UMTS access network (UTRAN) will need large investments in infrastructure. The existing access network infrastructure will have to be replaced (at least at the beginning). The UMTS core network will include two subsystems, the legacy, *voice* call domain and the IP *packet* domain. The IP-based packet domain of 2.5G will not need to be replaced; yet it will need to be enhanced in order to provide QoS guarantees for supporting real-time multimedia applications. Through the deployment of UMTS, future mobile communications will combine personalized and universal services. UMTS is expected to enable the creation of a virtual home environment. This is defined as a universal and portable personalized service environment across network boundaries and across terminals (Huber, Weiler & Brand, 2000).

Mobile operators and ISPs are the key players for the initial deployment of UMTS networks. Mobile operators have already bought UMTS licenses in many European markets. They own the mobile network infrastructure, which will be used for developing the access network (UTRAN). In addition, the UMTS core network will incorporate an expanded GPRS packet domain. Mobile operators will be able to exploit UMTS technologies and mobility in order to provide new value added services such as multimedia messaging services (MMS) and location-based services (UMTS Forum, 2000). With respect to voice communication, UMTS enables the provision of rich voice services (i.e., videophone, multimedia communications). These factors may increase the mobile operators' competitive advantage over any new entrant in the UMTS market.

ISPs own or lease the fixed Internet access network. UMTS technology will expand their business scope, by providing them enhanced technological infrastructure to develop value added services and establish a strong position in the new market. They are expected to play a key role in the interconnection between fixed and mobile networks, as well as supply the backbone infrastructure for the UMTS core network. Concerning telecommunication service provision, ISPs may exploit Voice over IP or similar technologies to offer lower cost long distance voice services to mobile customers and enter a new market segment by squeezing revenues from mobile operators.

The *Content Providers* are Internet companies providing information services or usage of specific applications by customers. When focusing on network service provision, they are considered customers. UMTS technologies will facilitate new business services by providing the infrastructure improvements (e.g., higher data rates, QoS, etc.) necessary for their provision. In addition they will be able to explore new business opportunities by creating value added services based on mobility and universality that UMTS provides. A new breed of content provider will enter the market, developing mobile portals that will focus on particular personalized services for mobile customers. These services will include access to selected content, based on established partnerships or agreements with other information providers.

During the initial phase of UMTS adoption and market penetration in the network services segment, the role of mobile device manufacturers will be critical. Moreover, a device that gives users access to the full range of applications and services, and the ability to roam between networks–in spite of operators' desire to limit access to their own portals/services–is likely to be attractive to consumers. Such devices will challenge UMTS operator strategies.

Various scenarios on UMTS business relationship models are expected to emerge (UMTS Forum, 2000). We present three generic ones:

*   The *fragmented* business relationship model (Figure 2): In this scenario each key player remains a separate business entity and provides the same services as prior to UMTS deployment. This model may appear during the initial phase of a new market, where existing technologies of the mobile network infrastructure will still be used (e.g., GSM, GPRS) and new entrants will be limited,

leaving space for key legacy players. Agreements would then be needed between market players involved in UMTS service provision. The user might have business relationships with various different entities. The UMTS operator will provide mobile access; the Internet service provider will offer network services, and the content or the applications provider, information services and personalized content services. The Internet will be the interconnection network between the different operator and provider domains. Therefore, the parties involved are able to choose completely different ways of handling mobility, QoS and security. The market segment for transmission and connectivity services—the Internet—will only be controlled by backbone providers, being outside the reach and control of UMTS operators and content providers. Thus, new pricing mechanisms will be needed to give backbone providers sufficient incentives to cover the various QoS requirements of UMTS operators and content providers for end-to-end service delivery. In the current Internet connectivity market, end-to-end QoS is not provided (Kano, 2000).

- The *cooperative* business relationship model: In this scenario key players will develop long term co-operation agreements in order to minimize initial high costs of fixed investments and transaction costs for the provision of services to the end user. Collaboration between network service providers for network infrastructure and resources sharing is expected to develop, and between the network service provider and information provider on the bundled service they deliver to the end user. In this scenario, the key players will be enabled to exploit the market opportunities that will appear as user demand will increase with respect to Value Added Services (personalization and localization) and new applications (UMTS Forum, 2000).

- The *ownership* business relationship model (Figure 3): In this scenario one business entity will provide both communication and information services. In this scenario one business entity provides bundles of services including access, connectivity and content services. In this case there is complete control both of the UMTS access, as well as on the IP side, on transmission and connectivity services. This business entity can decide autonomously the solutions for mobility, QoS and security control that are best suited to its business relationship model, since all the nodes and networks involved are under its own control. From the end-user's perspective, this model restricts the selection of mobile operators, and content/application providers. The end-user has to accept the services offered within the services bundle. Whether the end-user will be allowed to access additional services will depend on the market power of the key player in the ownership business relationship model. In a competitive market any key player may increase its market share by exploiting economies of scale and scope, and/or acquire small players to establish a leading position.

Such scenarios might bring into direct competition mobile operators and ISPs that currently have large overlapping customer bases. They will either develop the same infrastructure to provide multiple services to end users, or share the common infrastructure and resources. As technology evolves to GPRS and UMTS, mobile devices will have direct access and instant connectivity to an IP network. In this case, the end user may decide to ask for content from a variety of information providers, thus only use the UMTS provider for network services. In such case the value added that the UMTS provider expects to generate from its customers will decrease or even virtually disappear. Therefore, the UMTS provider will need to devise a new strategy in order to maintain customer lock-in. In a simple example its strategy may be to guarantee QoS for its customers in its own services and delay or even disrupt services from other providers. For customers that demand services from other providers, it may just offer best effort services. Another option involves different pricing mechanisms for services that are provided and delivered within its network versus out of its network. However, these activities may be against consumer protection laws and activate government intervention.

Figure 2: The Fragmented Business Relationships' Model

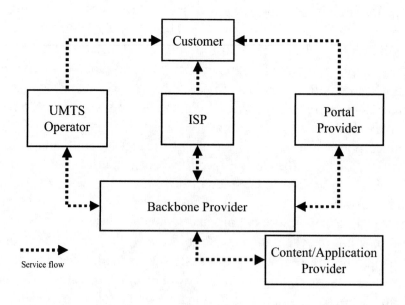

Figure 3: Ownership Business Relationships' Model

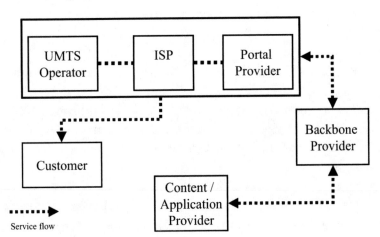

BUSINESS RELATIONSHIP MODELS IN 4TH GENERATION TECHNOLOGIES

The systems enabled by 4G technologies will combine mobility with broadband services on converging future networks. A broad range of access systems will be offered to the subscriber in order to cover a variety of requirements. Hence different access systems will have to be integrated with the backbone network. Convergence and ubiquitous networking are going to be key concepts. Technological evolution will lead to a seamless network where the customer will be able to access his/her application, from any access infrastructure, terminal or user interface. Wireless networks will evolve towards higher data rates, flexible bandwidth allocation in any part of the assigned spectrum and the ability to efficiently handle asymmetric services.

End-to-end IP connectivity over wireline and wireless networks will support multimedia applications. Consistent mobility, QoS and security are of strategic importance for any player involved and must be offered at the link, transport and application levels in a coordinated manner. The same services are expected to be available in all environments using intelligent application layer adaptation technology to cope with widely variable bit rates (Wireless Strategic Initiative-IST Project, 2000). Customers will be able to use any service of any third party without being limited by exclusive arrangements or other exclusionary tactics of the access provider. From a business perspective, the objective is to provide cost effective bandwidth to mobile customers while focusing on increasingly individualized, content and commercial applications.

"Fourth Generation" Impact on Access Networks

The introduction of 4G technologies will have critical implications on access networks. Aiming at the development of a globally integrated access network and the provision of "seamless service," a layered structure of the access technologies is expected to appear. This can be compared to hierarchical cell structures in cellular mobile radio systems. This concept facilitates an optimum system design for different application areas, cell ranges and radio environments, since a variety of access technologies complement each other on a common platform. In this structure the degree of support for mobility and the cell sizes increase from the lower layer to the top layer.

The *Broadcasting* layer contains emerging digital broadcasting (or distribution) systems such as Digital Audio Broadcasting (DAB), Digital Video Broadcasting (DVB), High Altitude Platforms (HAP) and satellite systems that have a global coverage and support large cells, full mobility, as well as global access. Individual links are not necessarily needed for broadcasting services. This technology can be used as a broadband downlink channel to provide fast transfer of Internet content. Other access systems may be used as return channels for data requests and acknowledgment signaling in highly asymmetric services.

The *2G and 3G* layer enables a high system capacity in terms of customers and data rates per unit area. It will consist of 2G and 3G mobile radio systems for data rates up to 2 Mbps. The systems on this layer provide full coverage, full mobility and global roaming. The 2G and 3G layer is well suited for small to medium bit rate multimedia applications and supports individual links.

The *LAN* layer is intended for very high data rate applications. It should be employed in "hot spots" such as in company campus areas, conference centers and airports. This layer contains WLAN (Wireless Local Area Network) type systems. These systems are flexible with respect to the supported data rates, adaptive modulation schemes and support asymmetric services. In contrast to 2G and 3G systems, this layer contains systems that are characterized by a shorter range and provide mainly local coverage with local mobility. Where global roaming will be required, however, full coverage is not expected.

The *Personal Area Network* (PAN) layer will mainly be used in office and home environments. Various "information appliances" (laptops, printers, personal digital assistants, etc.) and (traditional) appliances (video cameras, TVs, refrigerators, toasters, washing machines, smart sensors, etc.) can be connected to each other to provide short-range communication via systems such as Bluetooth. These systems can also be used to connect the equipment directly to the medium access system or to multi-mode terminals that can also communicate on one of the other network layers and are, of course, also equipped with a short range connectivity system. This facilitates an efficient interconnection between the devices as well as a connection from these devices to the public network. PANs may not support mobility (Wireless Strategic Initiative-IST Project, 2000).

The *Fixed network* layer contains fixed access systems such as optical fiber, twisted pair systems (e.g., xDSL) and coaxial systems (e.g., CATV). Furthermore, fixed wireless access or wireless local loops can be included in this category. Fixed access systems do not support mobility. However, portability with global roaming is feasible and might be supported. These systems of the fixed network layer are characterized by high capacity and relatively low cost.

The seamless network will ensure inter-working between these systems on the common platform by horizontal handover within an access system and by vertical handover between different access systems, as presented in Figure 4. Vertical handover takes place between different layers of the common platform. Vertical handover is combined with service negotiations to ensure seamless service, because different access systems support different data transfer rates and service parameters. Inter-working between systems, mobility management and roaming may be handled via the IP based core network and the medium access system.

However, this prospect depends on ISPs, access providers and other telecom carriers to agree to open their networks to common standards (other than IP) that

Figure 4: Layered Structure of Future Seamless Network of Complementary Access Systems

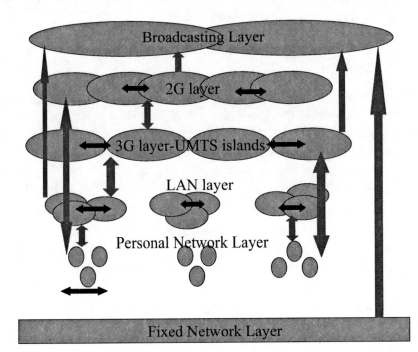

enable transparent network services and interconnection. This may be hard to achieve, particularly during early stages of 4G development when technological innovation (e.g., in security or personalization) will be a critical competitive differentiator. Market players will face significant short-term incentives to differentiate their offering through non-interconnection in order to lock-in their customers and extract higher margins. Having said that, all these stakeholders will also face long-term incentives to collaborate on building interoperable networks, to the extent that end users will derive more value from seamless and ubiquitous (rather than differentiated) service.

Interconnection Issues on 4G Networks

Interconnection will be key to the formation of 4G networks. Both mobile communications and the Internet market have established various types of interconnection agreements to ensure connectivity between networks. An interconnection agreement ensures bilateral exchange of traffic between two networks according to specific conditions.

Based on the description of the access systems that will be included in the future network, various scenarios for interconnection agreements can be envisaged. In this context, we will briefly present existing types of interconnection agreements and consider their applicability to the future network from two perspectives: the vertical, as an integrated network that provides services to the customer, and the horizontal, as a structure of complementary access networks that need to be interconnected with each other and the core network.

When considering interconnection in mobile communications, where services used to be provided through circuit switched networking, agreements were more straightforward and related to the total amount of traffic and the peak amounts exchanged between networks. However, the future seamless network will be based on packet switched technologies. The Internet market provides a suitable metaphor for analysis. Internet interconnection agreements are broadly classified into two categories: peering that involves exchange of traffic free of charge, and transit that involves usage based pricing. The new seamless network will have an IP based core network managed by Internet legacy key players such as backbone providers. Therefore, interconnection agreements, at least at the initial phase of future network development, will be Internet driven. However, issues related to mobility and roaming may be better handled by mobile operators that have already developed such core competences in circuit switched networks (Laffont & Tirole, 2000).

The new integrated market includes many players coming from the legacy markets of telecommunications and the Internet. Communication and network services between various networks will tend to become commodities, as technological innovation will lead to low cost provision. However, the various autonomous systems will need to collaborate and communicate closely in order to increase the overall efficiency of the future network and provide services to the customers. Interconnection between the various networks will be necessary. In order to minimize inefficiencies observed in the Internet (free riding, asymmetric information), a common framework for interconnection agreements is needed.

The main objective is to enable connectivity and universal access while mitigating adverse effects. Peering agreements may be suitable for networks of similar size whereas transit agreements may be more appropriate for networks of different sizes. However, experience has shown that these types of agreements are not sufficient for the Internet anymore (Huston, 1999). It is reasonable to expect that the complexities of 4G will soon render contemporary peering and transit agreements obsolete (Huston, 1999). Other forms of contracting for specific service levels will be needed. There is some initial evidence of this direction, especially when considering vertical interconnection between access networks with different characteristics. The value added of the future network comes from ubiquitous service and mobility support, and from the ability to handle asymmetric services efficiently. Therefore, providers that are able to offer this type of service to other networks through interconnection may charge premium prices.

When interconnecting, access providers face conflicting interests, which provide a basis for opportunistic behavior. In addition, access providers have incomplete and asymmetric information regarding traffic conditions on each other's networks. All this, in combination with uncertainty about the future, complicates matters when it comes to negotiating interconnection agreements. In order to facilitate coordination and ensure collaboration for seamless service provision to the customer, access providers will have to devise novel incentive compatible contracting schemes.

However, the various access networks may not be viable if they cannot generate sufficient revenue. This scenario would lead to horizontal and vertical mergers. Given the strong economies of scale and the externalities of 4G networks, a market structure involving local monopolies and an oligopoly of global backbone interconnection is quite likely. In addition, the high fixed costs associated with developing, managing and upgrading an access network may lead mobile operators to open their financial position through borrowing. UMTS licensing has already led to such outcomes with significant uncertainty regarding payoff periods. In this highly dynamic environment, regulators will have to rise to the challenges by intervening in order to mitigate the risks of monopolistic deviations and short-termism in investment.

Impact of Spectrum Cost in 4G Networks

The cost of spectrum is a key difference between mobile and wireless networks as compared to wireline networks. Consortia of mobile operators in Europe have already spent very large amounts on acquiring their licenses. At the same time, wireline networks are being upgraded in order to provide higher data rates and QoS. Such upgrades seem less costly than building the UMTS infrastructure from scratch.

In the 3G and 4G environments, wireless and wireline networks will compete to some extent for the provision of network services. However, the higher cost of mobile and wireless networks is expected to lead to higher service prices. In order to avoid price competition, network operators will be pursuing differentiation strategies by bundling services with different technical specifications and by introducing content personalization.

In 4G, Wireless LANs (WLANs) and other unlicensed spectrum local connectivity solutions will compete as substitutes to mobile networks (which are attempting to provide ubiquitous service). WLANs will provide an alternative access network in the framework of a 4G seamless (inter-)network. WLAN range covers small areas (e.g., a building or a campus). WLAN access can be envisioned as involving no usage cost since the use of the spectrum will be free; the main cost would be the cost of deployment (and secondarily maintenance). Depending on the evolution of Internet pricing schemes, similar schemes could be adopted as well, particularly in order to ensure particular QoS levels. However, the cost of access using WLANs is expected to be much lower than that achievable by mobile operators operating in wide areas using licensed spectrum.

The initial deployment of WLAN is expected to focus on access from customers within specific areas that it covers, the "hot spot" e.g., as a corporate network. The comparatively low cost of implementing WLAN technologies in specific "hot spots" may increase competition in the market for network access, thus putting more pressure on prices. Furthermore, in order to exploit network externalities, WLAN operators may co-operate to create a wide coverage access network based on WLAN islands. This raises several issues on internal pricing and interconnection between the various local "access networks."

CONCLUSIONS

The evolution of business relationships models in mobile networking indicates the leading role of mobile operators. However, the ability of manufacturers to supply infrastructure and handsets according to the launch timeframes of new technologies is critical. In addition, technological innovation enables the provision of new value added services that create new business opportunities for Internet players. Ultimately, the proliferation of wireless data and multimedia is likely to be driven by the extent to which applications simplify or add value to peoples' lives and by the usability of the devices they run on.

The increasing importance of information services for mobile customers, along with the decreasing margins from communication services provision, suggests that mobile operators should reconsider their strategy, in order to maintain the leading position in the future seamless network.

REFERENCES

Bettstetter, C., Vögel, H.-J., & Eberspächer, J. (1999). GSM Phase 2+ General Packet Radio Service GPRS: Architecture, protocols, and air interface. *IEEE Communications Surveys and Tutorials*, 2(3).

DoCoMo Report. (1999). Internet-compatible "i-mode" cell phones captivate Japanese market. http://www.nttdocomo.com/release/press.html.

Fransman, M. (2001). Evolution of the telecommunications industry into the Internet age. *International Handbook on Telecommunications Economics*. Edward Elgar Publishing.

Huber, J., Weiler, D., & Brand, H. (2000). Mobile radio advances in Europe: UMTS, the mobile multimedia vision for IMT-2000: A focus on standardization. *IEEE Communications Magazine*, September.

Huston, G. (1999), Interconnection, peering and settlements. *Proceedings of the Inet'99 Internet Society Conference*.

Kano, S. (2000). Technical innovations, standardisation and regional comparison: A case study in mobile communications. *Telecommunications Policy*, 24(4), 305-321.

Laffont, J.J. & Tirole J. (2000). *Competition in Telecommunications*, MIT Press.

UMTS Forum. (2000). Shaping the Mobile Multimedia Future—An extended Vision from the UMTS Forum, Report No. 10, September.

Wireless Strategic Initiative-IST Project. (2000). The Book of Visions 2000: Visions of the Wireless World. http://www.ist-wsi.org.

ENDNOTES

1   This research was supported by the European Commission's Fifth Framework, IST Project MobiCom (Evolution Scenarios for Emerging M-Commerce Services: New Policy, Market Dynamics, Methods of Work and Business Models–IST-1999-21000).

Chapter VII

The Ecology of Mobile Commerce: Charting a Course for Success Using Value Chain Analysis

Andreas Rülke
PRTM, UK

Anand Iyer and Greg Chiasson
PRTM, USA

ABSTRACT

The convergence of the Internet with wireless telecommunications has profound and pressing implications for enterprises ranging from long-distance carriers to record labels to automakers. The fast-growing ability of wireless devices to handle a wealth of data content as well as voice transmission is opening the door to the creation of new products, services, markets, and revenue streams. But in what prevailing form will mobile commerce—the still-nascent effort to assemble and monetize the wireless Internet—emerge? How will the vast potential variety of data-based content be created, aggregated, and profitably delivered to both individual and business customers? The essential tool for approaching these still-open questions is value chain analysis. A value chain is a map of the entire set of competencies, investments, and activities required to produce, deliver, maintain, and reap the proceeds from a product or service. The profits and competitive advantages of participation in a given value chain reside dynamically within the chain, pooling at the positions of greatest value. (The returns to the different forms of participation in a value chain, particularly one as complex as mobile commerce, are anything but equal.) This chapter presents and analyzes an extended model of the unfolding m-commerce value chain. The goal is to provide an effective tool for planning and executing relevant business decisions in the face of such complicating factors as technology migration, the absence of market data, and inescapable constraints on organizational resources. The analysis and recommendations are supported by data from a survey with wide participation conducted by the authors.

INTRODUCTION

How does a breakthrough technology, or a breakthrough combination of formerly separate technologies, become a viable business? What are the necessary conditions, competencies, and organizing mechanisms? Which enterprises are in the best positions to provide the various competencies and to organize the new business? How will the new business unfold?

Such are the questions posed by mobile commerce, the still-nascent effort to monetize the Internet's convergence with wireless telecommunications. Businesses ranging from telecom service providers to automakers are grappling with these questions, and are betting heavily on their answers. The purpose of this chapter is to present a tool for understanding the ecology of mobile commerce: the very dynamic relationships among all the elements that are required to make it work as a business. The tool is the value chain model.

In the pages that follow, we'll examine the multiple elements of the mobile commerce value chain. The specific technologies, investments, and competencies required to execute each element will be made clear, as will the relationships among the elements. We'll also describe the key approaches by which companies can create positions of strength within the value chain, including the use of different partnership structures to create integrated m-commerce products and services. Our goal is to impart an understanding of how to use the m-commerce value chain as an effective tool for planning and executing business initiatives in the face of a host of complicating factors, including technology migration, globalization, the absence of market data, and organizational resource constraints.

DEFINING MOBILE COMMERCE

M-commerce is "simply" wireless electronic commerce. Just as e-commerce is a layer of applications on top of the Internet, m-commerce is a layer of applications atop the "Mobile Internet"—the relatively recent technological feat of a two-way link between the Internet, with its data-based content, and wireless telephones and other handheld communication devices. Let's take a look at the technologies involved.

The first manifestation of m-commerce is already fairly common: the use of handheld terminals, wireless phones, and personal digital assistants to receive brief text messages: stock quotes, weather conditions, sports scores, and so on. Today's wireless phones use digital technology, commonly referred to as Second Generation, or '2G,' which provides for transmission and reception using relatively limited bandwidth. Wireless Application Protocol (WAP) based on an open, global specification, is one means of enabling m-commerce capabilities on 2G phones. Given the small screen sizes and slow transmission rates of current mobile devices, however, WAP-based services have been considered disappointing.

More capable mobile Internet appliances, referred to as Third Generation, or '3G' devices, are under development at this writing. These devices combine high-speed, "always on" wireless access with Internet Protocol (IP) networking, thus accommodating many innovative forms of media-rich applications, including simultaneous voice and data communications. So-called 2.5G technology is an intermediate step that attempts to bridge the deployment gap between 2G and 3G systems. General Packet Radio Service (GPRS) is a so-called 2.5G technology. GPRS networks are up and running in Europe, and will gain momentum as GPRS handsets become more widely available in 2002. The "always on" feature, along with higher bandwidth, are what distinguishes 2.5 and 3G devices from 2G devices.

Handheld 2.5G and 3G devices will surpass the capabilities of today's digital wireless phones, providing instant and seamless access to the new, packet-based networks, both public and private, such as the Internet and virtual wireless networks. With data rates of up to 384Kbps, services based on GPRS will offer mobile users the high-bandwidth access to the Internet now available only over fixed wireline networks. Although voice will remain the primary application, 2.5G and 3G will open the door to applications ranging from navigational aid to video teleconferencing and streaming media.

In what prevailing form will mobile commerce emerge? At the end of 2001, the only substantial rollouts of wireless Internet businesses were in Japan. Some 27 million users signed up for NTT DoCoMo's "i-mode" mobile Internet service in the 30 months following its February 1999 launch. The service, which offers such features as email access, already accounts for a quarter of the revenues of NTT DoCoMo, Japan's largest mobile network operator. DoCoMo has just launched a full-fledged 3G service, called FOMA (freedom of mobile multimedia access), in the metropolitan Tokyo area. Initial market response was tepid.

While the idiosyncrasies of Japanese consumers—their love of gadgets, eagerness to adopt fashionable technologies, and so on—are routinely cited in connection with i-mode's success thus far, the most important factor may be that each of Japan's mobile Internet services is an independent, vertically integrated entity unto itself. The networks, devices, and downloadable content of i-mode and its two current competitors, KDDI and J-Phone, are proprietary and non-interoperable. This 'silo' arrangement would not work in the West, since North American and European network operators, application providers, and handset makers are independent businesses, operating in markets that will not accept a non-interoperable product or service. As independent businesses, all three types of entities are reluctant to complete their pieces of the 3G puzzle until the other two pieces are in place. None wants to arrive at the party before it starts, burdened with armloads of not-yet-performing assets (Rowello, 2001). DoCoMo also enjoys a geographic advantage. Operating in a relatively small country, it can afford to overspend on infrastructure in order to ensure that there are no "dead zones" in coverage. North American wireless-service providers deploy their networks over vastly larger geographies than do their Japanese counterparts, so they need to economize as much as possible on infrastructure costs. In short, the direct lessons that Western companies can draw from the DoCoMo example are limited. That is not to say, however, that DoCoMo's dominant-provider, volume-driven strategy could not be successfully deployed by a North American or European wireless-service provider. In addition to a basic monthly charge for voice service, DoCoMo charges on a per-packet basis for the data content used by its subscribers. DoCoMo returns those revenues to its content-provider partners, after deducting a 9% commission. The question of whether DoCoMo is giving away too much revenue to its content partners is a fair one, but the result thus far has been a dominant position in content delivery, which has driven growth in both network traffic and new subscriptions.

CHALLENGES OF DEVELOPING A MOBILE COMMERCE STRATEGY

A valid m-commerce business model must surmount a host of analytical challenges. In our view, these challenges take three forms: complexity, uncertainty, and disruption.

Complexity—Today's wireless phone networks are more complex than most people realize. Making a call involves the use of a handset, a radio tower and base station, and a wireline network. Add the transmission of a couple of lines of data from the Internet, such as a quick weather report, and the complications multiply. The content must be produced, aggregated, and delivered by a combination of business entities working in coordination with one another and with the network, including a meteorological reporting service, a web hosting company, and an Internet portal.

Uncertainty—Whenever a new technology is commercialized, the market is an unknown quantity. What will the initial reception be? How will the adoption curve play out? What products and services based on the technology will customers gravitate toward? How will different consumers in different markets respond to the different product and service offerings? The sheer variety of products and services that the mobile Internet makes possible compounds these uncertainties. But the uncertainties extend well beyond the marketplace. Which technical standards will prevail? One such struggle has already taken place, between TDMA and CDMA, particularly in North America. A similar contest is now taking shape between Wideband CDMA (W-CDMA) and CDMA2000. Uncertainties also exist in the regulatory realm. At present, for example, there is heightened uncertainty over how much additional bandwidth the U.S. Federal Communications Commission will allocate for commercial use, and how much it will insist on reserving for military and other uses.

Disruption—For 75 years, the phone industry as essentially comprised vertically integrated carriers with uncontested monopolies over their markets. Phone companies manufactured voice-transmission equipment, sold it, installed it, serviced it, and billed for it. This business model, which served the phone industry so well for 75 years, was rendered obsolete with the disaggregation of the telecommunications industry. Into the stable world of the giant telcos came independent long-distance services, and then Internet services. The arrival of the mobile Internet has accelerated this disaggregation. How quickly? Consider the plight of Internet service providers. If Internet access is available through mobile phones, will customers be willing to pay a monthly fee to an ISP? Will wireless operators become the largest ISPs, as has occurred in Japan?

An important parallel phenomenon to disaggregation is its opposite—convergence. This is the ability of single devices or applications to perform multiple functions that formerly required multiple applications or devices. Convergence can create new industry segments. The telematics industry arose in order to bring mobile telephony into automobiles. Convergence can also create new competitors. Cable television companies have begun offering Internet services. Just as PCs can now serve as radios and CD players, 3G devices will serve as PCs, telephones, handheld computer games, music players, and even video cameras. The opportunity/hazard implications for electronic equipment manufacturers are obvious.

In his book, *Clockspeed* (Fine, 1998), MIT Sloan School of Management professor Charles Fine argues that technology and competition have combined to create economies, industries, and markets that are changing at unprecedented rates of speed. Fine has categorically stated that the rate of change makes all forms of competitive advantage temporary. That will be a truism of mobile commerce, a sector that will be continually reshaped by fast-evolving technologies, companies, and markets. A valid and useful m-commerce business model must explicitly account for all the complexities, uncertainties, and disruptions that will characterize the sector. Which brings us back to the value chain.

EMPLOYING VALUE CHAIN ANALYSIS TO CREATE A BUSINESS MODEL

A value chain is a map of the entire set of competencies, investments, and activities required to create, produce, deliver, maintain, and reap the proceeds from a product or service, and the relationships among those investments and activities. The profits and competitive advantages of participation in a given value chain reside dynamically within the chain, accumulating at the positions of greatest value. The enterprises that hold these positions have a great deal of control over how the chain operates and how the benefits are distributed (Rülke, 2000). Harvard Business School professor Michael Porter is well known for helping to popularize value chain analysis, beginning with his 1983 book, *Cases in Competitive Strategy*. Porter contended that understanding the structure of an industry is the key to strategic positioning. As the wireless Internet has gathered momentum, various academic institutions and businesses have published depictions of mobile commerce value chains, including INSEAD (2001), Goethe University (2001), and Intuwave (Jeremy Burton, 2000).

Let's look at the mobile commerce value chain's multiplying modes of participation since its birth in the mid-1980s with the first commercial deployments of cellular phone service.

The first commercial cell phone services, based on analog, or so-called '1G' technology, appeared in the mid-1980s. These services involved just three business elements: a wireless service provider that erected and operated the radio towers

Figure 1: First-Generation Value Chain: Cellular Voice Service

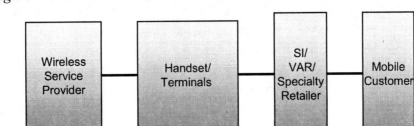

Figure 2: Second-Generation Value Chain: Digital Voice and Data

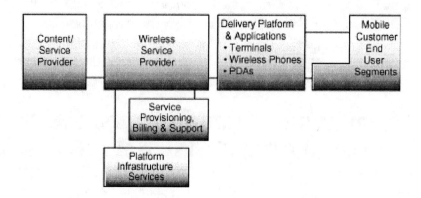

that distributed the signals; manufacturers of the terminals and handsets used by customers; and the system integrators, value-added resellers, and specialty retailers that installed the terminals and handsets (see Figure 1). Due to the weight and bulk of the equipment and the limited lives of the batteries, most of the early cell phone installations were in customers' vehicles. Due to the high costs, the market was generally restricted to business users. It is interesting to note that AT&T initially rejected the technology, but later reversed its position, buying McCaw Communications in order to form AT&T Wireless Services. Motorola, Nokia, Ericsson, and Siemens dominated the early market for handsets and terminals, and remain the market leaders today.

Digital voice and simple data services appeared in the early/mid 1990s. The value in the chain became more distributed, due to both the outsourcing of service and infrastructure providers' services and the emergence of new businesses within the chain that provide data-based content and services (see Figure 2). A particularly popular data-based application is simple text messaging (Short Messaging Service, or SMS). Today, billions of these messages are generated every month, creating substantial revenues for network operators. In terms of the value chain, SMS does not require any additional elements. It is noteworthy that SMS capabilities have been available since the early 1990s, but exploded in popularity only in the last couple of years. This says much about the unpredictable nature of consumer adoption curves. It is worth noting that with the growing segmentation of any value chain comes the demise of some technologies.

It was not until the arrival of the WAP protocol in the late 1990s that data-based services such as mobile banking became commercially viable. Despite WAP's general failure, its advent marked the addition of content and service providers to the mobile commerce value chain. But value chain evolution can mean

Figure 3: Next-Generation Value Chain: The Wireless Internet

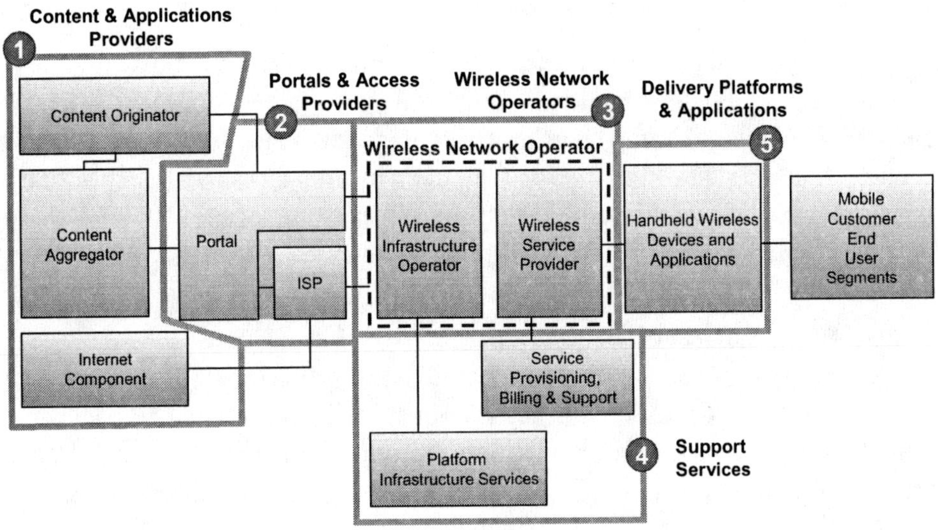

subtractions as well as additions. Formerly important products may disappear: the pager industry is rapidly shrinking because of the proliferation of wireless phones, for example. Ongoing value chain analysis can help companies anticipate the displacements of technology elements, and shift their modes of business participation accordingly.

ELEMENTS OF THE MOBILE VALUE CHAIN

Our model of the unfolding '3G' mobile commerce value chain groups the participants into five major elements.

Examples of current participants in each value chain element:

Element 1	Element 2	Element 3	Element 4	Element 5
Content & Application Providers	Portal and Access Providers	Wireless Network Operators	Support Services	Delivery Platforms & Applications
Bloomberg	Yahoo	Sprint PCS	Spectrasite	Nokia
MapQuest	AvantGo	Vodafone	Convergys	Palm

This value chain is highly horizontal, reflecting the multiplication of the required investments and competencies (see Figure 3). The sections numbered 1–5 highlight the paths and supporting capabilities required to consummate mobile commerce: to create, aggregate, sell, and deliver content. Note that the model includes both traditional and nontraditional companies. We will address the implications later in the chapter.

Barring the arrival of some massively disruptive new technology, we believe that the value chain depicted in this chart will be valid for purposes of industry analysis and strategy setting for at least the remainder of the first decade in the 21st century. Although telecommunications equipment providers have already begun to discuss 4G technology, with anticipated transmission rates of up to 10MB/s, we do not believe that such technology could become launch-ready before 2010.

Element 1: Content and Applications Providers

Content Originators—These are the businesses that create the vast number of highly specific types of content that is variously enhanced, combined, packaged, transmitted, and sold to customers. Content originators can run the gamut from a record studio (a specific song) to a pharmaceutical data publisher (the contra-indications of a specific drug) to a financial news service (the price of a specific stock).

Content Aggregator—These are the businesses that transform individuated content into specific and customer-tailored forms. For example, an aggregator might purchase detailed city maps from a variety of publishers, and obtain data on construction delays from local traffic services. It can then produce accurate maps of where the construction delays are in multiple cities, and suggest alternative routes. It now has a viable m-commerce product to sell.

Internet—The Internet component of Element 1 consists of the web-hosting companies, where the information for the web pages resides, and the wireline transport companies, which route the information from the web-hosting servers.

Element 2: Portals and Access Providers

Portals—A portal offers the consumer a single, convenient point of access to all the products and services produced by the content originators/aggregators. Designed to be individually customized by subscribers, portals are more or less synonymous with the Internet itself in most people's minds. At present, there are numerous competing portals. In Europe, for example, the list includes T-Motion, the joint venture between Deutsche Telekom's T-Mobile and T-Online; 02 (formerly Genie), the portal to British Telecom's 02 (formerly BT Cellnet); and OrangeWorld, the portal to France Telecom's Orange. Vodafone and Vivendi have jointly articulated a vision for a "super portal," called Vizzavi, accessible through any networkable device, wireless or wireline.

Internet Service Providers—ISPs provide the hardware that connects customers to content and applications providers, usually for a monthly fee. Using a dial-up modem connection with a local number, customers connect to a computer at a regional POP (Point of Presence). The POP then connects the user to the Internet, enabling access to all public sites on the World Wide Web. ISPs differentiate themselves through their ability to provide on-demand access, their connection speeds, and their pricing structures.

Traditional ISPs will face strong competition as wireless Internet services come into their own. As wireless network operators begin functioning as ISPs, giving their customers Internet access through handheld wireless devices, the market for conventional ISP services will likely diminish. Only the ISPs that provide content as well as access, such as AOL, will be in a position to counter this challenge.

Element 3: Wireless Network Operators

These operators provide the communication channels—the highways over which content is transported from providers to consumers. Wireless networks, which can reach customers anywhere, are an important alternative to today's wireline networks. Building and operating wireless networks is very expensive and complex, and requires a large organization with very substantial resources. In contrast to Internet highways, on which users can travel for free, the owners of wireless networks bill their customers. The wireless network operator element consists of the following:

Wireless Service Providers—These are the customer-facing elements of wireless networks--the services whose quality is perceived by customers, speed of connection, clarity, and so on. Wireless service providers can buy or rent capacity from network operators. They base their strengths on brand name or customer channels. Virgin Mobile, for instance, which buys wholesale airtime and resells it to end users, benefits from the high brand recognition of the Virgin name.

Network Infrastructure Operator—These are the network-facing elements of wireless networks, which provide the software and hardware that enable online communications. Customers judge network infrastructure operators according to how long it takes to obtain a connection, the quality of the signal, and the frequency of lost connections during calls ("call drops"). These characteristics reflect the quality of a network's management.

Element 4: Support Services

Service Provisioning, Billing, and Support—Various individual elements of customer service may be outsourced by wireless service providers, depending on their business focuses and competencies. The printing and mailing of customer invoices is being increasingly outsourced, for example. The billing function itself is generally kept in-house, however, since billing information captures customers' movement and usage patterns and is thus important for marketing and sales purposes.

Platform Infrastructure Services—These entities provide aspects of the physical network on an outsourcing basis. Examples of companies in this category include SpectraSite and Crown Castle, which own portfolios of telecommunication towers in various countries and rent space on the towers to network infrastructure operators.

Element 5: Delivery Platforms and Applications

This element is the realm of the handheld wireless device makers, of which there are three major groups of players. The first group is made up of the large and predominantly Asian consumer electronics firms. The second consists of manufacturers of handheld computing devices and personal digital assistants. Mobile handset manufacturers make up the third group. Handset makers are also active in the development of in-vehicle telematics units. This delivery platforms/applications element is currently a bottleneck in the proliferation of next-generation wireless Internet services in North America and Europe.

THE ADVANTAGED POSITION OF WIRELESS OPERATORS

Value chain analysis is predicated on the fact that some forms of participation in horizontally distributed businesses are more advantageous than others. Because wireless network operators are in a position to leverage their existing relationships with their mobile phone customers into m-commerce relationships, they currently occupy one of the strongest positions within the mobile commerce value chain. Moreover, incumbent network operators have cleared the very high capital-investment hurdle facing would-be competitors. Market history underscores the importance of seizing these two advantages. Many providers of wireline Internet services were slow to press their advantage in developing Internet offerings, only to see their customers migrate to new entities such as Yahoo! and AOL.

Wireless operators' third current advantage within the m-commerce value chain is their information on their subscribers' whereabouts. In the U.S., this information is becoming fairly accurate with the advent of 'E911' emergency location-determination requirements. Knowledge of subscribers' locations is invaluable in directing contextual advertising messages, and in providing services such as directions to and phone numbers of nearby businesses. This information provides more than a source of advantage over other m-commerce participants; it also affords a significant advantage over wireline ISPs, which don't typically track users' whereabouts.

Wireless operators have other advantages as well. They have extensive billing systems in place, which are generally flexible enough to capture m-commerce as well as access charges. This is of particular advantage in areas of the world where credit cards are less common, or where there is greater reluctance to use them for online transactions. The microbilling system used by NTT DoCoMo's i-mode service to aggregate charges from approved sites is a substantial factor in the service's success.

Value chain analysis also reveals how the positions of advantage within a chain may shift. The key challengers to wireless operators will be the Internet and 'dotcom' companies that are part of the World Wide Web. These companies include many thousands of ISPs, business portals, content providers, and other software companies. These entities tend to be extremely quick to react to—or even to create—market changes. They're in the business of being first to market with a product that works. They're also in the business of continuous product improvement, iteratively building customer solutions that are very much on target.

As traditional and nontraditional wireless enterprises converge to form content/delivery partnerships in the mobile commerce space, the positions of advantage within the industry's value chain will shift according to still-emerging patterns of consumer demand and preference. How will the distribution of revenues and profits change over time? What types of service offerings and business structures will be required to manage these changes? How can companies best implement the necessary changes?

EXPECTATIONS FOR NEXT-GENERATION MOBILE COMMERCE

In 2001, our firm, management consultants PRTM (www.prtm.com), conducted a survey in order to get a sense of the wireless industry's expectations associated with the rollout of next-generation wireless technologies over the next five years. We used our value chain model as the basis for the survey.

A total of 91 respondents, representing more than 80 companies, participated in the survey. Wireless infrastructure manufacturers, terminal and handset makers, wireless operators, content and applications providers, portal companies, and providers of a variety of specialized wireless-related services were all represented. A brief overview of the findings was published in the October 15, 2001, issue of *Telephony* magazine.

Figure 4: Expected Wireless Services Revenue Distribution 2001–2005

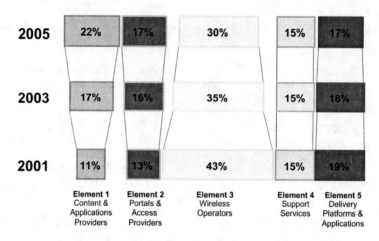

The survey results were quite consistent with the expectations stemming from our value chain model and analysis. The first finding was of immediate interest, given the recession in the telecommunications industry at the time: respondents collectively expected next-generation wireless networks, based initially on 2.5G technology, to be operational in their primary geographies by early in 2003. The findings are presented here.

Integrated Next-Generation Offerings—Companies are seeking to capitalize on or compensate for shifts in revenues and profits among the various elements of the wireless value chain through integrated next-generation offerings. Respondents expected a gradual revenue shift away from wireless network operators and toward content and applications providers as the locus of value shifts from transport toward content.

New location-based wireless services, coupled with an improved ability to charge for content (through microbilling, ASP models, etc.), are the key factors behind the anticipated reapportionment of revenues within the wireless industry (see Figure 4). At the same time, wireless operators face both mounting competition and the commoditization of their offerings, much as we saw with land-line long-distance operators.

While the largest share of revenue will continue to accrue to wireless operators over the next five years, the distribution of revenue across the value chain elements will become more equitable. Respondents expect content and applications providers' annual revenues to grow the fastest, doubling from 11% to 22% of the total. This makes sense, given the coming growth in data-based content and the growing consumer willingness to pay for it. Today, most of the revenue is generated through voice, and that revenue belongs to the wireless operators. In the future, portals will give customers access to graphics-rich websites via mobile handset screens. To limit their decline in revenue share, wireless operators need to avoid becoming mere fungible pipelines.

Respondents expect the wireless industry as a whole to become more profitable over the next five years. Net profit margins are expected to grow, with the largest increases in profitability accruing to content and applications providers that are able to take advantage of economies of scale by spreading their fixed costs over wider customer bases. Wireless operators are expected to hold net profit margins constant by focusing on their most profitable customers, and through increasing efficiencies.

For the wireless industry's expected gains to materialize, companies will need to develop offerings that span multiple value chain elements. The large majority of respondents (78%) report that their next-generation business plans focus on multiple value chain elements. Furthermore, companies with superior relative rates of revenue growth attach the greatest importance to integrated offerings. Traditional wireless companies will look to share in the growth and profitability of the new content- and applications-based developments, while companies new to wireless will need to offer solutions that combine delivery with content. Recent telematics ventures such as Wingcast and OnStar are excellent examples of nontraditional players (automotive OEMs in this case) partnering with existing wireless value chain participants to bring new integrated offerings to end consumers.

Multi-element participation in the value chain, both direct and indirect, is expected to increase over the next five years. The largest shift is expected from "one element" to "two element" companies. The number of companies participating in only one value chain element is expected to decrease from 39% in 2001 to 28% by the end of 2005.

Partnerships Preferred—Partnerships will be the preferred means of integrating across the next-generation wireless value chain. The desire to provide integrated offerings, and the expectation of participating in multiple value chain elements both indicate the transition away from transaction-oriented interactions among value chain participants in favor of more closely coupled business structures. Although the development of cross-chain capabilities in-house is the ultimate in integration, survey respondents generally prefer to partner with holders of existing capabilities. Figure 5 shows how respondents expect to obtain each of the five elements of next-generation capability. Note that the content and applications providers element, which is expected to enjoy the greatest revenue and profitability growth, is expected to see the most partnering activity. Wireless operators expect to form, on average, 21 partnerships with content and applications providers by 2003. Conversely, content and applications providers expect to form, on average, two partnerships with wireless operators and three partnerships with delivery platforms and applications by 2003.

Figure 5: Partnering vs. In-House Development for Integrated Offerings

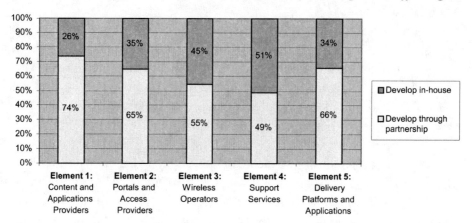

Why is partnering so evident today in the industries with the fastest rates of change? *Clockspeed* author Charles Fine sees it as hedging behavior. "They're hedging against this or that part of the chain becoming more important than their part of the chain. They're worried that they'll be shut out, so they think, 'maybe I need an alliance with somebody.' Everyone wants to be at the pinch point of the chain. The faster the clockspeed, the more uncertainty about where the next pinch point will occur"(Cooper, 1999).

We surveyed respondents about a range of anticipated partnership types, including alliances, minority investments, joint ventures, and acquisitions, with varying degrees of coupling and tightness of integration. While the types of partnerships preferred depend on respondents' particular locations within the value chain, strategic/product alliances are the most likely approach to creating partnerships in all value chain elements.

Overall, respondents expect to form "tighter" partnerships (i.e., acquisitions, joint ventures, minority investments) with wireless operators and delivery platforms/applications providers. "Looser" partnerships (i.e., non-exclusive strategy/product alliances and marketing alliances) are favored with content and applications providers, and with portals and access providers. Verizon Wireless, AT&T Wireless, Sprint PCS, and Palm all have alliances with Yahoo Mobile, for instance, and are listed as official partners on Yahoo's website.

The two most important reasons cited for forming partnerships are to exploit capabilities not available in-house, and to gain time-to-market advantage (see Figure 6). Contributing reasons for partnering are to obtain a cost advantage over in-house capabilities, and to leverage the brands or customer bases of potential partners. Brand and customer leverage were particularly cited by content and applications providers as reasons for partnering.

Figure 6: Primary Reasons for Forming Partnerships

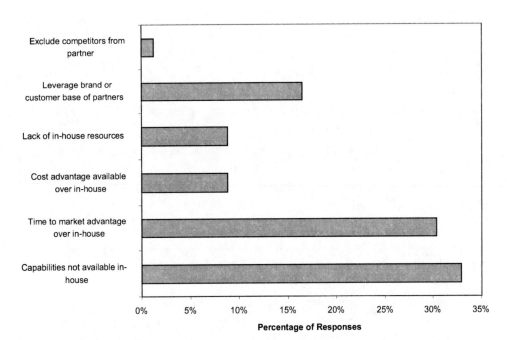

Partnering Best Practices—Partnerships will be most successful when formed and managed according to "partnering best-practices." Clearly, companies face challenges in partnering with new elements of the value chain with which they have no familiarity or relationships. The most-cited challenge in forming partnerships is in making the right connections and introductions, particularly as traditional wireless companies and non-traditional companies try to make connections. Beyond forging these connections, wireless participants then face the challenges of sharing customer information, integrating processes and systems, and aligning business models. Those hurdles are not insurmountable. For instance, NTT DoCoMo has formed a partnership with Coca-Cola to trial "intelligent" soft-drink vending machines that disseminate brand messaging to consumers in conjunction with the i-mode network.

Through its engagements with clients in the wireless services industry, PRTM has identified a set of seven partnering best practices. We asked respondents about their application of those practices.

Most companies use some sort of structured approach to determining whether a partnership is warranted, selecting the best partner, and then building an effective partnership. Over two-thirds (70%) use two or more of the partnering best

---

Partnering Best Practices

- Clear partnership goals are mutually identified and communicated.
- The due diligence process begins with an exacting evaluation of partnership needs, followed by partner selection based on explicit criteria.
- A single senior management sponsor speaks for the company both before and after the partnership.
- Partnerships are analyzed to determine which functions should be integrated and which should remain separate.
- Potential conflicts among the partners' operational functions are identified and controlled prior to partnership formation.
- Partners create a network of interpersonal relationships from executives through middle management.
- Partnerships are managed to be seamless from the customer perspective.

---

practices, and 40% use four or more. The best practices most commonly used by respondents are a due diligence process, senior management sponsorship, and clear identification of common goals. Those companies that report greater past success at forming partnerships are more likely to use a greater number of the partnering best practices. Those same companies are also more likely to enjoy higher profit margins.

CHOOSING A MODEL FOR YOUR CRITICAL PARTNERSHIPS

The highly horizontal, distributed nature of the mobile commerce value chain makes one fact plain: no single enterprise has the wherewithal to provide a true end-to-end solution that simultaneously optimizes business, shareholder, and customer objectives. But if partnership is vital, what is the right partnership model? Value chain analysis can guide you to the answer.

Each value chain element can be broken down into its constituent links. For instance, delivery platform applications consist of the key hardware, software, operating systems, standards, etc. that comprise the technological heart of the solution. Support services can be broken down into installation, provisioning, customer care, and billing. Each of those links can, in turn, be subdivided into layers. For example, the hardware link of delivery platform applications can be decomposed into cellular receivers, GPS receivers, processors, logic, and the like.

To illustrate how this link-and-layer decomposition of the value chain can help a company determine its optimum modes of participation in a mobile commerce solution, we'll describe how one of our clients, a major automaker, used the mobile commerce value chain to chart its telematics strategy. When our client first decided to incorporate telematics services into its vehicles, it assumed that it was competent to provide two inputs: the vehicular component of the delivery platform, and a captive customer base—the drivers. But when value chain analysis was used to map out the elements, links, and layers of the telematics offering, the company saw that it could bring more to the table than just cars and drivers. It could deliver a high level of value by acting as a portal to aggregate automotive-specific content and present it to the driver in a safe and useful manner. In the telematics service envisioned by the automaker, diagnostic data on a vehicle's operational status (temperature, oil pressure, tire pressure, etc.) would be aggregated with information on the probable cause and seriousness of any current or impending problem. This combined information, when further aggregated with GPS location-based information and services (i.e., driving directions to the nearest service station), would deliver a new and compelling type of value to drivers, as dashboard indictor data are "upward-aggregated" into enhanced driver safety and security. This link-and-layer analysis of the telematics value chain allowed the automaker to leverage its deep competency in automotive systems diagnosis by creating a new content-aggregator role for itself.

Once a company has charted the complete set of elements needed to deliver the intended solution, it can begin to structure an appropriate business relationship with the complementary parties required. A fundamental question now arises: for which of the required elements can your enterprise offer best-in-class value? Will you deliver this level of value through minimum cost structure, maximum customer flexibility, price, or some other differentiator, such as technology advantage or customer intimacy? Companies often rush to "own" as many elements of the solution as they can, despite their limitations.

At present, there are telematics business models in which automakers have taken it upon themselves to provide all the solution elements. While these self-contained models have generated some awareness in the marketplace, they have not provided the business, its customers, or its shareholders with appropriate returns.

Alternatively, there are "distributed competency" telematics models. The client example just cited is one. Another is Wireless Car, the recent joint venture between Telia, Ericsson, and Volvo. In terms of our value chain model, Telia operates the venture's wireless network and provides support for the communications channel. Ericsson provides the delivery platform and applications, and Volvo provides the motor vehicle—also part of the delivery platform—along with its know-how in integrating automotive technologies. Wireless Car itself is the portal. Content providers, such as news, weather, financial services, and travel services, are brought in on an as-needed basis.

Once you've decomposed the value chain into its underlying elements, you can begin to associate elements with types or classes of candidate alliances. This brings us to an important principle: an alliance should be mutually exclusive, yet collectively exhaustive. In other words, it should consist of the minimum number of players (element suppliers) required to deliver the solution, with no overlaps and no gaps. And, of course, the fewer the players, the better: less administrative and governance complexity, and more margin to go around.

We'll use 'N' to represent this minimum number of mutually exclusive/collectively exhaustive players. Once you've determined your candidate list of N candidates, the options in terms of business arrangement can be depicted along a simple spectrum. At the far left of the spectrum is the consortium. At the far right is the formation of a new company ('newco'—see Figure 7). Along the way are intermediate forms of business arrangement. While there are no explicit variables that are intended to depict this continuum, one can envision that factors such as degree of management control, investment required, and governance complexity may increase from left to right.

In the context of mobile commerce, it has been our experience that the partnership forms on the right side of the spectrum have worked better than those on the left side.

A Standards Consortium, Consisting of N Participants—A standards consortium is a collective of companies that define the elements of a common solution that each company could implement, either on its own or through the use of partners within or outside the consortium. Very often, the key outputs of such consortia are standards or design rules that govern how solutions should be implemented. The result is a hig   aality and interoperability than would be achieved in the absence of the consortium.

The consortium also has been a prevalent form of business arrangement in designing and implementing complex new products and services. A good example is the CDMA Development Group, or CDG. In this case, intellectual property providers, chip manufacturers, device manufacturers, and wireless carriers were

Figure 7: Partnership Value Chain Models

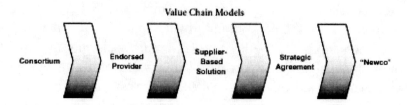

interested in how to best approach the commercialization of the IS95 standard for Code Division Multiple Access technology. The objective in this case was relatively straightforward: how to make the transition from analog cellular to the next generation of standards for digital voice communications. As a consortium, CDG has achieved its direct aim. But in the more complex case of mobile commerce solutions, where the objective spans voice, data, applications, content, and—very importantly—alternative business models, and calls for uniting competencies in design, fabrication, and delivery, the consortium is almost certainly the wrong form of business relationship.

Why? Despite the fact that the consortium may be the corporate lawyer's dream, it rarely achieves the goals of the partnership. It's often too slow and awkward. Mobile commerce partnerships typically involve participants with very diverse competencies and few, if any, prior relationships. The consortium approach to partnership poses the danger of combining enterprises that are structurally and behaviorally incompatible. Counterproductive competition among the partners is almost inevitable, and the governance challenges are apt to prove overwhelming. The endorsed provider solution certainly has the advantage of expediency. It's simple to arrange, but it means surrendering control of the customer solution to an outside party. The supplier-based approach represents business as usual. It won't maximize the collective potential of the participants because it won't align their interests.

An Endorsed Provider—In this model, a company may actually designate or endorse a specific vendor to provide the solution through a bidding or alternative selection process. The company itself may participate, either by providing a captive set of customers, participating in the definitions of solution requirements, providing sales support, or even providing co-branding services. However, the vast majority of the solution elements will be provided by the endorsed provider. For example, if telematics was defined as a set of position-enabled wireless services that follow a customer (as opposed to a vehicle), then various automobile manufacturers might endorse a wireless carrier (e.g., AT&T Wireless, Verizon, Sprint PCS, etc.) to be their telematics service provider in the U.S.

A Supplier-Based Solution—If your organization has the capability to provide many or most of the key elements of the solution, perhaps the entire solution can be best accomplished in-house, with the assistance of external suppliers. Nextel Communications' delivery of push-to-talk services, combined with PCS service, is an example of a supplier-based solution example, since Motorola is the sole telecommunications equipment manufacturer that supplies the ESMR-based solution.

A Strategic Joint Venture Agreement—A joint venture among the N players gives them the ability to enter new markets (vertical or horizontal expansion) or to combine their strengths in a market they are currently serving. An example of a recent joint venture is Sony Ericsson, which combined the wireless handset efforts of two secondary players into a more competitive entity.

A Newco–Formation of a New Company—A newco can take on several forms. While it typically involves some form of equity participation, it could also include equity sources from the N players, customers, or even suppliers of the N players. It can also include other, non-equity forms of participation, such as revenue sharing or voting seats on the newco's board. Examples of newcos include Covisint (between automotive OEMs) and Spain's Vodafone Airtel (between mobile telecommunications network operators).

A set of strategic and operational criteria can be applied to determine the best form for the business alliance from the options along the spectrum. The criteria include:

— Overall ability of the solution to meet customer goals
— Degree of up-front investment required
— Time-to-market requirement
— Degree of legal/regulatory compliance difficulty
— Degree of public acceptance of the solution offering
— Ability of the partnership form to absorb start-up costs
— Degree of brand leverage potential

In addition to leveraging multiple brands, the newco approach also leverages multiple core competencies, minimizing the new business's learning curve. The approach is well suited to combining best-in-class capabilities to produce a best-in-class offering. If a heavy infusion of capital into the partnership is required, the newco approach has the advantage of being able to draw capital from multiple sources. This approach can also be fastest to market, and has the greatest ability to absorb start-up costs. In addition, this approach has a major advantage when the time comes for market launch. If the newco brings together five companies, then it also brings together five sets of customers, creating a large and well-primed trial market.

In the case of our automaker client, the newco approach to creating and delivering the telematics solution provided all these advantages.

Our model and analysis are consistent with our survey findings. Partnering is and will remain the most productive approach to participating in the mobile commerce value chain. More specifically, well-chosen, well-constructed, well-managed partnerships will be the key to transforming the opportunities within the next-generation wireless value chain into profitable ventures.

VALUE CHAIN STRATEGY: WEAVING THE WEB

To describe mobile commerce as an emerging business opportunity, or even as a "business of businesses," would grossly fail to capture its extraordinary dynamism. M-commerce is a genesis in progress; a new and growing source of value at the confluence of two technological revolutions. The metaphor of a web is useful in that it conveys the idea of an intricate and purposeful pattern, but the pattern of m-commerce is anything but fixed or final. Its patterns are being woven today, at the interconnections of the many diverse technologies, competencies, investments, and business models that comprise the mobile commerce value chain. As we have emphasized in this article, m-commerce participation opportunities are value-chain (or "value-web") participation opportunities, and the key to participation is partnership.

The value chain model we have presented, along with the guidance we have offered on selecting the most appropriate form of partnership, will help companies set appropriate participation strategies, execute their strategies effectively and efficiently, and revise their strategies as circumstances change and opportunities arise. We have identified some clear trends in revenue distribution within the m-commerce value chain: for example, the trend toward a more equitable distribution of revenue across the elements of the chain over the next five years, notably favoring content and applications providers.

The emergence of mobile commerce, in all its complexity and flux, calls some basic business assumptions into question. For instance, to the old question, "Who owns the customer?", m-commerce value chain strategy poses the question, "Who *are* my customers in this web of entities and partnerships, and to what degree can I own the relationships with value chain partners and end users?" The possibilities for positioning and branding within—and across—the value chain appear boundless. Mobile commerce will become even more complex in the future. The companies that begin proactively carving out their positions now will be best able not just to cope with the value chain's mounting complexities, but to shape the chain's evolution to their advantage.

REFERENCES

Cooper, V. (1999). Industry clockspeed is getting faster: An interview with MIT's Charlie Fine. *PRTM's Insight*, (Winter), http://www.prtm.com/insight/, accessed January 2002.

Fine, C.H. (1998). *Clockspeed: Winning Industry Control in the Age of Temporary Advantage*. Reading, MA: Perseus Books.

Rowello, R. (2001). The Internet unplugged: Consumer electronics in a 3G world. *PRTM's Insight*, (Spring), http://www.prtm.com/insight/, accessed January 2002.

Rülke, A. (2000). Prospering in the age of mobile commerce. *PRTM's Insight*, (Fall/Winter), http://www.prtm.com/insight/, accessed January 2002.

Chapter VIII

The Wireless Application Protocol: Strategic Implications for Wireless Internet Services

Stuart J. Barnes
Victoria University of Wellington, New Zealand

ABSTRACT

Individually, the Internet and mobile telephony have witnessed extraordinary growth during the last decade. However, only recently have these two areas of technological development begun to converge. The result is the availability of wireless data communications on remote devices, enabling an array of applications tailored for consumer mobility. In this new era, one standard has been hailed as the entry platform for creating mobile Internet services – the Wireless Application Protocol (WAP). Whether WAP will become a key platform is unclear, but it has provided an interesting starting point for the emergence of mobile data services. This paper explores the dynamics of the emerging market for WAP services, examining the role of the consumer, suppliers, substitutes, new entrants and rivalry among the players. The paper concludes by examining some of the key strategies for WAP service provision, making some predictions regarding the future of strategic Internet service delivery.

INTRODUCTION

The growth of the Internet during the last decade has been phenomenal, as witnessed by the massive surge in users and connected computers; in 2002, the estimated number of Internet users stood at 529.9 million, and is expected to rise to 709.1 million by 2004 (eMarketer, 2002). Not only do the efficient services themselves attract people, but also the convenient way of accessing them via an Internet browser. Under most circumstances the same services can be used all over the world—as long as the user has access to an appropriately configured personal computer and access to the Internet (AU System, 1999).

Evidence now suggests that growth in Internet use is likely to emerge from a new channel—mobile devices. Throughout the 1990s, mobile telephony has undergone impressive technological development, and alongside, the saturation of mobile phones and other mobile handsets such as personal digital assistants (PDAs) has continued unabated. From a penetration of only 8% in 1995, more than half of the UK population now owns a mobile phone. Similar patterns can also be seen in Japan, the US, and many other countries. In some places, such as some parts of Scandinavia and Hong Kong, the saturation of mobile phone ownership is now in excess of 80 per cent (Fernández, 2000). Recently, the inevitable convergence of wireless and the Internet has occurred—bringing 'the Internet in your pocket' for which the potential applications are many and varied, including shopping, banking, news feeds, and e-mail.

Under the present technological constraints of low bandwidths and high latency in wireless networks, as well as the low power and small screens of handheld devices, a key standard has emerged for Internet service provision—the Wireless Application Protocol (WAP). WAP provides the means for bringing the Internet and a range of services to the wireless consumer. The emergence of WAP has created a whole new set of dynamics in the wireless industry driven by this new era of value-added service provision. During 2002, data is predicted to account for 20 to 30% of all wireless network traffic, and by 2005, there could be more mobile phones connected to the Internet than PCs (Logica, 2000). Further, the value of commercial transactions made over this channel could be worth more than $200 billion during that period (Strategy Analytics, 2000).

The objective of this chapter is to analyze the strategic implications of the WAP platform for the provision of wireless Internet services. It begins by providing a brief overview of the development of WAP. It continues with a detailed analysis of the WAP service industry, including the role of customers, suppliers, rivalry, new entrants and substitutes. The main focus for this chapter is business-to-consumer mobile commerce—currently the fastest-growing sector (Datamonitor, 2000). The chapter synthesizes and analyses some of the key issues, culminating in an original strategic framework for examining the development of WAP service provision. The chapter ends with some conclusions and predictions for the future of wireless Internet services.

THE WIRELESS APPLICATION PROTOCOL– BACKGROUND

Until very recently, the Internet and the mobile phone have appeared to be largely separate. However, since the mid-1990s, mobile technology providers have been working on a way to bring convergence between these two worlds to provide the wireless Internet to customers. In 1995, Ericsson initiated a project to develop the Intelligent Terminal Transfer Protocol (ITTP) to provide a standard for value added services in mobile networks. Similarly, in 1996, Unwired Planet launched the Handheld Device Markup Language (HDML) and Handheld Device Transport Protocol (HDTP), which respectively describe content/user interface and transaction protocols for wireless devices. Later, in 1997, Nokia introduced its Short Message Service (SMS) and a language called Tagged Text Markup Language (TTML).

With a multitude of concepts, there was substantial risk that the market could become fragmented—a development that the involved companies did not relish. Therefore, all the major players agreed upon bringing forth a joint solution. The outcome was the Wireless Application Protocol (WAP), and the industry group involved is called the WAP Forum (www.wapforum.org)—a group with over 200 members dedicated to enabling sophisticated telephony and information services on handheld wireless devices (Logica, 2000). In essence, WAP could roughly be described as "a set of protocols that has inherited its characteristics and functionality from Internet standards and standards for wireless devices developed by some of the world's leading companies in the business of wireless telecommunications" (AU-System, 1999); building on previous efforts at standardization and lessons learnt from well-known Internet technology, WAP scales a broad range of wireless networks and has the potential to become a global standard for wireless Internet.

WAP is a universal standard for bringing Internet-based content and advanced value-added services to wireless devices such as phones and personal digital assistants (PDAs). In order to integrate as seamlessly as possible with the Web, WAP sites are hosted on Web servers and use the same transmission protocol as Web sites, that is Hypertext Transport Protocol (HTTP) (WAP Forum, 1999). The most important difference between Web and WAP sites is the application environment. Whereas a Web site is coded mainly using Hypertext Markup Language (HTML), WAP sites use a similar but more streamlined formatting language—Wireless Markup Language (WML). WAP data flows between the

Web server and a wireless device in both directions. A wireless device will send a request for information to a server, and the server will respond by sending packets of data, which are formatted for display on a small screen by a piece of software in the wireless device called a microbrowser (Durlacher, 1999). Figure 1 provides some examples of WAP services—in this case advertisements—on WAP-enabled mobile phones.

Figure 1: Examples of WAP Pages (emulated WindWire ads)

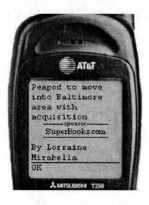  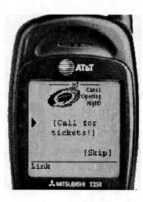

| a. Simple text ad | b. Rich ad | c. Interstitial ad |

The first WAP services and devices were launched in 1999. However, despite high expectations, WAP adoption by consumers is both patchy and limited. As of July 2001, the use of WAP phones has been disappointingly low; just 6% of Finnish and US mobile phone users access the Internet using their phones, compared with only 10% in the UK and 16% in Germany (eMarketer, 2001). Predictions are much better for some parts of the Asia-Pacific (Dataquest, 2000). In Japan, the success of WAP services has been greatest, with 6 million subscribers to the EZWeb WAP service in July 2001 (Mobile Media Japan, 2001).

While the impact of WAP has not been insignificant, in most countries the expectations of consumers have not been met and WAP has been considerably oversold. Part of the problem is the limitation of technology and the non-subtractive nature of services; WAP is not a replacement for the wired Internet and involves an important trade-off between richness and reach in providing data services (Wurster and Evans, 2000). Furthermore, whilst proponents argue that WAP is scalable and extensible enough to endure (Leavitt, 2000), many see WAP as a stopgap until 3G phones. In particular, critics point to the primitive nature of WAP, which is too closely aligned to the current generation of mobile phones, and the possible control of material by cellular operating companies, which will stifle creativity (Goodman, 2000). Key problems include security (Korpela, 1999), the high cost (until networks become packet-switched and the pricing model changes) and limited infrastructure (from networks and devices) (Barnes et al., 2001). Notwithstanding, WAP is recognized industry-wide as an important stepping stone on the path to the wireless Internet. The next section examines the strategic impact of WAP in the provision of wireless Internet services.

A STRATEGIC ANALYSIS OF THE IMPLICATIONS OF WAP AND THE WIRELESS INTERNET

In order to understand the industry segment associated with provision of WAP services, we need a comprehensive strategic framework encapsulating all of the major industry players. Porter (1980) provides such as framework, arguing that economic and competitive forces in an industry segment—such as the WAP service industry—is the result of five basic forces: a) positioning of traditional intra-industry rivals; b) threat of new entrants into the industry segment; c) threat of substitute products or services; d) bargaining power of buyers; and e) bargaining power of suppliers.

Table 1: Strategic Analysis of the WAP Service Industry

Force	Key pressures
Rivalry (central force)	• Falling average revenue per user (ARPU)
	• Cost of 3G technology
	• Consolidation via merger, acquisition and strategic alliance
New entrants	Emerging service content providers:
	• Web portals
	• Handset vendors
	• Retailers
	• Independents
Buyer power	• Sophisticated needs of consumers
	• Established buyer relationships (e.g., operator billing)
	• Pressure for customer-centric mobile offerings
Substitutes	Competing service platforms:
	• I-mode and i-appli service platforms
	• Java platforms, e.g., MExE and J2ME
Supplier power	• Market power of infrastructure providers, e.g., handsets
	• Developer technical competencies in service provision
	• Consolidation via merger, acquisition and strategic alliance

This section aims to provide an analysis of the WAP service sector from a strategic viewpoint. The purpose of this analysis is to provide some understanding of the key forces impacting on the ability of WAP to succeed in provision of mobile Internet services. Table 1 summarizes Porter's framework, highlighting some of the key elements for each of the five forces. Let us examine the framework in more detail.

Rivalry

The network operators are powerful players in the WAP service industry. Traditionally there has been a reasonably high concentration of players in mobile telecommunications. Recently, rivalry has been exacerbated by developments in service pricing and future service provision. With the implementation of 3G transmission technologies on the horizon, network operators have been clambering for licenses to provide services. The next generation of technologies promise transmission speeds of up to 2 megabits per second, opening the door to a raft of high bandwidth services and multimedia. However, the cost of access to such applications, in terms of the frequency licensing arrangements, have not been cheap: in the UK, the cost of 3G License B purchase soared to over £20 billion, and similar figures were seen in other European countries (e.g., Germany). Such costs will inevitably need to be passed on to the consumer.

On the other side of the coin, the mobile market has been squeezed in terms of consumer pricing arrangements. As the network operators have sought to increase the volume of mobile telecommunications and to provide differentiated packages to the customer, profit margins have fallen. Generally speaking, the Average Revenue Per User (ARPU) has declined steadily over the last 10 years and is now an estimated 77% lower than that of 1990 (Barnett et al., 2000).

The industry response has been a global consolidation as operators try to deal with their high up-front investments for 3G and decreasing ARPU. This consolidation trend, along with control of access and direct ownership of the customer, make the operators the most powerful players in WAP services (WireFree-Solutions, 2000c). However, as we shall see below, legal issues reduce the power of operators to restrict access to content, and a decrease in barriers to entry increases competition from mobile Internet content providers.

All of the other forces in Porter's framework—new entrants, substitutes, buyers and suppliers—further contribute directly to rivalry. Let us explore each of them in turn.

New Entrants

The predicted revenues from wireless data services are enormous and have provided an attractive impetus to the entry of new players to the WAP service industry. However, entry to this market is not without its problems. Not least, the incumbent operators, suffering competitive pressures, have used their control of the network infrastructure to try and lock-in potential value; by presetting their subscribers' telephones to make themselves the default Internet access provider and blocking unauthorized services, operators have the opportunity both to charge application providers for access to their subscriber base and to build their own branded services (Barnett et al., 2000). Nevertheless, where an industry is driven by consumer choice and varied access to services, such a strategy may not prove to be effective in retaining customers into the longer term.

The dominance of operators in the area of wireless content delivery is by no means assured. Mobile carriers' ability to extract maximum value from subscribers will depend upon their success in combating the threat represented by a host of other players—including both established and new Internet players, platform vendors, terminal vendors and other third parties such as banks (Yankee Group, 2000). This threat is demonstrated in Figure 2.

The key business-to-consumer market makers on the mobile Internet are mobile portals (or m-portals), revenues of which are predicted to be $42 billion by 2005 (Ovum, 2000). Literally, the word 'portal' means a doorway or gate; mobile portals are high-level information and service aggregators (Ticoll et al., 1998) or intermediaries (Chircu and Kauffman, 2000) that provide a powerful role in access to the mobile Internet. Their main aim is the provision of a range of content and services tailored to the needs of the customer, including: *communication*, e.g., e-mail, voice mail and messaging; *personalized content* and *alerts*, e.g., news, sports, weather, stock prices and betting; *personal information management* (PIM), e.g., 'filofax' functions; and, *location-specific information*, e.g., traffic reports, nearest ATM, film listings, hotels and restaurant bookings.

As such, mobile portals are usually characterized by a much greater degree of customization and personalization than standard Web-based portals in order to suit

Figure 2: Players in the Mobile Portal Market

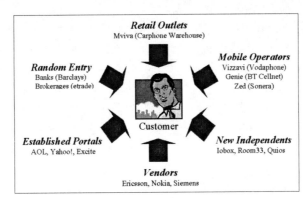

the habits of the consumer (Durlacher, 2000). The current technology restraints dictate that this should be necessary: whilst a standard Web page may have an average of 25 links to other sites or pages, on a WAP phone the average is only 5 links. Therefore, whilst three-clicks on the Web might provide access to 25^3 ($=15,625$) core sites or pages, on the mobile Web this falls to just 5^3 ($=125$) pages (Wappup.com, 2000). As a result, the mobile portal must be suitably tailored to the user's needs so as to present the right information at the right time.

More than 200 WAP portals have been launched in Europe alone since Autumn 1999 (Bughin et al., 2001). Players have attempted to build on existing brands, competencies and customer-relationships to develop a subscriber base. The key players in this market include:

- *Mobile operators*. Players include BT's Genie, Sonera's Zed and Vodaphone' Vizzavi portals. Although experience in content provision is limited, such players have a strong brand and existing customer relationships. Before the evolution of m-commerce, mobile operators were the key market makers in wireless telecommunications, controlling customer billing and the information and services offered.
- *Technology vendors*. For example, Nokia, Ericsson and Motorola have all developed portal services. Whilst new to this area, they have significant technological expertise.
- *Traditional Web portals.* Major players in the Web portal market have developed core mobile offerings, such as Yahoo! Mobile, AOL, MSN and Mobile Excite.
- *Retail outlets.* Some retailers, noticeably those associated with mobile handset sales, are ideally positioned as a front-line portal brand. For example, Carphone Warehouse strongly advertised its Mviva portal in the UK in 2001.
- *Random new entrants.* Mobile portal ideas are also being derived from the most unlikely of places. In particular, the financial services (such as banks and brokerages), who are already developing significant mobile content, are quite well positioned to draw customers to their niche market services. Barclays Bank is one example in the UK.
- *New independents.* This final breed of mobile portals tends to be very flexible and niche-oriented. Whilst generally not experienced in content or partnering, they have creative ideas and are able to position themselves to market segments. In the UK, such players include Room33, Iobox and Quios.

Given time, one might expect the portal market to consolidate, although the potential role of niche players appears much greater than the traditional Web portal market.

Substitutes

In addition to WAP, and as a result of consumer apathy, attention is now being drawn to a variety of other standards for wireless Internet provision, either under development or in use. One of these is the i-mode standard in Japan, which is based on compact HTML (cHTML). The growth and success of i-mode provides considerable food for thought for WAP proponents. Launched in February 1999, i-mode has a subscriber growth rate of nearly 1 million per month, standing at 30 million in December 2001 (Mobile Media Japan, 2002). This is nearly four times more than the competing WAP service, EZWeb. NTT DoCoMo, the owner of the i-mode brand and service, are now planning to 'export' this model to the US and Europe. Through a strategy of partnering, NTT DoCoMo hopes to emulate its earlier success (Associated Press, 2001; Business Week, 2001).

Analysts attribute the success of i-mode to a number of reasons, including (Funk, 2000; Kramer and Simpson, 1999; WireFree-Solutions, 2000a):

- Internet penetration in the home was low and expensive in Japanese homes (at 13%).
- NTT DoCoMo is a leading mobile operator (with 60% market share) and has a strong position and brand in the mobile value chain, being vertically integrated into chip, handset and infrastructure research and development.
- DoCoMo put in place a packet data overlay on their network, allowing for relatively fast (28Kbit/s), cheap, efficient, 'always-on,' push-based services.
- The Japanese culture has a strong tendency towards uniformity and mass acceptance of technological innovations. This is particularly the case for small devices.

Clearly, the development of i-mode is very different to WAP. In some ways, the Japanese i-mode example is unique and perhaps unlikely to be emulated in very different markets such as the US (Diercks and Skedd, 2000). However, there appear to be some important lessons that can be gleaned. I-mode is very definitely a brand and stands for key concepts like simplicity, functionality and meeting consumer needs (WireFree-Solutions, 2000a). I-mode is a market consolidator, adding value by bringing together a great variety of providers in an easy-to-use environment. In this respect, WAP has some way to go to catch up with i-mode; WAP is a bundle of technologies and protocols, which on its own does not deliver value to the end-user.

Another possible alternative to WAP are standards based on Java—a 'write once, run anywhere' programming language—to provide a full application execution environment. These include Java 2 Micro Edition (J2ME) (Newsbytes, 2001) and the Mobile Station Application Execution Environment (MExE) (Durlacher,

1999). These standards are primarily aimed at the next generation of powerful smartphones. MExE, for example, incorporates some advanced features to provide intelligent customer menus, voice recognition and softkeys, as well as to facilitate intelligent network services. Although these standards are not well known or well understood at the present, they have the opportunity to develop into a key role in the technologically superior devices and networks of the future.

Customers

Network operators dominate the wireless market as a key intermediary. Nevertheless, the signs are that this situation will change very quickly and may, to some extent, mirror the business model of Internet Service Providers (ISPs) in the traditional Internet market (Mobilocity, 2000). The key driver here is service provision: the operators, in order to increase their ARPU, have to provide services that increase the customer's willingness to pay. Whereas there are currently very few services linked to cellular telephone companies (cellcos), estimates suggest that by 2004, around 75% of wireless revenues will be from the provision of services (KPMG, 2000).

The convergence of mobile telecommunications and the Internet leads to more personalization and customer empowerment (Arthur D. Little, 2000; Barnett et al., 2000; WireFree-Solutions, 2000b). Although segmentation is important, e.g., focusing on specific market segments with designated products or specific high quality products for various segments, there is a strong recognition that one size does not fit all on the wireless Internet (Arthur D. Little, 2000). After all, a customer who is not fully satisfied can move to another service provider immediately. Three key features enable the personalization of wireless Internet: the 'always at hand' nature of the mobile phone; the unique identifying nature of the phone; and the ability to detect a user's location (Barnett et al., 2000). Using 'intelligent' personalization tools, such information can be used to enhance the richness of the user's service experience, anticipating customer needs. For example, data-based marketing is likely to develop from the current event-based customer relationship management era to a new time-and-place problem and solution management paradigm, based on previous transactions and preferences (Mobilocity, 2000).

Presently, the dominant business model for service provision involves mobile operators aggregating content and services from third-party partners and providing these services directly to their subscribers. Here, mobile (m-) businesses and portals are obliged to reach customers through proprietary networks. However, as the diffusion of WAP accelerates and consumers begin demanding access to m-commerce offerings independent of the wireless carrier, a model similar to that of ISPs is likely to emerge; under this model, wireless service subscribers will have access to any mobile site, and the open-access system will spur companies' development of their m-commerce presence (Mobilocity, 2000). However, new entrants will be severely challenged by the incumbents that have already built strong customer relationships. For example, users of mobile devices will generally have billing relationships with very few service providers (because of the convenience it offers). Therefore, unlike the Web, where customers traditionally maintain multiple accounts, companies in the WAP service industry will be highly dependent on network operators and first movers that control billing relationships. In this sense, the customer-centric model will not prevent companies from creating partnership consortiums to provide the entire range of value-added services to customers (Barnett et al., 2000).

Suppliers

Suppliers of hand-held mobile devices exert a powerful influence on the WAP service industry. In the smartphone market, as in the PDA market, the brand and model are the most important part of the purchase decision; the service provider or network provider is less important (Peter D. Hart, 2000). One of the reasons for this is the importance of 'image' and 'personality' to young customers as associated with specific mobile phones. This places a lot of power in the hands of smartphone producers, who also decide which technologies are incorporated into the end products. These producers must continue to innovate and support leading edge technologies and services in their new products if the wireless Internet is to prosper (Financial Times, 2000).

A stable oligopoly of four smartphone suppliers set prices to sell what they can produce; there is no omnipotent force pressuring prices, and little evidence that low cost strategies win market share (Kramer and Simpson, 1999). Players such as Nokia have proven that barriers to entry in the handset market are substantial, with the cost of branding, production capacity and R&D deflecting considerable competition and making high margins sustainable. The next wave of consolidation in the wireless industry will most likely involve handset vendors strengthening their position prior to the new wave of sophisticated wireless services (Kramer and Simpson, 1999). Simple strategies of 'safety in numbers' will not address the deep impacts of deploying next-generation networks and services; access to leading-edge competencies in software and services is likely to be more important than being the largest supplier of a given element of the network.

Mobile network operators—such as Mannesmann, Telia and Vodaphone—are an important part of the transport process. Notwithstanding, these players are now leveraging their infrastructure advantages in transport to enable movement along the value chain towards mobile services, delivery support and market making. Typically, these operators control the billing relationships and SIM (Subscriber Identification Module) cards on mobile phones and are ideally positioned to become mobile Internet service providers (MISPs) or portals, thereby establishing a transport pipeline for content services (Durlacher, 1999).

Apart from the infrastructure suppliers who are driving technological progress, a host of other suppliers are important, such as those who handle financial transactions, software application developers, content packagers and content providers. The simple value chain that is mostly controlled by the network operator and heavily influenced by handset vendors is being transformed into a complex value network where alliances play a key role. In this new digital economy, consumer online services demand that diverse inputs must be combined to create and deliver value. No single industry alone has what it takes to establish the online digital economy; success requires inputs from diverse industries that have only been peripherally related in the past (Schleuter and Shaw, 1997; Tapscott, 1995). As a result, co-operation, collaboration and consolidation have been the key watchwords, as arrangements are struck between companies in complementary industries. Noticeably, companies in telecommunications, computer hardware and software, entertainment, creative content, news distribution and financial services have seized opportunities by aligning competencies and assets via mergers and acquisitions, resulting in a major consolidation of information-based industries (Symonds, 1999).

A FRAMEWORK FOR STRATEGIC WAP SERVICES

The WAP service industry, itself only a few years old, is in a state of flux. Driven by this new platform of value-added services, rivalry has begun to develop. Pressured by falling revenues, operators are seeking to build on important relationships with mobile customers, such as billing, to extend their portfolio of offerings. However, the demands for services from customers are unlikely to be met from a sole company; with the opening of WAP channels to the customer, partnership is one key trend in an area where pressure is mounting from players entering the market either directly or from adjacent competencies. In other parts of the value chain, the power of suppliers, such as handset vendors, is also distorting the market.

Figure 3 provides a simple framework for visualizing some of the key strategies in the WAP service industry. In particular, this shows how strategies are likely to change over time, driven by the increasing trend towards an open, customer-centric industry model and full service provision. The matrix has two axes: market focus and channel access. In the framework, market focus can either be *broad*, as the WAP service provider aims to be a portal, or *niche*, as the WAP service provider aims to target a specific segment. In terms of channel access, this can either be *closed*, where control falls to the network operator as an intermediary, or *open*, where access to the customer is direct and partnerships are likely to play a role in the provision of a range of services.

In the early days of WAP—from late 1999 to the present—services have been content focused, as indicated by the leftmost cells on the grid; the provision of WAP services is based more on a supplier 'push' than a customer 'pull.' Typically, the network operator has played the role of content enabler, providing a range of selected content services to its subscribers. It is the controlling faction in exclusive alliances. Such players include BT, Mannesmann and Sonera. Other companies—content providers—supply focused, niche-oriented digital content to the network operator portal. Examples include Kizoom, the travel information provider, and BBC News Online, via its WAP news site.

As the WAP service market becomes more open, the user is likely to become the key focus; in this new era, WAP services become demand-led by the ever-sophisticated needs of the consumer. As the channels to the consumer become more accessible, other players will enter the increasingly lucrative portal market, attempting to gain a share of increasing service revenues. In a market driven by personalization and consumer choice, alliances provide an important way to give the full range of consumer-demanded services. Participants in such alliances—service enablers—are an integral part of service offerings. Those who provide the 'front-end' of these offerings—service providers—are far less dominant than in the closed-channel era. Nonetheless, network operators are still likely to be central players in the early stages of open-channel access due their knowledge of the customer and pre-existing relationships – particularly via billing.

Figure 3: Strategic Framework for WAP Service Provision

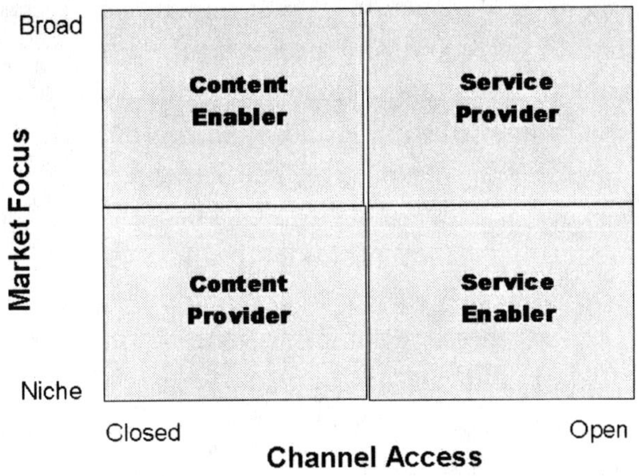

CONCLUSIONS

Early attempts to introduce wireless Internet services based on the WAP platform have been somewhat mixed and apathetic. Notwithstanding, in spite of some misgivings, the wireless Internet is now firmly on the map, aided by the platform provided by WAP. The WAP standard provides the first steps in a path towards mobile Internet, taking stock of the current limitations of wireless networks and mobile devices. WAP creates an interesting and powerful set of dynamics for the industry of mobile Internet service provision, with competition and collaboration coming from a variety of avenues. Mergers and acquisitions have been rife as players in wireless, IT and media industries have attempted to reposition under the increasing threat of competition. As well as the transformation of incumbents, the lure of service revenues and the drive to serve the customer brings many new players from adjacent and even unrelated markets.

In order to understand some of the strategic implications of WAP for companies involved in m-commerce, this chapter has applied Porter's model of industry structure to an analysis of m-commerce issues. This has proved to be a valuable approach to examining strategic issues pertaining to WAP, and is likely to prove equally useful for future analysis of other m-commerce applications and technologies. Building on the analysis, the chapter has provided a matrix to chart some of the key market strategies for WAP service provision. Future research is aimed at a more detailed investigation of the development of business models for WAP service provision, particularly in relation to changing market dynamics. Also of interest is the impact of other technologies and applications on the provision of wireless Internet services.

The implications of the above analysis for firms involved in WAP service provision is clear; successful offerings must be demand-led rather than supply-driven. Successful wireless Internet offerings are likely to be those combining content, infrastructure and services in a seamless way, attempting to be relevant and personal to the mobile phone user. I-mode provides a pertinent example of how this can be accomplished, and it is successful for entirely these reasons. This can only be achieved by greater openness and inclusiveness in the WAP service industry, emphasizing the importance of strategic co-operation and alliance to achieve market share. From a marketing perspective, the core consumer market for mobile Internet services is likely to be users under age 35, whose trade-off between reach and richness has proved most favorable (Wurster and Evans, 2000).

Whether WAP continues to thrive into the medium-term is uncertain. The implementation of the next generation of transmission technologies will enable a new breed of high bandwidth mobile networking that will stretch the abilities of WAP. Alongside, mobile devices are becoming more powerful—combining the capabilities of a mobile phone and small computer into a PDA. Whether WAP is extensible enough to cope with the possibility of rich multimedia and 'always-on' connection remains to be seen. WAP will always exist as a technology alternative, but the strengths of other application protocols such cHTML-based i-mode and those based on Java provide attractive replacements. Such replacements are built for a world where complex interactivity is paramount. Notwithstanding, in the absence of more advanced infrastructure, WAP provides the de facto standard for the wireless Internet; WAP will most likely endure into the short term, spearheading initial attempts at wireless data services for business-to-consumer markets. During 2001, growth of WAP services and sales of WAP phones in some parts of Europe and Asia were surprisingly buoyant, although this was less so in North America.

REFERENCES

Arthur D. Little. (2000). *Serving the Mobile Customer*. Retrieved November 15, 2000, from http://www.arthurdlittle.com/ebusiness/ebusiness.html.

Associated Press. (2001). *NTT DoCoMo to Offer iMode in Europe Later this Year*. Retrieved January 18, 2001, from http://www.anywhereyougo.com/ayg/ayg/wireless/Article.po?id=1034.

AU System. (1999, February). *WAP White Paper*. Retrieved August 15, 2000, from http://www.ausystem.com/.

Barnes, S., Liu, K., & Vidgen, R. (2001). Evaluating WAP news sites: The WebQual/m approach. *Proceedings of the European Conference on Information Systems*, Bled, Slovenia, June.

Barnett, N., Hodges, S., & Wilshire, M. J. (2000). M-commerce: An operator's manual. *The McKinsey Quarterly*, 3, 163-173.

Bughin, J., Lind, F., Stenius, P., & Wilshire, M. (2001). Mobile portals: Mobilize for scale. *The McKinsey Quarterly*, 2, 118-127.

Business Week. (2001). *America Next on DoCoMo's Calling Card*. Retrieved January 15, 2001, from http://www.anywhereyougo.com/ayg/ayg/imode/Article.po?id=36786.

Chircu, A. M., & Kauffman, R. J. (2000). Digital intermediation in electronic commerce—the eBay model. In Barnes, S. and Hunt, B. (Eds.), *Electronic Commerce and Virtual Business* (pp. 45-66). Oxford: Butterworth-Heinemann.

Datamonitor. (2000). *US Mobile Market Worth USD1.2 Billion by 2005*. Retrieved August 28, 2000, from http://www.datamonitor.com/press/.

Dataquest. (2000). *Asia-Pacific Mobile Internet Service Dominated by WAP in First Quarter of 2000*. Retrieved September 18, 2000, from http://gartner6.gartnerweb.com/dq/static/about/press/pr-b09182000.html.

Dierks, B., & Skedd, K. (2000). *Global Demand for Wireless Internet on the Upswing—Carriers Must Structure Services Towards Specific Markets.* Retrieved October 18, 2000, from http://www.instat.com/pr/2000/md2004md_pr.htm.

Durlacher. (1999, November). *Mobile Commerce Report.* Retrieved January 15, 2000, from http://www.durlacher.com/research/.

Durlacher. (2000, June). *Internet Portals.* Retrieved August 15, 2000, from http://www.durlacher.com/research/.

Emarketer. (2001). *Wireless Web Growing Around the World.* Retrieved September 10, 2001, from http://www.nua.ie/surveys/index.cgi?f=VS&art_id=905357175&rel=true.

Emarketer. (2002). *Global Online Population Still Growing.* Retrieved February 15, 2002, from http://www.nua.ie/surveys/index.cgi?f=VS&art_id=905357630&rel=true.

Fernández, B. A. (2000). The future of mobile telephony. Paper presented at *Mobile Telephony and Communications*, Madrid, Spain, May, 22-23.

Financial Times. (2000). *Global M-Commerce Standard in Development.* Retrieved April 19, 2000, from http://www.nua.ie/surveys/index.cgi?f=VS&art_id=905355728&rel=true.

Funk, J. (2000). *The Internet market: Lessons from Japan's I-mode system.* Unpublished White Paper, Kobe University, Japan.

Goodman, D. J. (2000). The wireless Internet: Promises and challenges. *IEEE Computer*, (July), 36-41.

Korpela, T. (1999). *White Paper of Sonera Security Foundation v. 1.0.* Helsinki: Sonera Solutions.

KPMG. (2000). *Wireless Report.* London: KPMG.

Kramer, R., & Simpson, B. (1999). *Wireless Wave II: The Data Wave Unplugged.* London: Goldman Sachs.

Leavitt, N. (2000). Will WAP deliver the wireless Internet? *IEEE Computer*, (May), 16-20.

Logica. (2000). *The Mobile Internet Challenge.* London: Logica Telecoms.

Mobile Media Japan. (2002). *Japanese Mobile Net Users.* Retrieved February 15, 2002, from http://www.mobilemediajapan.com/.

Mobilocity. (2000, May). *Seizing the M-Commerce Opportunity: Strategies for Success on the Mobile Internet.* Retrieved July 15, 2001, from http://www.mobilocity.net/.

Newsbytes. (2001). *RIM Founder Lauds J2ME as Common Platform for Wireless.* Retrieved July 15, 2001, from http://www.ayg.com/j2me/Article.po?id=1575716.

Ovum. (2000). *Wireless Portal Revenues to Top USD42 Bn by 2005*. Retrieved from http://www.nua.ie/surveys/index.cgi?f=VS&art_id=90525588&rel=true on July 5, 2000.

Peter D. Hart. (2000). *The Wireless Marketplace in 2000*. Washington DC: Peter D. Hart Research Associates.

Porter, M. E. (1980). *Competitive Strategy*. New York: Free Press.

Schleuter, C., & Shaw, M. J. (1997). A strategic framework for developing electronic commerce. *IEEE Internet Computing*, 1(6), 20-28.

Strategy Analytics. (2000). *Strategy Analytics Forecasts $200 Billion Mobile Commerce Market by 2004*. Retrieved July 10, 2000, from http://www.wow-com.com/newsline/press_release.cfm?press_id=862.

Symonds, M. (1999, June 26). Business and the Internet: Survey. *Economist*, 1-44.

Tapscott, D. (1995). *The Digital Economy*. New York: McGraw-Hill.

Ticoll, D., Lowy, A., & Kalakota, R. (1998). Joined at the bit—the emergence of the e-business community. In Tapscott, D., Lowy, A., and Ticoll, D. (Eds.), *Blueprint to the Digital Economy*. New York: McGraw-Hill.

WAP Forum. (1999, February 12). *Wireless Application Protocol—Wireless Transport Layer Security Specification*. Retrieved January 15, 2000, from http://www.wapforum.org/.

Wappup.com. (2000). *Fighting for Mobile Internet Freedom of Choice*. Paper presented at WAP Wednesday, London, July.

WireFree-Solutions. (2000a). *WAP vs. i-mode—Let battle commence*. Retrieved June 15, 2000, from http://www.wirefree-solutions.com/.

WireFree-Solutions. (2000b). *Wireless applications—Where is your opportunity?* Retrieved June 15, 2000, from http://www.wirefree-solutions.com/.

WireFree-Solutions. (2000c). *An overview of the wireless Internet with WAP*. Retrieved June 15, 2000, from http://www.wirefree-solutions.com/.

Wurster, T., & Evans, P. (2000). *Blown to Bits*. Boston: Harvard University Press.

Yankee Group. (2000). *Mobile Portals: Carrier Positioning Strategies*. Boston: Yankee Group.

Chapter IX

Mobile Business Services: A Strategic Perspective

Jukka Alanen
McKinsey & Company, Finland

Erkko Autio
Helsinki University of Technology, Finland

ABSTRACT

Mobile business services are attracting increasing attention and they promise a multibillion dollar market whose characteristics are quite distinct compared to mobile consumer services. Competitive activity among players keen on tapping into this opportunity is increasing rapidly. In this article, we look at mobile business services from a strategic business perspective. We chart the mobile business services landscape and discuss the underlying market drivers and potential end-user benefits. Additionally, we describe the competitive landscape and discuss the relative positions of the primary player groups.

INTRODUCTION

Corporations and business users have traditionally been early adopters of telecommunications solutions. During the past few years, however, it is the personal rather than the business market that has driven mobile service innovation in many leading-edge markets. Mobile chat, mobile games, and downloadable mobile handset icons and ringing tones provide good examples in Europe and Japan. In the U.S., on the contrary, the evolution of mobile services has been driven more by the corporate sector as seen in the mobile fleet management systems of FedEx and

UPS, for instance. Such applications, however, represent only early precursors of what promises to become a significant industry with real value potential. Although consumer services have attracted the greatest media attention so far, corporations are now becoming more active in deploying new mobile business solutions to achieve tangible business benefits. Moreover, in the mobile service development and provisioning industry, business solutions are attracting increasing interest as consumers have not rushed to use mobile business-to-consumer (B2C) services as enthusiastically as expected.

This chapter sheds light on the mobile business services opportunity space from a strategic perspective. First, we discuss the underlying market drivers, current obstacles, and potential benefits of mobile business services. Second, we discuss the differences between mobile consumer and business services, and map the opportunity landscape of mobile business services. Third, we discuss the competitive landscape, the value chains, and the relative positions of various player groups in this industry.

A SILENT REVOLUTION: GRADUAL CHANGE IS LEADING TO FUNDAMENTAL TRANSFORMATIONS

Over time, enterprise IT solutions have evolved from the mainframe and client-server solutions to e-business solutions such as customer relationship management (CRM) and supply-chain management (SCM) that facilitate information flow and interaction within and between organizations. Mobile business solutions represent the next wave of this evolution, further extending connectedness and enhancing interaction. But should the emergence of the mobile data medium be considered predominantly as a new access channel to current enterprise IT applications, will it add a new functionality to these, or does the mobile data medium represent a more fundamental shift in the way companies operate?

The mobile business services sector is driven by both demand- and supply-side factors. According to the Yankee Group (1999a), among large U.S. companies (with more than 5,000 employees), 20% of the workforce is already mobile, with the share of mobile workforce set to increase constantly in the foreseeable future. The adoption of mobile business services is thus driven by an increasing need for mobility, but also by technical opportunities to streamline business processes and enhance interactivity. The decreased time and place dependency of many business processes enable appealing value propositions to many kinds of organizations. According to the Gartner Group (2001a), over 80% of European corporations consider mobile devices and applications as very important for their business. A Forrester Research survey (2001a) found that approximately 70% of the Global 3,500 are considering implementation of mobile data, and 20% have already piloted with such applications. Viable mobile services are also facilitated by recent developments in enabling technologies, for instance by the notably improved network coverage and quality in the U.S. Broader bandwidth and packet-based networks enable new services and promise reduced data transfer costs. New devices with more sophisticated user interfaces as well as mobile data protocol and security improvements are transforming the feasibility of mobile data services in business applications. Driven by these factors, it has been predicted that two-thirds of the Global 3,500 will roll-out mobile applications by 2003 (Forrester Research, 2001b).

So far, five major external obstacles have prevented a more widespread adoption of mobile business services: high costs, an unclear business case, technological uncertainty, the complexity of deployment options, and the complexity of coordination between service providers. The implementation of data services over circuit-switched networks is costly. This hinders service deployment especially as companies feel uncertain about the magnitude of benefits that can be realized from such investment. The lack of empirically validated reference cases thus creates a chicken-and-egg problem. The lack of dominant designs is also holding back adoption, as companies do not want to invest in solutions that may need to be replaced soon. Technological uncertainty is also manifested in immature security: a recent Gartner Group survey (2001a) identified mobile security and privacy as a top concern, with 76% of European CIOs worried about it. Technological uncertainty is amplified by the high technical complexity of deployment created by multiple devices, networks, protocols, and architectural options. According to the Gartner Group (2001b), companies will have to support a minimum of 50 different mobile device profiles and 10 different mobile network interfaces. Finally, the implementation of an end-to-end solution requires the involvement of many different service and technology providers. The uncertainty of their roles and the complexity of coordination of their activities may deter companies from implementation efforts.

The above external issues are complemented with internal inertia and the lack of know-how. To be able to capture maximum benefits from mobile data applications, companies need to redesign their business processes in order to fully embrace the value potential of mobility. Doing so will likely require a major change effort and a set of skills that extends beyond technological know-how.

Because of the forces of inertia such as the ones listed above, the adoption of mobile business services will not happen overnight. The paradigm shift will likely take years. However, the fundamental benefits of the mobile medium are compelling enough to drive gradual adoption. Corporations deploying mobile data services want to realize the full value of their mobile workforce—through time savings, faster decision making, and improved coordination for business anytime, anywhere. Further opportunities are created as corporations learn to leverage mobility for greater agility, flexibility, responsiveness, and innovation. Through a combination of business-to-employee (B2E) solutions and new types of mobile machine-to-machine (M2M) solutions, mobile data services may well herald a discontinuity in the way corporations organize their activities. Thus, as there is potential for fundamental but gradual transformations, the impact of mobility on businesses could be characterized as a silent revolution. On the consumer segment, mobile services will be more visible to the public, and therefore more likely to capture widespread attention.

The primary benefits from mobile business services and their underlying drivers are illustrated in Figure 1. At the most basic level, the plain mobile access enables rapid access to current business applications and databases irrespective of time and place. More sophisticated mobile functionality is represented by location-specific information and real-time push notifications, for instance. The most radical impact is generated when mobile data is used as an enabler of reconfigured business processes instead of simply providing an additional access point. Process flows and user interactions, e.g., in field service management, can be redesigned to fully leverage the fundamental characteristics of the mobile medium: immediacy, ubiquity, high-end personalization, and location awareness. The three driving features illustrated in Figure 1 enable productivity gains through increased efficiency and

Figure 1: Benefits from Mobile Business Services and Their Underlying Drivers

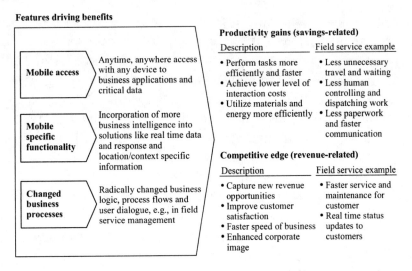

Source: McKinsey research

lower interaction costs. Additionally, they contribute to a competitive advantage and value creation by enabling enhanced value propositions to customers as well as greater agility and faster decision making.

As an example, a mobile field service solution can support productivity gains in several ways. First, it can increase the productivity of a service technician by reducing her need for travel, reducing idle time, and increasing the availability of problem-specific information at the service site, thereby enabling a more efficient service execution. Further time savings may arise from the automation of time sheets, for example. Second, a mobile field service application can improve dispatchers' productivity due to reduced manual work and more efficient communication. Third, it can also help streamline back-office processes and cut costs, e.g., in invoicing and administration. Besides productivity gains, field service applications can enhance revenue generation by enabling service companies to handle a larger service volume and cater to more demanding customers.

Interaction costs (i.e., search, monitoring, and coordination costs) account for approximately half of the labor costs in industrialized economies (e.g., 51% in the U.S.) and also for a substantial share of the labor costs in developing countries (e.g., 36% in India). Information technology can tremendously improve the interactive capabilities of individuals and organizations and thus bring huge productivity gains and savings (Butler, et al., 1997). As an added functionality of IT systems, mobile business services can significantly further enhance interactive capabilities. We estimate that mobile business services could thus generate at the global level, as a very indicative approximation, a return of tens of billions of U.S. dollars by 2005 through savings in interaction costs (see Figure 2).[1] The Gartner Group (2000), for instance, claims that mobile technology investments will likely increase the productivity of mobile workers by up to 30%. On top of interaction cost savings, further savings can be achieved in materials, supplies, and energy costs through mobile remote monitoring and materials management systems, for example. The aforementioned new revenue opportunities and improved customer satisfaction can bring economic benefits that are additional to cost savings. According to Forrester Research (2001c), firms can realize over 260% ROI on mobile business solutions over five years. To realize this value potential, an increasing number of companies across industries and geographies are working to determine what kind of services to deploy and when and how to deploy them.

Hardly any industry will be left untouched by the mobile medium. For example, mobile technology is affecting the entire energy industry including the process of drilling wells, meter reading, and energy trading (the Yankee Group, 2001a). The most significant impact is likely to be produced in industries where: (1) the degree of mobility or remoteness of activities, personnel, or materials is high; (2) the interaction (search, monitoring, and coordination) intensity is high; and (3) the

Figure 2: An Indicative Estimate of the Value Potential Arising from Mobile Business Services

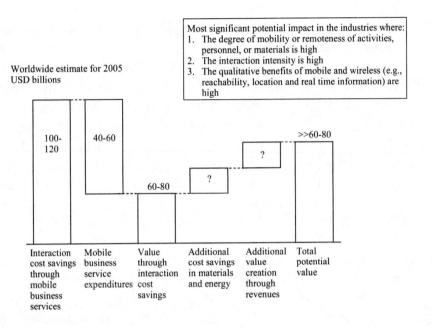

qualitative benefits (e.g., mobile or remote reachability, real-time and location-specific information) offered by the mobile medium are high. The largest opportunities naturally arise when these criteria are met in a large market (e.g., in terms of employees). The transportation industry provides an illustrative example of a large potential market significantly touched by mobile business services such as mobile fleet management and field force solutions.

THE OPPORTUNITY LANDSCAPES OFFERED BY THE MOBILE CONSUMER AND BUSINESS SERVICES DIFFER SUBSTANTIALLY FROM ONE ANOTHER

For the mobile services landscape, the main demarcation line runs between the consumer and business markets. The business market differs significantly from the consumer market in several ways. The needs of businesses are centered on supporting their mobile workforce and enhancing business processes, whereas consumers are focused on fun and convenient ways to stay in contact with their friends, spending their idle time, and finding rapidly relevant content such as local restaurants. In the business market, the greatest benefits will be created through the integration of mobile data services into business processes, e.g., through stream-lining field service operations with mobile dispatching, as well as through enhancing employees' personal productivity. Such differences in emphasis will naturally be reflected in mobile service offerings. Consumer services are generally targeted at the masses and are typically very generic. On the business side, there are also some horizontal services, but a large portion of them is quite vertically oriented (i.e., industry-specific) or tailored to the needs of a single enterprise. Business applica-tions are characterized by complexity, customization, integration with legacy information systems, and requirements for support and maintenance. In short, consumer services often build more on telecom services, whereas business services typically look more like extensions of corporate IT systems. Also, the terminals are somewhat different. The business market will have a stronger concentration of high-end devices (e.g., smartphones, integrated PDAs), specialized devices (e.g., rugged field terminals), and embedded machine-to-machine communication de-vices. Consumers are likely to use more generic, voice-dominant devices for years to come. In addition, the customer interfaces in the consumer and business markets will be occupied by very different sets of players. Mobile consumer services will be dominated by mobile operators and content providers, such as media houses. Mobile business services, on the contrary, will often be simply added into application providers' and systems integrators' current solution offerings.

Figure 3: A Generic Categorization of Mobile Services

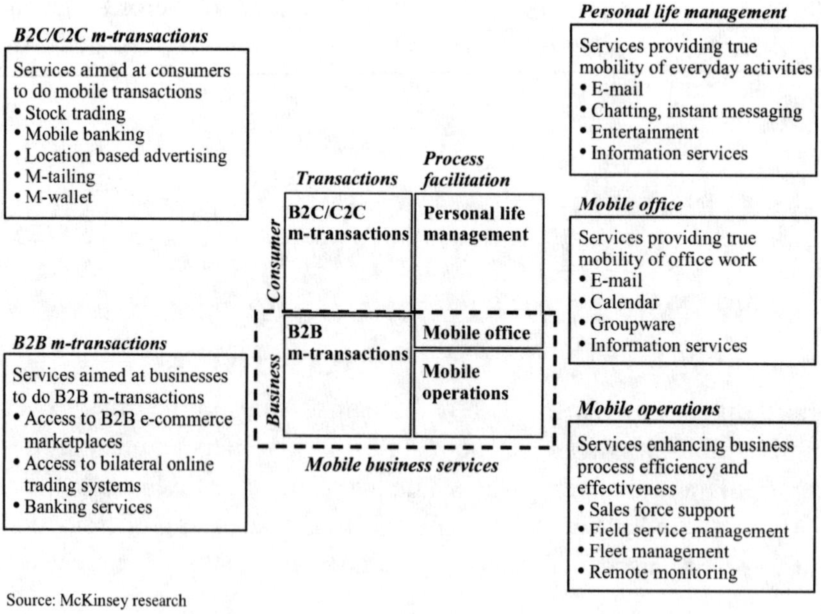

Personal life management

B2C/C2C m-transactions

Services aimed at consumers to do mobile transactions
• Stock trading
• Mobile banking
• Location based advertising
• M-tailing
• M-wallet

Services providing true mobility of everyday activities
• E-mail
• Chatting, instant messaging
• Entertainment
• Information services

	Transactions	*Process facilitation*
Consumer	B2C/C2C m-transactions	Personal life management
Business	B2B m-transactions	Mobile office
		Mobile operations

Mobile business services

Mobile office

Services providing true mobility of office work
• E-mail
• Calendar
• Groupware
• Information services

B2B m-transactions

Services aimed at businesses to do B2B m-transactions
• Access to B2B e-commerce marketplaces
• Access to bilateral online trading systems
• Banking services

Mobile operations

Services enhancing business process efficiency and effectiveness
• Sales force support
• Field service management
• Fleet management
• Remote monitoring

Source: McKinsey research

Another key demarcation line for types of mobile services runs between commercial transactions and process facilitation. Together, the two demarcation lines serve to map the mobile service landscape, as illustrated in Figure 3.

Opportunities in the mobile business services market can be further divided into three categories: mobile office, mobile operations, and mobile business-to-business (B2B) transactions. Mobile office services share some similarities with personal life management services, especially in the area of messaging. The primary role of mobile office services is to enhance personal efficiency and effectiveness. Mobile operations services are geared for efficiency and effectiveness in business operations. Of the various mobile business services, mobile office (a horizontal, industry-spanning service) and mobile operations (a group of quite vertical, industry-specific services) represent the major overall opportunities for value creation. On the other hand, we consider the size of the opportunity offered by mobile B2B transactions to be quite limited due to the typically low relevance of mobile-specific advantages in B2B transactions.

Mobile Operations

Mobile operations services can create significant cost savings through reduced interaction costs. However, significant value potential can also be found in new ways of enhancing the customer value proposition and managing the customer interface for improved customer loyalty. For example, providing field sales force with mobile access to product and inventory information can significantly enhance customer service.

One way to categorize mobile operations services is by distinguishing between vertical, industry-specific services (e.g., remote monitoring in oil fields) and horizontal, function-specific services (e.g., field sales management). The emphasis traditionally has been on tailored and quite experimental company-specific vertical services. These have focused on applying the wireless functionality in some specific areas, such as production process monitoring. However, fuelled by developments in mobile technologies, services of a more horizontal nature have strongly gained ground and are emerging with wide adoption across industry sectors.

Logistics, sales and marketing, and after-sales service represent the most immediately feasible functional areas for mobile operations services. These sectors typically comprise a significant amount of remote and field activities, and their needs are specialized enough not to be served adequately through generic mobile office services. Figure 4 presents examples of application areas in mobile operations services. Field service dispatch, sales force support, and fleet management have already stood out as the most widespread applications, and the user base for these types of services is in constant steady increase (the Yankee Group, 1999b).

Figure 4: Examples of Application Areas in Mobile Operations

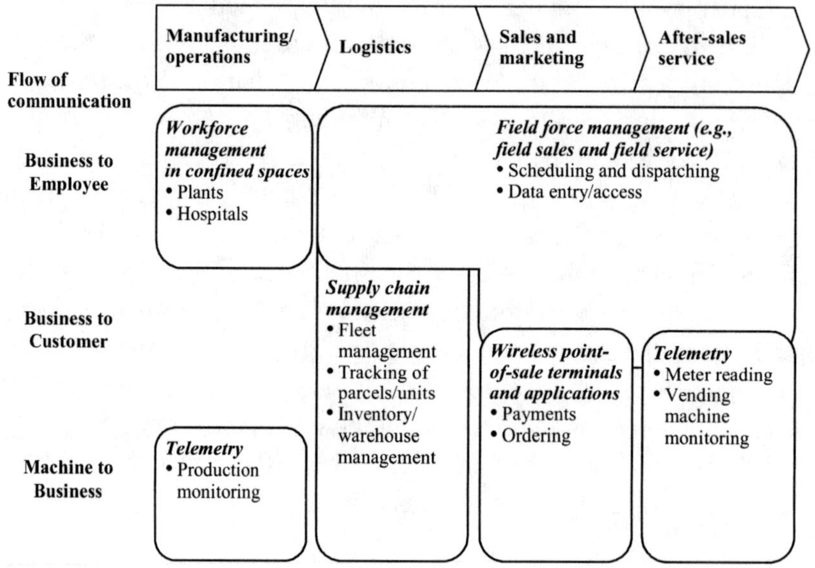

Source: McKinsey research

Many application areas presented in Figure 4 can be quite industry-specific. Therefore, there is demand for vertical services targeted at specific needs. Telemetry services, for example, are typically specific to each industry. The Yankee Group (2001b) expects U.S. mobile telemetry (e.g., mobile asset tracking, point-of-sale, vending equipment) revenues to grow to USD 3.4 billion by 2006. On top of this, there can be a vast market for fixed wireless telemetry solutions. For instance, the 240 million gas, water, and electricity meters in the USA represent a potentially large telemetry opportunity (Forrester Research, 1998). One example of wireless telemetry in the energy industry is the recent partnership between Aeris.net and American Innovations, who joined forces to offer remote monitoring capability covering two million miles of pipelines throughout North America. Data such as level of flow, temperature, and pressure is set through a wireless telemetry hub and then routed to the Internet. Wireless telemetry is also being utilized for meter reading and consumer monitoring of energy consumption (the Yankee Group, 2001a).

Although the market for industry specific mobile services is fragmented into numerous segments, the combined value potential is likely to be substantial. An illustrative example of a vertical niche solution is CRF Box's application for mobile collection and management of data from clinical trials of experimental drugs in the pharmaceutical industry. Given the huge R&D costs associated with the development of new drugs, any applications that enable more efficient product development are likely to offer considerable savings potential. These kinds of targeted solutions have potential to significantly transform key processes in their respective industries. However, one of the questions concerning the future development of industry-specific niche services is to what extent more generic and horizontal solutions could displace them in the future as they become more powerful and versatile. Increased versatility, e.g., enabled by more intelligent mobile terminals, can create opportunities to provide multi-industry solutions that may require only light configuration for each specific industry.

Mobile Office

The possibility to improve personal efficiency offers significant potential value. Enhanced communication and collaboration as well as management of personal information are already driving the roll-out of mobile office services. According to several surveys and estimates, mobile office is the most widely utilized mobile business solution with the global user base potentially exceeding 100 million in 2004 (Ovum, 1999; Yankee Group, 1999b; IDC, 2001a, 2001b). Mobile office services can entail communication and collaboration services (e-mail, unified messaging, groupware messaging), personal information management (PIM) services (e.g., calendar, task lists, calculator, word processor), and information access services (e.g., personal files, corporate databases, access to external business information services). Figure 5 illustrates an example of mobile office services. These personal process facilitation services are increasingly becoming incorporated in a mobile corporate portal, but they can also be partially situated on the

Figure 5: An Example of Mobile Office Services

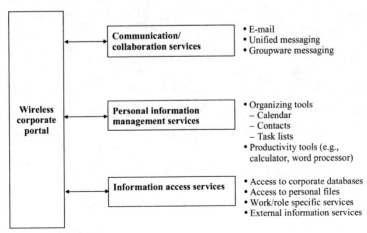

device. The same portal interface often also functions as the overall access channel to all the mobile services available to the business user, including mobile access to the public Internet.

Mobile B2B Transactions

Mobile devices are quite well suited for consumers to do small transactions, receive and review localized advertisements, and monitor bank accounts. However, both the applicability of mobile devices as well as the relevance of mobility is significantly lower in business-to-business m-transactions. A mobile terminal typically offers merely an access point to B2B trading, instead of adding significant new functionality or changing transaction processes fundamentally. The main value from the mobile medium is not in improved transaction processing. The revolution in B2B transactions occurred already with digitalization and the Internet, taking out most of the cost savings potential. But mobile will expand the reach of current e-marketplaces and enhance their place independence. Blackberry's mobile terminals are quite widely used for distributing financial information, for example. Also, the mobile medium may facilitate mobile field transactions, like in transportation capacity exchange. Using mobile technology, truckers would be able to access capacity requests while on the road and market their own excess capacity. For e-marketplaces, mobile can also offer added functionality through notifications and pre-defined alerts in auctions (a time-critical form of trade that consists of standardized simple transactions) and exchange trading, which lead to faster responsiveness. Moreover, some limited opportunities are emerging in one-to-one m-transactions and in mobile banking. The opportunities arise mainly for repeated and routine transactions where the number of transaction parameters is small, like in accepting scheduled bank account transactions or ordering regular spare parts. A potential example is direct field sales of timber where mobile can enhance integration of suppliers' and buyers' processes, and facilitate interaction in transaction acceptance and delivery.

Solution Types in Mobile Business Services

Solutions for mobile business services differ in terms of their sophistication of mobile functionality. Many of the solutions offered today provide simple mobile access (e.g., e-mail access) instead of leveraging new mobile-specific functionalities and business processes (e.g., fleet management solutions with location data). Another differentiating dimension is the scope and the level of integration of the solution to corporate IT systems. There are very focused (single-purpose), stand-alone solutions targeted for specific mobile purposes, whereas some solutions serve broader needs and integrate more deeply into existing systems. Solution types in

Figure 6: Solution Types in Mobile Business Services

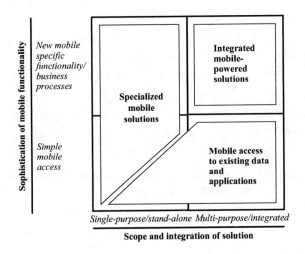

Source: McKinsey research

mobile business services can be categorized along these two dimensions as illustrated in Figure 6.

The vast majority of solutions seen today either take the form of mobile access to existing data and applications, or that of specialized mobile solutions. Companies that offer solutions enabling mobile access to existing data and applications include Microsoft, Fenestrae, Infowave, and Smartner Information Systems, for instance. Their solutions typically consist of middleware that supports connectivity with various device types, content formats, and networks while also offering interfaces for accessing business applications and databases. Essentially, they adapt content from existing systems and deliver it in a suitable format to mobile devices. Specialized mobile solutions, on the other hand, typically offer a stand-alone, end-to-end solution that supports focused mobile-driven business processes such as fleet management. Truck24's packetized fleet management solution for SMEs provides an example. Aether Systems is one of the prominent providers of specialized mobile solutions, having started from financial services and expanded to several other verticals.

Integrated mobile-powered solutions are capable of combining the business benefits of the two aforementioned solution types. They are solutions designed to interface tightly with a broad range of mobile and non-mobile legacy applications within an organization and to leverage the unique advantages of the mobile medium. Similar to specialized mobile solutions, integrated mobile-powered solutions also enable changed business processes, but they do so in a broader scope and in an integrated way. They are typically based on new application designs, or radical additions to existing applications with mobile-specific features, mobile-enabled process logic, flow, and user dialogue. There are very few examples seen today. One example could be FieldCentrix's mobile field service system that, besides dispatching messages, also covers online equipment catalogs and diagnostics, spare parts ordering, time sheet automation, completion e-mails to customers for review, customer acceptance signatures, and integration with accounting and invoicing as well as web-based dispatching systems. Mobile-powered construction project portals with project management applications, collaborative tools, supplies ordering, and job sheets and dispatching could provide another potential example.

Providers of mobile business services are developing their solutions predominantly using two alternative approaches and paths. Several players such as large established IT providers (e.g., Microsoft and SAP) have started from the lower left-hand corner by adding mobile access to single or very few existing applications. They are moving to the lower right-hand corner by extending the number of applications and databases that can be accessed through the mobile. The next large step is to incorporate more mobile functionalities and intelligence into these access solutions and thus move towards the top of the diagram, that is, to the integrated mobile powered solutions. Another group of more vertically focused players such as Aether also started from the lower left-hand corner by providing specialized solutions that can be accessed through the mobile. By incorporating more mobile functionalities to the design, they are moving toward the upper left-hand corner. A key question for further expansion is whether to stay in specialized solutions, or to move towards broader, integrated solutions by extending scope and tying separate specialized solutions together through leveraging common elements in these solutions.

THE COMPETITIVE LANDSCAPE: A MIXTURE OF IT AND TELECOM VALUE CHAINS

Corporations' increasing interest in mobilizing their workforce and business processes stimulates activity among players who try to develop and provide mobile services. The billions of dollars spent in deploying and utilizing mobile services will be divided between technology providers, solution providers, network operators, and business content and transaction service providers as illustrated in Figure 7. Technology providers are moving into the mobile space and mobile-enabling existing applications or creating brand new ones. A number of mobile network operators seek to expand their role beyond simply acting as the pipeline through

Figure 7: The Competitive Landscape in Mobile Business Services

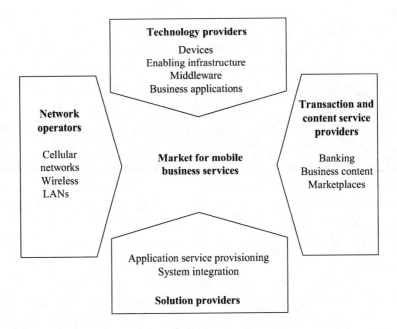

which the mobile traffic is channeled. Providers of transaction and business content services, such as financial information providers and marketplaces, are eyeing their current offerings to identify where mobile could offer a feasible value proposition. In addition, solution providers, including both the large established providers and new up-and-coming competitors, strive to be coordinators of the pack and owners of the corporate customer interface.

The competitive landscape for mobile business services is effectively a mixture of the traditional IT value chain (e.g., application software developers and system integrators) and the telecom value chain (e.g., network operators and handset vendors). Telecom players are traditionally strong in the mobile consumer market, whereas IT players carry significant weight in enterprise solutions. This also highlights one of the operators' challenges in the business market: how to develop the capability to actively participate in the IT-driven activities of the value chain instead of being just a passive data pipe. In mobile services, operators have better chances to establish a foothold in value-adding services, as the commoditization of data traffic has not yet proceeded as far as in the wireline and Internet operations. Because the capabilities, needs, and strengths of the different players are different, there is ample opportunity for partnering. IT houses, for example, largely see mobile carriers as valuable solution or distribution partners. As an example, IBM's current approach is to enable operators to better cater to the corporate sector.

When examining linkages between the various players in the market, the limits of the traditional value chain concept soon become evident. As illustrated in Figure 8, there typically does not exist a single value chain, but actually at least three distinct value chains linked to one another. The single value chain representation, illustrated at the top of the figure, is too simplistic, as the players in this market do not operate in a sequential mode, and there is no single identifiable value flow from the left to the right. Instead, we can identify three primary value chains in action in mobile business services. There is: (1) service development value chain (i.e., activities in building a technological solution), (2) service provisioning value chain (i.e., activities in operating and using the solution), and (3) device value chain (i.e., activities in providing end-user devices to enterprises). In mobile services these three chains are superimposed over one another, and the result is a value net in which several different players cooperate and compete with the objective of establishing a dominant position. This is the reason why we are observing a high level of interaction between different players and why we are seeing some players extend their position to cover several activities (or roles) in the value net.

When the three value chains are meshed together, the result is a value net. Figure 9 shows an illustration of the mobile business services value net and some of the primary linkages between players within it. In this simplified example, a system integrator works with the enterprise customer so as to tailor a solution that suits the customer's needs. A middleware provider and an application provider bring key elements to the overall solution that is put together by the system integrator. The

Figure 8: Examples of Value Chains in Mobile Business Services

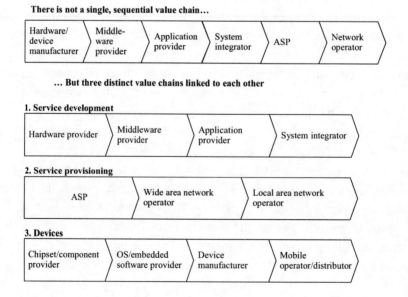

middleware provider may have established cooperation with a device manufacturer to incorporate support for the features and functionalities of the mobile terminal. An application service provider (ASP) takes the responsibility for hosting and operating the solution on behalf of the enterprise customer. The end-users use the ASP service through a mobile operator's network. This is only one example of many possible value nets, and not an all-purpose generalization. Roles, linkages, and power structures may vary depending on the specific service and the situation.

The value net provides a good model with which to analyze and anticipate the strategies of different players as the competitive landscape is under constant change. Because it clarifies the roles of different players, it can be used to analyze the strengths and potential complementarities between these. Thus, the value net model can be used to uncover and anticipate the alliance strategies of the different players as these jockey for position in the shifting landscape. Since the evolution of the mobile services industry is still in an early phase, the players' domains and roles are not clearly defined as yet. However, competition is already intensifying, and first movers are achieving significant scale. This is forcing players to sharpen their strategic focus and more clearly select and communicate the roles they want to pursue in the mobile business services sector. The initial period of broad strategic exploration and experimentation is ending, and more heated and focused battles are starting. Next, we briefly discuss the strategic positions of four player groups: technology providers, solution providers, transaction and content service providers, and network operators.

Figure 9: Key Linkages Between Roles in an Example of a Value Net

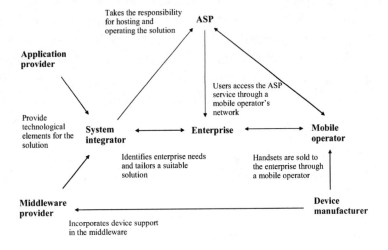

Technology Providers

Technology providers, such as middleware and application developers, are primarily leading innovation in mobile business services. Because of the need to integrate with corporate IT systems, business solutions tend to require more tailoring than average consumer services. This tends to emphasize the role of technology providers, as many services cannot be easily managed by the network operator. There is a significant market for software developers to mobile-enable traditional enterprise application software, such as sales force management systems, and to create new mobile services, such as fleet management applications. The scale of the opportunity depends partly on the solution types and their evolution. If the prevailing solutions are those offering basic access to existing applications and data, their relatively low complexity, functionality, and intelligence may limit the technology providers' upside. On the other hand, the emergence of specialized or integrated mobile solutions will also help strengthen the position of technology providers. To capture the opportunity, technology providers need to overcome the challenge of building distribution channels. To get their technology to the market, technology providers may need to offer basic integration services and even manage the new service in the beginning. At some point, however, technology providers need to decide whether they want to focus and excel in productizing technology, and partner with system integrators, ASPs, and network operators to get the products to enterprises. Some technology areas such as middleware platforms are by nature quite invisible to the end-customer, which makes their own complementary products or partnerships with other players a necessity.

From a device manufacturer's perspective, a central question in mobile business services is what capabilities should be incorporated in the device, and where the focus should lie between specialized and generic devices. Mobile office solutions can build on generic devices, but mobile operations services, especially some vertical applications, may require specialized or rugged devices designed for specific purposes. It may take time, though, until these vertical segments are large enough to merit specialized devices. As device manufacturing is largely a volume-driven business, there is a trade-off involved between high volumes with low unit costs and the need to better serve specific segments. Another key question is where and to what extent device capabilities should be integrated with server-side systems to form end-to-end solutions. RIM's initial success with the BlackBerry e-mail device was partly due to combining applications, device, and infrastructure in the right way. RIM directly addressed enterprises' key concerns: integration, provisioning, security, and maintenance from server through to device. Device manufacturers need to consider whether they should take a broader role in facilitating mobile business services, as they stand to significantly gain from faster adoption. There is a continuum between stand-alone devices and devices closely bundled with partner- or self-provided enterprise solutions.

Solution Providers

In the IT world, the corporate customer interface is largely managed by solution providers who know the corporate IT infrastructure and can deal with legacy systems. Mobile solutions are broadening in scope, and they are becoming increasingly integrated to other channels and systems, which implies a need for system integration skills. According to an IDC survey (2001b), 80% of companies prefer an integrated mobile solution to a standalone system. IDC (2001c) also reports that 91% of U.S. companies considering the deployment of a mobile solution indicated they would consider using an external service provider. Driven by the needs and preferences of companies, integration services are expected to capture clearly the largest share of the software-related mobile spending (Ovum, 2001). The key question for solution providers is how to ensure the ownership of the corporate customer and become the hub in mobile business service provisioning. In the case of incumbent players, this may call for significant knowledge and skill upgrades. Besides specific technical skills, the ability to educate organizations about mobile opportunities is an important ingredient to strengthening corporate customer ownership. One of the incumbents' most important assets is broad corporate IT understanding, as many mobile business services also require a wireline dimension. Also the end-to-end solution provisioning capability can be in incumbents' favor. However, there may be room for challengers, especially in specialized mobile solutions. These require specific skills and may be too niche oriented, at least initially, for most large system integrators. The stand-alone characteristics of specialized solutions also require less integration and legacy know-how, thus weakening incumbents' starting position. Especially in new, untested, and niche-oriented specialized mobile solutions, technology providers may have to initially take the role of the system integrator to themselves, as established integrators often wait for market proof and substantial growth before they commit resources to it. When this happens, acquisitions of specialized technology providers may provide a rapid entry to the market.

Many system integrators, like IBM, have traditionally offered hosting services to complement their integration business. In addition to incumbents, also focused challengers and mobile network operators are targeting the managed services opportunity. Hosting and other ASP (application service provisioning) operations can be even more important in mobile business services due to the higher complexity, the low initial strategic importance of the mobile medium, the limited internal mobile know-how of enterprise IT departments, and the willingness to undergo trials before broader adoption. Players like Aether have developed easy-to-use service offerings that reduce complexity in specific vertical areas, like financial services. By offering a fully managed service with a hosted set of applications, Aether managed to build a solid position in providing mobile financial services and, from that basis, started to expand into other vertical services.

Transaction and Content Service Providers

Transaction and content service providers belong to a key group of players in the consumer market, along with mobile operators. In the business market, these players have a clearly weaker relative position. Although businesses may be more willing than consumers to pay a 'mobile premium' for timely and relevant content, the majority of the content and applications in mobile corporate use is internal to the corporation or belongs to its business network. Corporate customers are largely owned by solution providers, such as system integrators, and providers of external services such as financial information agencies are often pushed to a peripheral position. External service providers can typically only offer simple mobile access, instead of mobile-driven functionalities and improved business processes. Therefore mobile can typically generate only limited, incremental value to business content and transaction service providers.

Banks and transaction providers are experimenting to offer their services to corporations and business users through the mobile medium, but many still struggle, for instance, with security and quality issues. For content providers it is critical to customize their service offerings—such as intelligent information alerts—for well-defined business needs. Services often tend to lack distinctive differentiation and thus fail to build a loyal user base. The critical issue is to understand where the mobile medium really adds value to business content or transactions and to distinguish true market opportunities from mere technical possibilities.

Network Operators

The business segment has been the main source of value for telecom operators in the past, but the profitability is currently at risk, as mobile is becoming bundled as part of the overall data service offering. As new technologies, like mobile LANs, become practical alternatives for providing high-speed mobile data access especially in hot spot areas, new players utilizing these technologies may enter the network operator market. All this is likely to contribute to gradual commoditization of mobile data capacity provisioning, leading to continuing price pressure and intensifying battle for market share in communication services.

The scenario of weakening data traffic margins, combined with the fundamental differences between the business and the consumer services, implies a degree of urgency for mobile network operators, who must think through their approach to mobile business services. To recoup the high 3G license costs and network investments, mobile operators may have to turn to the business market and take a more aggressive posture. Some operators, like Nextel Communications, are already using mobile business services as a way of profiling themselves against competitors. It is naturally in operators' interests to accelerate the adoption of mobile solutions as operators are one of the primary beneficiaries of increased data traffic.

Potential operating layers from a network operator's perspective are illustrated in Figure 10. In the lower, communication-focused layers, operators may have a corporate market opportunity in wireless LANs and VPNs, for instance. If operators aspire to a broader role and want to strengthen their ownership of the corporate customer, they can consider expanding to wireless system integration and managed services. This can, however, be a major stretch for most operators as they typically lack knowledge of corporate legacy IT systems, integration skills, and software vendor partnerships as well as software business mindset and culture. NTT DoCoMo, for instance, has primarily relied on partnerships with technology providers (e.g., Lotus) and system integrators (e.g., IBM and NEC). On the other hand, Sonera, the Finnish mobile operator and service provider, has taken multiple roles in its fleet management solution by developing the technology itself, and working simultaneously in the roles of system integrator, ASP, and data traffic provider. Moving beyond the traditional scope of operations can be difficult, and expansion efforts require considerable strategic prioritization based on existing key

Figure 10: Potential Operating Layers from a Network Operator's Perspective

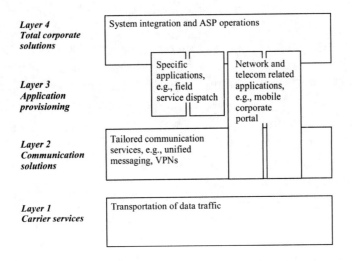

Source: McKinsey research

assets as well as several partnerships. Besides the potential vertical expansion, alone or through partnerships, another key issue is to which customer segments operators should put emphasis on. The SME (small and medium-sized enterprise) market is largely neglected by large system integrators who typically do not have a profitable model for serving it. Most operators already have SMEs as their customers and could be in a position to provide packetized, low-end solutions to them through the ASP model. However, the SME segment is quite heterogeneous, very cost-conscious, and poorly aware of the mobile opportunities, which does not make it an easy sell.

CONCLUSIONS

In this article, we discussed the drivers and obstacles for mobile business services and evaluated potential benefits of the mobile medium to enterprises. We briefly compared the business and the consumer markets and described the various types of mobile business services. Finally, we discussed the competitive activity, the value chains, and the starting positions of various types of competitors in the market. We summarize our observations in three key findings.

First, the impact of mobile on business can be seen as a silent revolution. There is potential for fundamental, transformative benefits through improved efficiency and effectiveness of both business-level and personal-level processes. This level can be achieved when enterprises go beyond plain mobile access and shape their operations using the full set of mobile and wireless features. However, the adoption of mobile in enterprises will be gradual and slowed down by technical, business, and organizational factors. Deployments may happen behind the scenes and capture clearly less media attention than consumer services.

Second, mobile opportunities in the business world differ significantly, in a number of ways, from the consumer market. Business services respond to enterprises' needs to support their mobile workforce and enhance business processes. Therefore, business solutions tend to be complex, customized, and require integration as well as support and maintenance. The two primary, large-scale opportunities within mobile business services are mobile office (e.g., e-mail and PIM) and mobile operations (e.g., field service, fleet management).

Third, the competitive landscape is a mixture of the IT and the telecom value chains. The strength of IT houses builds on their extensive expertise and relationships in the traditional corporate IT market and their legacy system integration know-how. On the other hand, the position of mobile network operators still appears quite central as IT houses prefer to partner with operators instead of bypassing them. The primary player groups in the market are technology providers, solution providers, and network operators, each having their own assets and roles. Linkages between companies are not best represented with a sequential value chain. Instead, we can use three distinct value chains (service development, service provisioning, and devices), or value net mapping to illustrate and analyze players' strategies and alliance behavior.

ENDNOTES

1   The methodology is partly based on Autio, Hacke, and Jutila (2001). The interaction cost savings in Figure 2 include the labor cost savings in search, monitoring, and coordination activities that are enabled by mobile business services. The analysis utilizes the interaction cost database of McKinsey & Company (see Butler et al., 1997). The expenditures include those costs such as data traffic, device, and IT solution costs that are attributable to mobile business services. Predictions for the mobile business service expenditures vary significantly. For instance, Ovum (2001) predicts the total worldwide spending (excl. devices) to amount to USD 29 billion in 2006. Merrill Lynch (2001) forecasts the enterprise mobile Internet market (excl. devices and data traffic) to represent globally a USD 15 billion market in 2005. The Yankee Group (2001c) estimates that enterprise messaging and remote access data revenues will total USD 17 billion in 2006 alone in Western Europe.

ACKNOWLEDGMENTS

The authors wish to thank present and former colleagues at McKinsey & Company whose work has contributed to the creation of this chapter. Specifically, we would like to name Vesa Jutila, Marcus Hacke, and Per Hansson.

REFERENCES

Autio, E., Hacke, M. & Jutila, V. (2001). Profit from wireless B2B. *McKinsey Quarterly, 1*,20-22.

Butler, P., et al. (1997). A revolution in interaction. *The McKinsey Quarterly, 1,* 4-23.

Forrester Research. (1998). *Telemetry's Time is Coming.*

Forrester Research. (2001a). *Mobile Data Finds Niche in Risk-Tolerant Firms.*

Forrester Research. (2001b). *The Real Cost of Mobility.*

Forrester Research. (2001c). *Winning Enterprise Mobile ROI.*

Gartner Group. (2000). *How to Build a Wireless Office: The Next Wireless Revolution.*

Gartner Group. (2001a). *European CIOs are Bullish about M-Commerce but Will Not Pay Extra for 3G.*

Gartner Group. (2001b). *Gartner Advises Businesses on Wireless Application Gateways Selection as Market Consolidates.*

IDC. (2001a). *Desired Mobile Enterprise/Corporate Applications in the U.S.*

IDC. (2001b). Wireless ASPs: *Where's the Demand?*

IDC. (2001c). *Going Mobile: A Look at the End Users' Needs.*

Merrill Lynch. (2001). *Mobile Internet/Wireless Data.*

Ovum. (1999). *Third Generation Mobile: Market Strategies.*

Ovum. (2001). *Mobile Intranets: Towards the Wireless Enterprise.*

Yankee Group. (1999a). *YC200 Survey.*

Yankee Group. (1999b). *Wireless/Mobile Data Applied Vertically: A Business Segmentation Model.*

Yankee Group. (2001a). *Wireless Technology in the Energy Industry.*

Yankee Group. (2001b). *Wireless Mobile Telemetry Forecast: Can Flora Grow on an Icy Glacier?*

Yankee Group. (2001c). *European Cellular Market Forecast, Part 2: Maintaining Growth through Mobile Data.*

Chapter X

Mobile Portals:
The Development
of M-Commerce Gateways

Irvine Clarke III and Theresa B. Flaherty
James Madison University, USA

"[Mobile Internet] is going to be the most fantastic thing that a time-starved world has ever seen."
                    Jeff Bezos, CEO, Amazon.com (KPMG, 2000, p. 4)

ABSTRACT

The proliferation of mobile Internet devices is creating an unparalleled opportunity for mobile commerce. Factors composing a productive M-commerce portal development strategy are investigated to improve a company's strategy. Also explored are the nonpareil benefits of mobile applications to introduce a five-step approach for developing an effective mobile portal strategy.

INTRODUCTION

In this new decade, marketing is poised to witness an unprecedented explosion of mobility, creating a new domain of mobile commerce (Kalakota & Robinson, 2002). Mobile commerce, or M-commerce, is the ability to purchase goods anywhere through a wireless Internet-enabled device (e.g., cellular phone, pager, PDA, etc.). M-commerce refers to any transaction with monetary value that is conducted via a mobile network. It will allow users to purchase products over the Internet without the use of a PC. Once nonexistent, M-commerce is now the buzzword of the marketing industry (King, 2000). "Within five years, individual E-commerce services will be primarily delivered by wireless, and the wireless terminal will become the window of choice to the transactional E-world," says Neil Montefiore, executive of Singapore mobile operator, M1 (Hoffman, 2000, p. 2).

Over the past few years, the marketing of E-commerce has become increasingly reliant upon portals to attract and retain users. Portals are the preferred starting point for searches that provide the user easily customizable architecture for finding relevant information. Portals provide the valuable gateways for getting users to their desired destinations. About 15% of all web page-view traffic goes through the top nine portals, making them some of the most valuable land on the web (Monohan, 1999). This heavy traffic flow gives the web-based portal a unique position in the corporate E-commerce strategy with even greater potential influence for mobile applications. For mobile devices, these portals take on even greater significance, as consumers are unwilling to spend long periods "surfing" on these inherently less user-friendly wireless devices. By the year 2006, 25 million people are expected to be dedicated wireless portal users (Carroll, 2000). Therefore, the success of M-commerce may be partially dependent upon the successful development of effective consumer-oriented mobile portals.

As M-commerce success will likely depend upon maintaining consumer utilization of these gateways, the companies that leverage the unique characteristics of wireless devices will gain exploitable advantages in the mobile marketplace. Due to current technological limitations, and varying mobile consumer behavior patterns, portals developed for mobile devices must emphasize differing characteristics than traditional web-based portals. As such, many traditional portals may be unsuited for application in the mobile world.

> *"Traditional portals are not providing information that is specific enough for the user of a mobile portal. They are not able to incorporate location-specific information nor do they have the data and knowledge of each customer that the mobile operator has."*
> (Durlacher Research, 2000, p.65)

Despite tremendous interest in the melioration of M-commerce, there is little, if any, research that examines how to develop and integrate portals into a comprehensive M-business strategy. Utilizing conventional portals may be insufficient in the mobile wireless world. Therefore, the primary purpose of this chapter is to explore the factors that compose a productive mobile portal strategy. The nonpareil benefits of mobile portal applications are investigated and a process is introduced for effective mobile portal strategy development. An enhanced understanding of mobile portals could acutely improve a marketer's ability to utilize this emerging mobile technology.

BACKGROUND

Mobile devices have been the fastest adopted consumer products of all time with more mobile phones shipped than automobiles and PCs combined (de Haan, 2000). By 2003, there will be 1.4 billion mobile phones worldwide and half will be Internet-enabled (Zabala, 2000). The proliferation of wireless capability has created an emerging opportunity for E-commerce providers to expand beyond the traditional limitations of the PC. Fueled by such enabling technologies as 3G broadband capability, eXtensible Markup Language (XML), Compact HTML (CHTML), Wireless Markup Language (WML), Wireless Application Protocol (WAP), General Packet Radio Services (GPRS) and Internet ready mobile terminals, E-commerce is now poised to take advantage of the current accretion of mobile devices. "The wireless world is a parallel universe almost as large as the Net, and the two are beginning a fascinating convergence," said Swapnil Shah, director of Inktomi Europe, a search engine and caching solutions company (Rao, 2000, p. 1).

Commerce will transpire as organizations induce new methods to employ these mobile devices to communicate, inform, transact and entertain using text and data via connection to public and private networks. It is predicted that this emergence of m-commerce will happen even faster than the development of E-commerce—in roughly the time between the invention of the first Web browser and now (Schenker, 2000). "If you look five to 10 years out, almost all of E-commerce will be on wireless devices" says Jeff Bezos, chief executive and founder of Amazon.com (McGinity, 2000, p.1). Consequently, within the next five years, one-quarter of all e-commerce will take place through wireless devices (Zabala, 2000). Forecasts estimate the wireless web to be as large as the wired web of today and worldwide m-commerce exceeding $200 billion by 2004 (M-commerce Times, 2000; Shaffer, 2000).

The potential of M-commerce is considerable for those willing to develop mobile-specific business models. However, as M-commerce matures, current mobile operators will rely less upon usage fees and increasingly derive revenues from content and services. Additionally, M-commerce is going to bring about a massive change in the way users consume products and services.

"It is key that commerce companies recognize M-commerce as a completely unique service. Cell phone users are more impatient than Internet users. The paradigm here is not surfing; all services for the mass market have to be pitched at users in such a seamless way that

they need not even be aware that they are accessing the Net."
Cindy Dahm, European director for Phone.com (Rao, 2000, p. 2)

Those best able to provide value-added user experiences, through content aggregation and portal development, will achieve long-term success. Merely extending the current Internet presence will not be enough. "Mobile Internet customers will be more demanding. They will want personalized service to meet their individual wants and needs and will no longer be satisfied with being a mass market" (KPMG, 2000, p. 2). Providers must take advantage of the characteristics which make M-commerce distinct from E-commerce to develop truly unique and compelling services rather than replicating current E-commerce models.

What Is a Portable Portal?

The word portal is derived from the Latin *porta*, or gate, through which something will pass, in an effort to get to another place. In the traditional sense of the word, the portal is not the desired end-state. Rather, a portal is a necessary, or convenient place one must go to get to the desired location. For example, the airport is not the desired location for most people, rather a necessary portal through which they must pass to obtain transportation to another location. Similarly, portals assist by directing the transport of the web-user, to the ultimate location of their choice. The portal is intended to be the beginning point of a consumer's Internet experience. For example, sports fans may start at the ESPN portal site [http://espn.go.com/] to find the latest sporting news and information. Busy women looking to share news and advice can begin at the iVillage.com portal [http://www.ivillage.com/]. Portals further differ from traditional web pages in that they "provide seamless access to a variety of goods and services via a single interface based on a predefined profile of preferences" (Lazar, 2000, p. 52). Portals serve as a gateway or entry point to other content and feature customizable architecture that allow users to integrate data from disparate sources.

Mobile portals, sometimes referred to as "portable portals," are typically developed to assist the wireless user in their interaction with web-based materials. Today, most mobile portals are being formed by syndicating content providers into a centralized source of personal productivity information. Mobile portals are often modeled by aggregating applications (e-mail, calendar, instant messaging, etc.) and content from various providers in order to become the user's prime supplier for web-based information. Mobile portals differ from traditional web-based portals by a greater degree of personalization and localization than traditional web portals (Daitch et al., 2000). Representative objectives for such mobile portals may be to attract the desired viewers, and build valuable customer relationships, through value-added content and community services, to augment the overall wireless Internet experience and build long-term customer loyalty.

Established portal players, such as AOL, Yahoo! and MSN, have recognized the potential impact of the mobile Internet and have created mobile portals targeting U.S. subscribers. However, many of the traditional portal players are experiencing difficulties adapting to the mobile world. The mobile portals emerging today are, in many ways, a stripped-down version of traditional web portals, without an understanding of the special requirements of mobile users (Kobielus, 1999). Consequently, these offerings are unacceptable to mobile Internet users (Datamonitor, 2000).

> *"The mobile portal strategy of the traditional portal players often lacks in-depth understanding of national mobile markets and of the specific local dynamics involved in building businesses in a territory. In addition, the differences between a fixed and more mobile portal model are non-trivial and, as such, lessons learned in the past are not necessarily directly transferable."*
>
> (Durlacher Research Ltd., 2000, p. 69)

Current portal strategy is based on a traditional paradigm of consumers as passive receivers of communication efforts with the portal provider holding control of the "when" and "where" of information. With wireless Internet-enabled devices, consumers now have more discretion of "when" and "where" the information is available, creating a demand for a specialized portal offering.

Unique Characteristics of Mobile Portals

The mobility-afforded wireless devices will shape mobile portals into a disparate entity from conventional portals. The advantages of mobile devices provide a greater offering of *value-for-time* to users. That is, by accessing the Internet through mobile devices, users will be able to realize additional value allowances for any specified period of time, that fixed-line users will not be able to achieve. Information will now truly become available anytime, anyplace and on any wireless device. As identified in Figure 1 below, m-commerce portals differ from traditional e-commerce portals on four dimensions: ubiquity, convenience, localization and personalization.

Ubiquity: Mobile devices offer users the ability to receive information and perform transactions from virtually any location on a real-time basis. Thus, mobile portal users can have a presence everywhere, or in many places simultaneously, with a similar level of access available through fixed-line technology. Communication can take place independent of the user's location. The advantages presented

Figure 1: Unique Characteristics of Mobile Portals

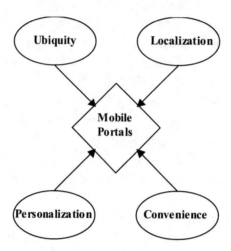

from the omnipresence of information and continual access to commerce will be exceptionally important to time-critical applications.

Mobile portals, for example, can leverage this advantage of ubiquity by providing alert notifications, such as for auctions, betting and stock price changes, which are specified by the user as an important part of relevant personal content. As such, the real-time, everywhere presence of mobile portals will offer capabilities uniquely beneficial to users. Industries that are time and location sensitive, such as financial services, travel and retail, are likely to benefit from portals exploiting this value-added feature of M-commerce.

Convenience: The agility and accessibility provided from wireless devices will further allow M-commerce to differentiate its abilities from E-commerce. People will no longer be constrained by time or place in accessing E-commerce activities. Rather, M-commerce could be accessed in a manner which may eliminate some of the labor of life's activities and thus become more convenient. For example, consumers waiting in line or stuck in traffic will be able to pursue favorite Internet-based activities or handle daily transactions through M-commerce portals. Consumers may recognize a higher level of convenience that could translate into an improved quality of life.

One opportunity to increase value lies in the fact that M-commerce capabilities allow the consumer to shop where they are not located. This ability to obtain information and conduct transactions from any location is inherently valuable to consumers. As such, M-commerce portals offer tremendous opportunities to expand a client base by providing value-added services to customers heretofore difficult to reach. By making services more convenient, the customer may actually become more loyal. Consequently, communication facilities within the mobile portal are key applications for the delivery of convenience. Consumers may seek M-commerce portals which can deliver functions such as: sending and receiving e-mail, voice mail forwarding, conference calling, faxing, document sharing, instant messaging and unified messaging, as well as transactional-based activities.

Localization: Knowing the geographical location of an Internet user creates a significant advantage for M-commerce over E-commerce. Location-based marketing, via cellular triangulation and global positioning technology (GPS), will soon be available in all mobile devices. Through GPS technology, service providers can accurately identify the location of the user so that M-commerce providers will be better able to receive and send information relative to a specific location. Since mobile devices, such as cell phones, are almost always on, vendors will know the location of their customers and can deliver promotions based on the likely consumer demands for that location.

Location-specific information leverages a key advantage a mobile portal has over a traditional web portal by supplying information relevant to the current geographic position of the user. Portable portal providers will be able to both push and access information relevant to the user's specific location. Portable portals may serve as points of consolidation of consumer information and disseminate the relevant information for a particular location. This can be based on profile data built on the user's past behavior, situation, profile and location. As such, real-time discounting may become the "killer application" for M-commerce. For example, each time a customer visits a local grocery store, Procter & Gamble can send, directly to a wireless device, through the designated mobile portal, tailored promotional information on their products.

Personalization: Mobile devices are typically used by a single person, making them ideal for individual-based and highly personalized target marketing efforts. Mobile technology offers the opportunity to personalize messages to various segments, based upon time and location, by altering both sight and sound. New developments in information technology and data mining make tailoring messages to individual consumers practical and cost-effective. For example, upon employing a mobile portal, advertising messages tailored to individual preferences can be provided. Relevance of material and the "de-massing" of marketing will become possible through the personal ownership of mobile devices.

Additionally, personalized content is paramount in using mobile devices, because of the limitation of the user interface. Relevant information must always be only a single "click" away, since web access with any existing wireless device is not comparable to a PC screen either by size, resolution or "surfability." Therefore, subscriber profile ownership is a key element in mobile portal success, as it will allow selectively targeted M-commerce and advertising. As such, the personalized database becomes central to all mobile portal activities. The mobile portal becomes the primary factor of M-commerce success by compiling personalized databases and providing personalized services. Personalized information and transaction feeds, via mobile portals, offer the greatest potential for the customization necessary for long-term success.

Value-added becomes maximized for those portals best able to implement the four distinguishing capabilities of ubiquity, convenience, localization and personalization. Mobile portals will become differentiated from traditional portals based upon their abilities to integrate and actuate the four advantages that are germane to mobile devices.

ISSUES, CONTROVERSIES AND PROBLEMS ASSOCIATED WITH MOBILE PORTALS

Since the success of M-commerce applications is dependent on the ease of use and the delivery of the right information at the right moment, *value-for-time* propositions will be key dynamics in determining the success of mobile portals. Consumers utilizing wireless devices for Internet access will come to expect improved value from their portal experience. Value-for-time will be accredited to those portals which leverage the unique advantages of mobility. Increased value-for-time propositions will enhance the success factors of mobile portals, with rewards to providers through increased user acquisition and retention.

Figure 2: Drivers of Mobile Portal Strategy

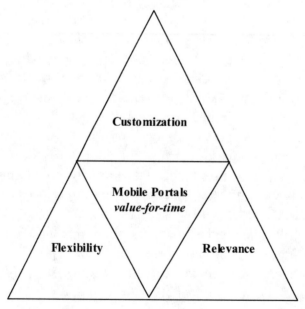

Three common themes are emerging that facilitate value-added accretion. These themes, applicable to mobile portals, are identified in Figure 2: customization, flexibility and relevance. Mobile portals should be able to increase the probability of success by incorporating these three major drivers into mobile portal strategies.

Portals must be easily customizable. The best mobile portals offer users a high degree of customization with organized and timely information in useful, related links to other websites. Value-for-time propositions will hinge on the ease-of-use and capabilities of the user to tailor the portal to their particular requirements and needs. Customization affords the user greater ease-of-use on a device that is inherently less friendly than full-sized computers. Also, as the portal becomes more customized to the specific user, the more likely the user will develop long-term loyalty to that portal service.

Mobile portals may be best able to achieve customization through personal information management systems (PIMs). PIMs include many of the functions of current PDAs, such as the maintenance of a personal address book; personalized calendar functionality; the management, writing and reading of notes and memos. PIMs can be synchronized with desktop applications (e.g. MS Outlook or Lotus Notes) to offer superior integration with other information technology tools. A PIM allows the user to maximize the value-for-time proposition while fully leveraging the advantages of mobility.

Portals must optimize flexibility. As might be expected, many companies have made substantial capital commitments to current technology and software to create a portal presence. But, users are not concerned with the delivery technology, rather the user wants the best available bundle of information, products and services. Mobile portals should be flexible enough to seamlessly adopt new technologies that enter new strategic areas without disrupting the viewers established usage habits. Successful portals will be able to adapt to the changing technological environment and incorporate into their platforms the value-added services that their audience desires. Mobile portals must be able to ensure anytime/ anywhere access to information, even as the company adapts to the ever-changing technical environment. An aperture in service will break the value-for-time compact with the consumer. For example, current wireless approaches center on technologies such as WAP and WML. However, the near future offers a new horizon of technological possibilities. As new technology is developed for mobile communication services, the most successful portals will be able to maneuver, without disrupting consumer patterns, into these new technical arenas. Consumers are not purchasing technology, rather, they are interested in the value achieved from the utilization of mobile technologies.

Portals must contain relevant content. Value-added content and relevant community services will become key elements for success in mobile portals. A good mobile portal develops a relationship with the user by helping to navigate through all the information on the Internet. Portals provide value-for-time services by continuously scanning the Internet, based on consumer profiles, for relevant and timely information. Not all web pages are equally valuable to all audiences. Therefore, successful portals will screen and prioritize the links pursuant to the demands of the target audience. Users will form a trusting relationship that the mobile portal is providing them with the best services possible. "The key is to provide information and how to make it relevant," says Glenn Gottleib, vice president of marketing for the wireless portal, AirFlash (Gohring, 2000). An opportunity exists for new mobile portals that can transform themselves into true content players and focus in particular market segments and niches, which are not currently addressed properly by portals designed for e-commerce applications.

For mobile portals, the technological limitations magnify these value-added concerns. It has been estimated that every additional click-through which a user needs to make in navigating through a commercial online environment with a mobile device reduces the possibility of a transaction by 50% (Durlacher Research, 2000). Providing the user with the desired, most relevant information without forcing a complex click-through sequence will significantly improve the effectiveness of any mobile portal.

SOLUTIONS AND RECOMMENDATIONS

The mobile portal development process will be a function of the complexity of the objectives for the portal. Since the mobile portal is another outward face for the organization, it should be mission driven. Too often, portable portals are developed as responses to short-term changes in the environment without long-term strategic considerations. The most successful portable portals will follow a systematic development process that supports the organization's overall strategic vision.

"A key difference between planning for a portal and planning for another internal software development is that eventually the portal will evolve to support not only the internal organization, but outside clients and customers as well."

(I/S Analyzer Case Studies, 2000, p. 2)

To decrease the risk for the organization, a methodical and thorough process for mobile portal development should be followed. To accomplish this mark, "Five Ds" of mobile portal strategy are proposed: (1) define, (2) design, (3) develop, (4) deliver and (5) defend. The Five Ds, each discussed below, offer a general blueprint for the creation of an effective portable portal strategy.

Figure 3: The Five Ds of Mobile Portal Strategy Development

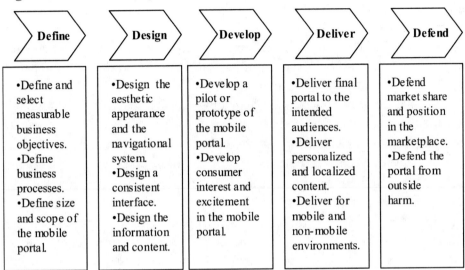

Define	Design	Develop	Deliver	Defend
•Define and select measurable business objectives. •Define business processes. •Define size and scope of the mobile portal.	•Design the aesthetic appearance and the navigational system. •Design a consistent interface. •Design the information and content.	•Develop a pilot or prototype of the mobile portal. •Develop consumer interest and excitement in the mobile portal.	•Deliver final portal to the intended audiences. •Deliver personalized and localized content. •Deliver for mobile and non-mobile environments.	•Defend market share and position in the marketplace. •Defend the portal from outside harm.

Define: First, as with most strategic endeavors, the starting point begins with the definition and selection of measurable business objectives. Some typical broad-based objectives for mobile portals might include the following: to conduct online transactions, to provide timely information, to increase sales, to improve customer service, to enhance customer relationships and to reach new market segments. In order to capitalize on the unique characteristics of mobile portals, strategists should consider how to incorporate the four unique characteristics of mobile portals (i.e., ubiquity, convenience, localization, and personalization) into their business objectives. For example, a mobile portal with the objective of "to allow consumers access to real-time stock quotes so they may conduct trades from anywhere and at any time," would be capitalizing on all four of the differentiators. The focal point of a successful strategy evolves from clearly defined objectives, which derived from the four unique characteristics of mobile portals.

Second, the business processes should be defined to encourage communication and provide organization to all participants involved in mobile portal development. Developers, strategists and other involved parties need a thorough understanding of the business processes that will be used in the development and maintenance of the mobile portal. Managers must define who will be involved in the project, the responsibilities of each party and the interrelationships required between the parties. Common models, such as a role interaction diagram or a workflow model, are extremely beneficial in organizing the business processes. If outsourcing is employed, those parties should be included in the process model as well.

Finally, a mobile portal strategy should define the initial, as well as future, size and scope of the portal. This includes defining the type of audience, the kind of information and the product/service mixes that will be provided. Through market research, understand which differentiators are most desired by customers (i.e., corporate channel partners, end-user customers, employees, etc.) since they represent the target market's primary needs.

Design: Design involves the aesthetic appearance and navigation of the mobile portal. Design can be considered a major barrier to entry; when consumers cannot maintain ease-of-navigation through the look, feel and organization of the mobile portal, it will not succeed. Traditionally, in non-mobile portals, layer approaches were often used for accessing data so that each page of information could be accessed through different angles and perspectives. These layered approaches often varied according to: (1) the experience of the user (novice, expert, manager), (2) the fields of interest and (3) resources of users. Current technological improvements will facilitate greater usage of layering in future generations of mobile portals.

However, with a mobile portal, other aspects of the design become more crucial to consider because of the smaller screen for the user interface and the slower download speeds that users experience. The mobile portal interface must be designed so that users can customize portal usability to fit their particular needs without taking an unreasonable amount of time. Additionally, the mobile portal display should be simple to understand and operate so that usage is as comfortable and effortless as possible.

Of critical importance is the development of a "consistent interface" when comparing the look and feel of the portable portal to the interface of a non-mobile portal. In order to maximize flexibility and minimize the amount of learning that consumers must endure, a consistent interface and similar look can provide a greater level of "stimulus generalization." Stimulus generalization explains why consumers tend to make the same response to only slightly different stimuli (Shiffman & Kanuk, 2000). In this case, the consistent interface allows consumers the ability to integrate mobile and non-mobile activities with ease when using either device to meet their needs. The key design issue is to avoid disrupting already established consumer habits any more than is absolutely necessary (Jin & Robey, 1999).

A final aspect of this phase entails designing the actual information and content for the portable portal. Focus groups and other exploratory marketing research methods should be conducted to assist in designing and selecting the most relevant content for the mobile portal. Strategic decisions need to be made in areas such as: (a) Which major categorical areas of content will be included in the mobile portal? (b) How many category options will be provided to users in order to maximize personalization? (c) With what frequency will the links be updated, altered and archived to provide the most user relevance? (d) What information is most attractive for localized target marketing opportunities? Amor (2000) contends that information should be organized into zones (sometimes called channels or guides) on general topic areas such as software, automobiles, travel and arts. Connections to relevant and customized content can be created from the zones. When defining the type of information and services that will be provided to the mobile portal audience, offerings should be tailored to user desires, mindful that more features are not always better. A successful mobile portal will have services offered which are tightly integrated with the overall purpose and objectives. Additionally, because portals can become quite large over time, the key is to define the area(s) that will provide the most value-for-time for the mobile portal audience. This can be accomplished by providing consumers with a solution to an unmet need or by addressing a problem that can be solved through the use of a mobile portal (e.g., last-minute shopping, in-store price comparisons, etc.).

Develop: The third stage involves developing a pilot or prototype of the mobile portal that can be tested on a small group of potential end users. Here, it is important to collect feedback, fine tune the portal and address problems revealed by the mobile users. A pilot portal can provide feedback in a number of areas such as user interface, system functionality, usability and workflow, general attractiveness and interaction with other systems (e.g., docking stations and cradles that sync with PCs). Mobile portal developers must make swift responses to customer feedback, and, if user-suggested modifications cannot be implemented, it is important to explain such limitations. Never discourage users from providing feedback, generating interest and feeling a high level of involvement in the mobile portal process.

The last aspect of the development stage calls for generating initial interest and excitement in the portal. Demonstrating mobile portal capabilities, announcing the launch date and offering product/service promotions can serve to generate attention. For instance, some companies have offered free activation, reduced set-up fees and free wireless modems as incentives for utilizing wireless portal services. Primary marketing efforts aimed at accruing awareness and interest will assist in the endeavors for the fourth stage—deliver.

Deliver: A portal is never "complete" because of the fluid nature of mobile technology. However, at some point a "finished" mobile portal must be delivered to the intended end-user audience. This stage involves the official launch of the mobile portal to intended end-user audiences. Mobile users will need to know that the portal exists, where and how it can be located and the value-added capacities of the service. Integrating the portal site into the domain of the mobile device may provide an initial patronage. Selecting a highly memorable, and easy-to-recognize, domain name is extremely helpful when portals cannot be directly integrated into mobile devices. In either case, it is imperative that the portal location and offerings are communicated effectively through as many channels as possible.

Content and other services must be delivered in both a personalized and localized fashion through the mobile portal. To illustrate, when working out of town, a traveling businessperson may wish to complete tasks such as changing an airline flight, making dinner reservations and finding driving directions to a client's location. These activities should be delivered through a portable portal with ease by effortlessly adapting to the end-user's location. At the same time, however, users need access to non-localized products and services through their portal in case that is what they need.

Portals furthermore should be deliverable in both mobile and non-mobile environments to maximize convenience for the customer. For instance, AOL mobile communicator users are able to login at both their desktop and handheld devices simultaneously using the same screen name. With the AOL mobile communicator, users are able to send and receive e-mail messages using their device. Although e-mail attachments cannot be sent and received, they are, however, able to see that the attachment exists so they can save the message to be retrieved later via their desktop. The overarching theme of this stage it to deliver the portal based on end-user needs, not necessarily technology-driven models.

Defend: All parties involved with the mobile portal need to ensure that the portal always remains a viable online entity. In other words, the mobile portal's position in the market, and in the minds of consumers, must be defended. This involves checking the competition regularly through strategic benchmarking programs. With the increasing number of organizations offering unique mobile devices, as well as new methods of engaging in M-commerce, it is relatively easy to benchmark the competition and respond quickly to market changes.

The portable portal should continuously promote input from users and allow modifications for improvements. This can be accomplished through a simple "feedback" or "contact us" button located in the mobile portal entry. If the portal is to remain the preferred starting point for the mobile user, it must continually seek out new ways to provide additional service. Ultimately, the portal is merely a mobile service provided to the target audience and must always seek means of improvement along these four dimensions.

Finally, appropriate mechanisms are required to defend the portable portal from outside harm. Measures are also needed to protect the privacy of consumers and help them feel secure in this environment. For the transactional aspects of mobile portals, it is imperative that the highest level of security possible is not only provided, but clearly communicated to the consumers. Encryption software that allows a website to accomplish secure tasks, such as receiving revenue from credit card-generated business, is a basic building block for building trust. Security is compulsory for any firm desiring to protect their customers and clients from "hackers" compromising the system.

CONCLUSIONS

This chapter provided some initial direction on how to develop and integrate portals into a comprehensive M-business strategy. Mobile portal development is likely to parallel the growth in overall M-commerce. New and even more innovative applications than those discussed here will arise as more people connect to the web through wireless devices and additional content becomes available. As other functionalities such as file attachments, faster network speeds, Bluetooth, speech recognition and E-commerce security features are added, more users will be attracted to this intriguing mobile marketplace. M-commerce is still not without its limitations. The problems it must overcome include: uniform standards, ease of operation, security for transactions, minimum screen size, display type and bandwidth, billing services and the relatively impoverished websites. In the short-term, portable portals must be mindful to operate within evolving technological constraints while providing the most amount of flexibility possible. In the end, these problems, much like the initial problems associated with E-commerce, will also be overcome. Long-term mobile portal success is likely to come from consumer-driven, rather than technology-based models.

Initially, mobile portals will also be limited by the dilemma of two different types of portal providers, (i.e., the traditional portals and mobile operators). Ultimately, for long-term success, the two portal types must join forces and merge as providers recognize that users do not want to access multiple levels of portals to help them manage their lives; they need only one gateway to the Internet world. Value-for-time will only be achieved when portals provide a single entry point for the relevant information to the consumer.

However, access to data essential to mobile activities, rather than the multimedia approach of fixed-lined portals, remains the key. Merely replicating traditional portals, or even "web-clipping" streamlined content from existing portals for delivery on PDAs, will not provide the value-added necessary for long-term success. Wireless content must be succinct, hyper-personalized and relate directly to the activities of being away from the PC.

In the end, mobile portals may prove to be fundamental to the success of M-commerce. For M-commerce to reach its full potential of information available—anytime, anyplace and on any device—portable portals must offer the user maximum effectiveness and value through leveraging the advantages unique to wireless technology. If this is accomplished, and E-commerce business models are adapted for wireless Internet technology, expect to see a second E-business revolution take place based on mobility of commerce. The marketers developing the best technological tools to operate in this emerging mobile environment, like mobile portals, will operate with a definite advantage as consumers seek increased value in their time-starved world.

ACKNOWLEDGMENT

The authors gratefully acknowledge the Commonwealth Information Security Center at James Madison University for support of their research.

REFERENCES

Amor, D. (2000). *The E-Business Revolution: Living and Working in an Interconnected World.* Upper Saddle River, NJ: Prentice Hall.

AOL Mobile Communicator. (2000). *Internet Product Watch.* Retrieved March 1, 2001, from http://ipw.internet.com/communication/tools/975595302.html.

Barnett, N., Hodges S., & Wilshire M. (2000). M-commerce: an operator's manual. *McKinsey Quarterly, (3),* 162-171.

Carroll, K. (2000). Portable portals. *Telephony, 238*(10), 10-11.

Daitch, J., Kamath, R., Kapoor, R., Nemiccolo, A., Sahni J. & Varma, S. (2000). Wireless applications for business: Business anytime, anywhere. *Kellogg TechVenture 2000 Anthology,* 88-120.

de Haan, A. (2000). The Internet goes wireless. *EAI Journal,* (April), 62-63.

Experts offer five keys to successful portals. (2000). *I/S Analyzer Case Studies,* 39(2), 2-4.

Hoffman, G. (2000). Start game. *Communications International,* (April), 18-22.

Kalakota, R. & Robinson M. (2002). *M-Business: The Race to Mobility.* New York: McGraw-Hill.

King, G. (2000, December 7). The m-marketing model dilemma: Push me or pull you? *Newsbeat ChannelSeven.com,* Retrieved January 15, 2001, from http://www.channelseven.com/newsbeat/2000features/news20001207.shtml.

Kobielus, J. (2000). Microsoft, Nextel need to rethink wireless portal strategy. *Network World, 16,*(23), 57-58.

KPMG. (2000). *Creating the New Wireless Operator.* (2000). Retrieved March 6, 2001, from http://www.kpmg.com.

Lazar, I. (2000). The state of the internet. *IT Professional, 2*(1), 52.

Lei, J. & Robey, D. (1999). Explaining cybermediation: An organizational analysis of electronic retailing. *International Journal of Electronic Commerce,* 3(4), 47-66.

M-commerce vital statistics. (2000). *M-commerce Times.* Retrieved March 4, 2001, from http://www.mcommercetimes.com.

McGinity, M. (2000). The net/wireless meeting of the minds. *Interactive Week,* (March 6), 2.

Mobile commerce report. (2000). *Durlacher Research Ltd.* Retrieved February 9, 2001, from http://www.durlacher.com/fr-research-reps.htm.

Monohan, J. (1999). Portal puzzle. *Banking Strategies, 75*(6), 148-158.

Rao, M. (2000). M-wire: E-commerce, M-commerce poised for rapid take-off in Europe. *Electronic Markets: The International Journal of Electronic Commerce & Business Media,* (April 6). Retrieved January 9, 2001, from http://www.electronicmarkets.org/electronicmarkets/electronicmarkets.nsf/pages/emw_0004_cell.html.

Schenker, J. (2000). Europe's high-tech start-ups have stolen a march on silicon valley by launching new mobile services. *Time Europe,* (February 7), 1.

Schiffman, L. & Kanuk L. (2000). *Consumer Behavior* (7th ed.). Upper Saddle River, NJ: Prentice Hall.

Shaffer, R. (2000). M-commerce: On-line selling wireless future. *Fortune,* (July 10), 262.

The race for M-commerce: Shifting paradigms in the world of mobile commerce strategy. (2000). *Datamonitor.* Retrieved April 29, 2001, from http://www.datamonitor.com.

Zabala, H. (2000). M-commerce, the next big thing? *Asian Business,* (June), 34-35.

Chapter XI

Factors Influencing the Adoption of Mobile Gaming Services

Mirella Kleijnen[1,2] and Ko de Ruyter
Maastricht University, The Netherlands

Martin G. M. Wetzels
Technical University, Eindhoven, The Netherlands

ABSTRACT

The current chapter focuses the adoption process of mobile gaming. After providing a brief introduction to the topic of m-commerce and m-services, several relevant adoption factors are highlighted. These factors have been researched empirically, via a conjoint study conducted in the Netherlands. The results illustrated a hierarchical importance of the factors identified, whereby perceived risk, complexity, and compatibility were identified as the three main regarded as the factors that are mainly influencing the adoption of mobile gaming applications. Based on these findings, we have provided several managerial implications.

INTRODUCTION

According to recent forecasts, the mobile services industry in Europe will be worth over 76 billion Euro by 2005 (Durlacher, 2001). Experts claim that wireless technology will usher in the next wave of electronic commerce—'mobile commerce' (m-commerce). The m-commerce industry provides unlimited opportunities for business growth, and forward-thinking companies are already integrating mobile commerce into their businesses to establish a vital competitive edge (http://www.mobileinfo.com). In order to accomplish this integration of the wireless Web in a successful way, consumers' acceptance of mobile commerce as a delivery channel is essential. Therefore, it seems critical to examine which factors influence customer adoption and diffusion of this new way of providing services.

Although there are numerous studies in the field of adoption and diffusion of marketing-enabling technology (Plouffe, Vandenbosch, & Hulland, 2001; Daghfous, Petrof, & Pons, 1999; Rogers, 1995; Holak & Lehman, 1990; Labay & Kinnear, 1981), previous work has mainly focused on the adoption of products and technology (Verhoef & Langerak, 2001; Au & Enderwick, 2000; Eastlick & Lotz, 1999; Davis, 1989). In contrast, the perspective on services and service-enabling technologies is considerably less pronounced. Despite the fact that several trend studies have been conducted regarding the potential of wireless technology and 3G services (Durlacher, 2001; UMTS Forum, 2001), there exists a need for more substantive, theory-based research, creating a more in-depth understanding of consumer behavior with regard to m-commerce. The current study aims to define critical factors in the adoption of mobile services and determine consumers' preferential structure with regard to this technology.

Our contribution to this encyclopedic book on m-commerce is structured as follows. First, we briefly introduce the field of m-commerce and m-services to set the scene. Subsequently, based on a literature review of adoption and diffusion theory, several success factors enhancing mobile services adoption are identified. Moreover, mobile entertainment services, such as playing games via hand-held devices, are used as a setting for our research. Consequently, we report on an empirical study that was completed among 99 consumers in The Netherlands. By using a conjoint measurement design, we are able to obtain a detailed insight into consumer preference structures regarding mobile gaming services. Interpretation of the analysis yields a hierarchy of importance concerning m-services adoption factors. Finally, the chapter concludes with a discussion of the results and theoretical as well as managerial implications of our study.

MOBILE COMMERCE

Frequently, m-commerce is viewed as the next frontier in the electronic market place. E-commerce adoption and diffusion has led to widespread acceptance of electronic transactions (May, 2001). It is argued that "m-commerce allows users to access the Internet without needing to find a place to plug in" (http://whatis.techtarget.com). As a result, it is "the effective delivery of electronic commerce into the consumer's hand, anywhere, using wireless technology" (http://

/www.gsmworld.com). Although m-commerce is regularly defined as an extension or next step of e-commerce, it should be acknowledged as a business opportunity with its own distinctive characteristics and functions (http://www.mobileinfo.com) resulting from unique advantages wireless technology holds over wired technology. First, the use of a wireless device enables the user to receive information and conduct transactions anywhere, at anytime, guaranteeing customers virtual and physical mobility (UMTS Forum, 2000). Second, the emergence of location-specific-based applications will enable the user to receive context-specific information on which to act. The combination of localization and personalization will create unique possibilities for reaching and attracting customers. Personalized services incorporate customized information, meeting users' preferences, and payment mechanisms that allow for personal information storage, eliminating the need to enter credit card information for each transaction (May, 2001).

A considerable range of mobile services is already available, only to be extended by so-called third generation services based on new wireless technologies such as GRPS and UMTS. In a taxonomy of wireless services, Durlacher (2001) identifies four main categories: communication, information, entertainment, and transaction services. Communication services are the foundation of mobile services. Voice-to-voice application is still the primary service in wireless technology (http://www.itweb.co.za). Further examples are person-to-person messaging and SMS-based services, such as mobile chat.

Information provision is a second fundamental service of m-commerce (Durlacher, 2000). This category includes information services, such as general news, sport news, financial news, and weather reports. Newell and Newell-Lemon (2001) mention SmartRay.com as a frustration-saver that delivers personalized flight updates to mobile appliances. Another example of information services is convenient access to product and price comparisons. Customers can use their mobile device to compare product characteristics while shopping in a brick-and-mortar store or mall.

A third category, entertainment services, seems to be the most promising application judging from the explosive growth of i-mode in Japan, which was mainly driven by entertainment services. Virtually every i-mode user consumes the entertainment services offered. Entertainment services vary from mobile music provision to mobile gambling in virtual casinos (Durlacher, 2001; May, 2001).

Finally, transaction services mainly consist of m-shopping, m-finance, and m-payment (Durlacher, 2001). Mobile shopping services supply customers with the possibility to purchase anything at any point in time. In parallel a "one-button purchase experience for mobile shopping" will be desirable for mobile customers (Durlacher, 2000). M-finance relates to mobile banking and brokerage. Banks provide mobile banking as an additional distribution channel to electronic banking (e-banking). In addition to m-banking, mobile brokerage is expected to become a major business driver (Durlacher, 2001). Professional as well as private traders can use mobile commerce solutions to access information about stock price development of personalized stock portfolios. Finally, with mobile payment, consumers circumvent the necessity to queue by simply transferring payments through their wireless devices (Tarasewich & Warketin, 2000).

Presently, the market is dominated by communication services. However, several other service types are predicted to become at least similarly important revenue sources. Although the digital distribution of information is growing extensively, entertainment services in particular will gain momentum from the end of 2002 onwards, being the most fruitful revenue opportunity in the B2C market. According to recent findings of the Arc Group (2001), the total number of 'mobile gamers' is set to increase from 43 million in 2001 to almost 850 million by 2006. Online mobile games are predicted to exhibit strong growth, rising from 21.8% of the total gaming market in 2001 to 43.2% in 2006 (http://www.arcgroup.com). Durlacher (2001) states that "it is expected that mobile games especially will become the number one service and generate annual revenues of around 8.1 billion by 2005."

Therefore, in this chapter we focus on mobile gaming. Mobile gaming services allow users to play interactive multi-player games (MPG) against other remote users (UMTS Forum, 2000) independent of time and location. M-gaming services serve as leisure time entertainment as well as time-killing activity (May, 2001). Existing mobile games frequently consist of simplistic single-player games (Durlacher, 2000) with poor graphical resolutions (Stone, 2001). Examples of existing games can be found for instance at www.wirelessgames.com or http://www.nokia.com/games/games_extra.html. However, due to increasing quality of wireless technology, future mobile gaming services will allow users an experience similar to the high-quality experience provided by existing PC solutions. After this elaboration on mobile commerce and an illustration of its potential impact, we will now continue our discussion by identifying the factors that influence the actual adoption of mobile services in consumers' daily lives.

CRITICAL FACTORS IN THE ADOPTION PROCESS OF MOBILE SERVICES

Although little empirical research has been conducted on mobile services so far, there is a broad range of theories and previous studies that may assist in setting up a systematic assessment of critical success factors. Many of the earlier adoption models investigate behavioral characteristics, like perception and attitude, and frequently they integrate innovation literature with other constructs to develop a new framework (Akkeren & Cavaye, 1999). Customer acceptance of mobile commerce can be identified as a technology adoption. Several theories have been developed to investigate technology adoption, of which the Technology Acceptance Model (TAM) (Davis, 1993,1989) is well established throughout the literature (Moon & Kim, 2001; Karahanna & Limayem, 2000; Lederer, Maupin, Sena, & Zhuang, 2000). Central in this model are the notions of ease-of-use and usefulness (Davis, 1986). At the same time, the narrow focus of these concepts prohibits us from examining other potential drivers of m-commerce adoption. Alternatively, adoption process theory may provide valuable insights for building a theoretical framework (Eastlick & Lotz, 1999). Landmark studies in this field are the work of Rogers (1962) on diffusion of innovations, and Bass (1969), who pioneered in developing the first analytical marketing models concerning adoption of innovations (Daghfous et al., 1999). Their work initiated the development of extensive research in this field (Daghfous et al., 1999). Several recent empirical studies have validated adoption theory in relation to a wide range of products (Rogers 1995; Holak & Lehman, 1990; Labay & Kinnear, 1981; Ostlund, 1973) and technology (Beatty, Shim, & Jones, 2001; Plouffe et al., 2001). A large number of studies have investigated the use of electronic commerce, but the field of mobile commerce has been left virtually unexplored. In the current study, the conceptual framework of Rogers (1995) has been expanded by several constructs that influence individuals' adoption decisions. Rogers (1995) has defined five factors that influence rate of adoption: (1) relative advantage; (2) compatibility; (3) complexity; (4) communicability; and (5) triability. In the remainder of this section, these factors—including the additional constructs that were identified—will be explained and applied to the context of mobile gaming.

Relative advantage is the extent to which an innovation is perceived as being better than the idea it supersedes (Rogers, 1995). Consumers are not likely to start using new technology just because it is there. The main advantage of mobile services is that they are accessible 'anytime, anywhere.' Time killing displays one of the goals of using mobile entertainment services. Playing m-games while traveling to school or work seems to be increasingly popular (http://www.i-moder.nl).

Compatibility relates to the fit between the innovation and the existing values, past experiences, and needs of potential adopters (Rogers, 1995). Contextually, this means addressing the issue of how well mobile services fit into the respondent's daily activities, comparable to the concept of perceived usefulness defined in the TAM model (Davis, 1989). Playing mobile games can for instance fit very well into the lifestyle of a student, who travels to school for an hour everyday and who has adopted the Internet in his daily activities already.

Complexity, also regularly referred to as ease-of-use, is the extent to which an innovation is perceived as relatively difficult to understand and use (Plouffe et al., 2001; Karahanna & Straub, 1999; Agarwal & Prasad, 1998; Rogers 1995; Davis, 1989). Complexity in our study relates to the use of m-services. This can relate for instance to ease of accessing a game, the amount of effort it takes to understand the rules of the game, and how easy it is to find somebody to play the game with.

However, there is another factor related to complexity that might play a role, which is very specific for mobile commerce. We have defined this variable as *navigation*. Design has been identified before as a critical factor in the success of mobile Internet (Dolan, 2000; Kaasinen, Aaltone, Kolari, Melakoski, & Laakko, 2000). In the present context, design is related to the hand-held device. Several mobile devices make use of touch screens; others use dual thumb navigation buttons. Especially in playing network games, navigation or maneuvering ergonomics is critical, since it influences the reaction speed of participants. This is a very specific feature that relates to the hand-held devices and not to the mobile services themselves and is therefore explicitly taken into account in this research.

Communicability refers to the extent to which the innovation lends itself for communication, particularly the extent to which the use of the innovation is observable by others (Verhoef & Langerak, 2001; Rogers, 1995). This factor resembles social influence, which has been identified as a critical factor in the adoption process by several authors (Karahanna & Limayem, 2000; Karahanna & Straub, 1999; Fang, 1998). Usage of innovation, apart from other factors, is often influenced by a social context (Karahanna & Limayem, 2000; Karahanna & Straub, 1999). Fang (1998) indicates that social influence and social pressure are strongly linked: social pressure refers to the service usage and choice as the result of influence from supervisors, peers, or others that are highly regarded. It signifies the extent to which an individual believes that an innovation will give him added prestige or status in his relevant community (Plouffe et al., 2001). Consumers might experience the need to play mobile games in order to feel accepted by their friends.

The last factor identified by Roger (1995) is *triability*, the degree to which an innovation may be experimented with on a limited basis, but without a large commitment. This is not a realistic factor when discussing mobile services. Considerable effort is necessary for trying mobile services. Consumers have to invest a substantial amount of time to familiarize themselves with WAP portals (the current technology providing mobile services via mobile phones), and to personalize the configurations according to their preferences, e.g., which type of games they like to play. Subsequently, consumers can set their preferences for a specific game. Furthermore, the consumer has to possess a hand-held device capable of mobile gaming. This requires significant monetary investment. Moreover, the mobile service (like network games) and time spent online is charged for. There are virtually no trial opportunities for network games. Only a limited number of single-player games, such as 'Snake,' are offered as free trial via the Internet. Therefore, the variable *triability* was not taken into account in the current research.

Frequently, the Rogers taxonomy has been extended to include *perceived risk* (Eastlick & Lotz, 1999; Ostlund, 1973). The use of highly personalized and context-based technology is particularly prone to consumer risk perception (Newell & Newell-Lemon, 2001). Perceived risk is defined as the extent to which risks are attributed to the mobile services. Risk can be recognized as total risk or as a specific type of risk (performance, privacy, or psychosocial risk). The opinions of others about the person adopting an innovation can be considered a psychosocial risk and therefore could be defined as part of perceived risk (Ortt, 1998). However, in the current study psychosocial effects are enclosed in communicability. Consequently, perceived risk will focus on performance and privacy risk. Performance risk of mobile games can encompass the breakdown of the operator network or the browser in the mobile device. Privacy reflects mistrust in mobile security, but also relates to customers fearing that their personal information will be misused (Sutherland, 2001). Since consumers have to reveal a substantial amount of personal information while setting their WAP configurations, privacy issues might play a critical role in the adoption process.

Furthermore, *critical mass* seems to play an important role in the adoption of mobile services. Critical mass theory states that individuals who have access to multiple communication media will generally use the medium most widely available within their communication community, even when it is not the medium they prefer (Fang, 1998). Critical mass is defined as the minimal number of adopters of an interactive innovation for the further rate of adoption to be self-sustaining. Especially interactive innovations (like mobile entertainment services) are dependent on the number of others who have already adopted the innovation (Mahler & Rogers, 1999). Excitement about mobile gaming is increased with the establishment of critical mass. The main entertainment service of i-mode is network gaming. These network games make it possible to compete against other players, either friend or stranger. As the number of consumers using mobile entertainment increases, the playground transforms into a greater, potentially global arena.

The final factor that is included in our study is *payment options*. One of the factors mentioned as a critical success factor for i-mode in Japan is a convenient billing system (Dolan, 2000). In the current marketplace, there are several payment options available. Consumers can be charged for the minutes they are online via their mobile device. Another option is that they are charged only for the amount of data they download regardless of the time needed to perform the service (e.g., i-mode). In this way, mobile devices can be used to support 'always-on' services (Newell & Newell-Lemon, 2001). When playing games via a hand-held device, costs could be very diverse. It depends, for instance, on the network zone, time of day, and kind of game (strategy games can take a few minutes or a few hours depending on the competence of the players). All these factors might lead to fluctuating prices. A flat fee on the other hand overcomes the aforementioned obstacles. Now that we have introduced our theoretical framework, the next part of the chapter will continue with an elaboration of the research design and the analysis of the results of the empirical study.

RESEARCH DESIGN

The main goal of this research is to identify a hierarchy of importance concerning the critical factors influencing the adoption of mobile services. To realize this research objective, conjoint analysis was seen as the appropriate statistical tool.

Conjoint Analysis

Conjoint analysis is a technique which allows a set of overall responses to factorially designed stimuli to be decomposed so that the utility of each stimulus attribute can be inferred from the respondents' overall evaluations of the stimuli (Green, Helsen, & Shandler, 1988). A number of (hypothetical) combinations of service elements can be formulated that will be presented to a sample of customers. According to Lilien and Rangaswamy (1997), the analysis comprises three stages. The first stage is concerned with the design of the study, where the attributes and levels relevant to the product or service category will be selected. In the second stage customers rate the attractiveness of a number of possible combinations of customer service elements. Finally, in the third stage these ratings are used to estimate part-worth utilities, i.e., the utility which is attached to the individual levels of each service element included in the research design. Consequently, an accurate estimate of customer trade-offs between services elements can be obtained.

We used "Adaptive Conjoint Analysis" (ACA) to conduct our conjoint study. ACA is a PC-based system for conjoint analysis. The term 'adaptive' refers to the fact that the computer-administered interview is customized for each respondent; at each step, previous answers are used to decide which question to ask next, to obtain most of the information about the respondent's preferences. The program allows the researcher to design a computer-interactive interview and administer the interview to respondents. The interview can consider many factors and levels, paying special attention to those the respondent considers most important. Questioning is done in an "intelligent" way; the respondent's utilities are continually re-estimated as the interview progresses, and each question is chosen to provide the greatest amount of additional information, given what is already known about the respondent's preferences (Sawtooth, 1985-87).

The dependent variable in our study was the intention to make use of mobile services. The eight independent variables were perceived risk (three levels: no risk, medium, high), relative advantage (three levels: no relative advantage, medium, high), compatibility (three levels: no match with current behavior/prior experiences, medium match, high match), complexity (three levels: easy, medium, difficult), communicability (three levels: communication with friends, colleagues, family), critical mass (three levels: nobody, some people, a lot of people), navigation (four levels: input via mini-keyboard, normal button system, dual thumb, touch screen), and finally payment options (three levels: minutes online, based on data bytes, flat fee per month).

Sample

By means of pseudo-random sampling, a total of 99 respondents were intercepted on the street of a mid-sized city in The Netherlands. Every third person that passed the data collection point was invited to participate in our study. A negligible amount of respondents that were approached refused to participate in the study (3.5%). Internal validation was achieved by an investigation of the correlation coefficient. This coefficient represents the correlation between the respondent's predicted and actual answers to the calibration concepts. A cut-off point of 0.5 was used, which led to a usable sample of 84 respondents. The sample can be described as follows: gender (female: 52.4%, male: 47.6%), age (18-25 years: 42.8%, 26-35 years: 28.6%, >35 years: 28.6%), and level of education (at least secondary level: 52.4%, higher level education: 26.2%, university: 21.4%). A complete overview of the sample characteristics can be found in Table 1.

Table 1: Sample Characteristics (n = 84)

Variable	Caterogies	%
Gender	Male	47.6
	Female	52.4
Age	18-25 years	42.8
	26-35 years	28.6
	> 35 years	28.6
Level of Education	At least secondary level	52.4
	Higher level education*	26.2
	University	21.4

* equivalent with polytech level

The adopted sampling procedure gave the researchers the ability to perform face-to-face interviews with the respondents, which was desirable since the conjoint method was used (Green & Krieger, 1991; Green & Srinivasan, 1990). The interviewer had the option to explain the technique carefully, and to make sure the data collection system worked properly (laptops were use to collect the data via the ACA system).

One of the major concerns when using street interview surveys is to ensure the sampling procedure is performed in a manner that the correct respondents are chosen (Bush & Hair, 1985). In order to avoid respondent bias, the research was completed over a 12-day period that included weekend days and weekdays, and different hours of the day. Respondents were asked if they were familiar with mobile services before they were invited to complete the survey. Through this procedure, we were assured that they understood the meaning of the factors presented to them in the right context.

Analysis and Results

As was stated before, conjoint analysis allows us to define a hierarchy of importance concerning the critical factors influencing adoption of mobile services. Based on the importance ratings, it can be concluded that perceived risk is the most important factor in adopting mobile services (20.69%), complexity is second in importance (15.19%), and compatibility ranks third (13.71%). Payment options (10.77%), navigation (10.73%), and relative advantage (10.50%) seem to be approximately equal in importance. Critical mass (9.86%) and communicability (8.51%) seem to have a weaker impact on the adoption decision.

Table 2: Importance Ratings (n = 84)

Variable	Importance Ratings	Ranking number
Perceived Risk	20.69%	1
Complexity	15.19%	2
Compatibility	13.71%	3
Payment Options	10.77%	4
Navigation	10.73%	5
Relative Advantage	10.50%	6
Critical Mass	9.86%	7
Communicability	8.51%	8

CONCLUSIONS

This study aimed to define significant factors in the adoption of mobile gaming services and determine consumers' preferential structure with regard to this technology. Based on the conceptual framework of Rogers (1995), relative advantage, compatibility, complexity, and communicability were incorporated as critical factors. Triability was excluded from the current study. Furthermore, additional constructs that seemed relevant in the present context were identified: navigation, perceived risk, critical mass, and payment options. Based on the analysis, the following conclusions can be drawn. Perceived risk, complexity, and compatibility were identified as the three main factors influencing the adoption of mobile gaming applications. This is consistent with earlier findings from the adoption literature (Verhoef & Langerak, 2001; Ruyter, Wetzels, & Kleijnen, 2001) in relation to the wired Web. In particular, risk is the most important reason that consumers avoid engaging in a wireless transaction (Newell & Newell-Lemon, 2001). Therefore, security and privacy issues should be considered in any effort to introduce new m-services. Complexity is another barrier. Although one of the primary benefits of m-commerce is to make life simpler for consumers (Koranteng, 2000), many new consumers are put off by the complexity associated with using m-services. Finally, as technology has become more personal, it needs to fit into a consumer's lifestyle. Critical mass and communicability seem to have the least influence. These factors relate to the social aspect of mobile services, but our results indicate that first and foremost m-commerce is regarded as a personal technology.

Limitations and Suggestions for Further Research

The limitations of this study provide directions for future research and point to several theoretical implications. First, in this study a limited number of factors were taken into account, which is inherent to the nature of conjoint analysis. Consequently, other relevant marketing mix variables such as actual price levels that might be of importance have been excluded. Furthermore, cross-sectional research provides a snapshot of the variables of interest at one point in time. Longitudinal analysis will possibly show a different emphasis on the importance of certain characteristics. Additionally, the study was conducted in The Netherlands, therefore the results might not be generalizable to countries where the uptake of mobile commerce has been demonstrated to be faster and more widespread. A further limitation of this research relates to the fact that our research was embedded in the context of mobile gaming. Consequently, the generalizability of our results is limited. Finally, the different scenarios were presented to the research subjects in a textual format. Testing in a more real-life experimental setting using, for example, different prototypes of mobile devices might provide further validation of results.

Managerial Implications

The findings of our study hold several specific implications for managers. Results illustrate the importance of perceived risk related to the adoption of mobile gaming services. Consequently, managers need to focus on diminishing perceived risk and increasing consumer trust in order to increase adoption rates. Issues currently under debate are concerned with the security of transactions and privacy associated with personal information. Golden (2000) acknowledges that securing information from unauthorized access is a vital problem for any network, wired or wireless, since imminent security gaps exist in the present security framework for mobile business. While wired connections using standards such as Transport Layer Security ensure a secure connection between PC and Web server, GSM and GPRS communication mainly provide data-encryption between mobile phone and transmitter. Several measures are currently being introduced, including securing WAP gateways, sophisticated encryption, digital signatures and Public Key Infrastructure (PKI). Furthermore, Hoffman et al. (1998) state that mistrust arises from consumers' perceived lack of control over the access third parties have to their personal information during the online navigation process. Companies can develop privacy statements or employ eTRUST certifications for example, ensuring consumers that their information will be used for identification purposes only.

Although conjoint analysis revealed a major emphasis on perceived risk as the most prevalent factor in the adoption process, other factors should not be ignored by marketers. Complexity was the second most important critical adoption factor. This illustrates a need for information that marketers can fulfill via alternative channels such as television or magazines, informing customers about m-services usage. Moreover, the use of relatively uncomplicated services should be encouraged, such as information services similar to SMS services. Customer acquaintance to this category of m-services will lower the barrier to exploit other m-services as well. Compatibility to daily lifestyle can be illustrated via different media as well. Consumers need to be educated about the possibilities of m-commerce and the convenience it can bring them by incorporating it into their daily routines. Advertising should also focus on the unique advantages of mobile services, thereby not only stressing ubiquity of m-commerce but also pointing out the opportunities of localization and personalization. The introduction of 3G technologies will provide numerous opportunities in this area.

An additional factor that will become more prevalent with the introduction of new technologies is payment options. Regardless of the service category customers exploit, m-services are perceived as relatively expensive. With the introduction of new technologies such as GPRS and UMTS, constant connectivity to the Internet will be offered. Payment will more likely be based on the amount of data-download than on the duration of time spent online. Another option providers can offer is a flat rate, which will stimulate the use of other mobile services as well. Nevertheless mobile billing will present a challenge to providers, as there is no standard yet concerning billing procedures. A clarification of who charges the consumers is needed—the service content provider or the technology provider.

Attention should be devoted to enhancing navigation systems, particularly in m-entertainment, but also in other service categories. Current devices typically provide cumbersome navigation via the standard buttons on the device. However, newly introduced devices have implemented enhanced features, such as dual thumb buttons, one-button access for mobile Internet, touch screens, and mobile keyboards. Specifically, M-gaming gadgets like mobile joysticks that can be clipped on the mobile phone have been introduced. A continuation of this trend is expected to create profitable opportunities for marketers.

Critical mass and communicability will most likely become more prevailing factors as mobile technology develops into a more social technology, for instance with the development of (more) sophisticated multi-player games. The visibility of mobile services can be increased by offering communities or buddy list options through which customers can alert their peers or invite them to join a game for instance.

ENDNOTES

1   Correspondence to Mirella Kleijnen, PhD candidate, Maastricht University, Faculty of Economics and Business Administration, P.O. Box 616, NL - 6200 MD Maastricht, tel. +31.43.3883819, fax: +31.43.3884918, e-mail: m.kleijnen@mw.unimaas.nl.
2   The authors would like to extend their thanks to Jeanien Werkman for her assistance to this study.

REFERENCES

Agarwal, R., & Prasad, J. (1999). Are individual differences germane to the acceptance of new information technologies? *Decision Sciences, 30* (2), 361-391.

Akkeren, J. van, & Cavaye, A.L.M. (1999). Factors influencing entry-level Internet technology adoption by small business in Australia: An empirical study. *Proceedings of the 10th Australian Conference on Information Systems.*

Au, A.K., & Enderwick, P. (2000). A cognitive model on attitude towards technology adoption. *Journal of Managerial Psychology, 15*(4), 266-282.

Bass, F.M. (1969). A new product growth model for consumer durables. *Management Science, 15* (January), 215-227.

Beatty, R.C., Shim, J.P., & Jones, M.C. (2001). Factors influencing corporate Web site adoption: A time-based assessment. *Information & Management, 38,* 337-354.

Bush, A.J., & Hair, Jr. J.F. (1985). An assessment of the mall intercept as a data collection method. *Journal of Marketing Research, 32* (November), 385-391.

Daghfous, N., Petrof, J. V., & Pons, F. (1999). Values and adoption of innovations: A cross-cultural study. *Journal of Consumer Marketing, 16*(4), 314-331.

Davis, F.D. (1986), *A Technology Acceptance Model for Empirically Testing New End-User Information Systems: Theory and Results.* Doctoral dissertation, Massachusetts Institute of Technology, Boston, MA.

Davis, F.D. (1989). Perceived usefulness, perceived ease of use, and usage of information technology: A replication. *MIS Quarterly, 16*(2), 319-339.

Davis, F.D. (1993). User acceptance of information technology: System characteristics, user perceptions and behavioral impacts. *International Journal of Man-Machine Studies, 38,*475-487.

Dolan, D.P. (2000). The big bumpy shift: Digital music via mobile Internet. *First Monday, 5*(12). Retrieved January 10, 2001, from the World Wide Web: http://www.firstmonday.org.

Durlacher. (2000). Mobile Commerce Report. *Durlacher Research Ltd.* Retrieved May 20, 2001, from the World Wide Web: http://www.durlacher.com.

Durlacher. (2001). UMTS Report. *Durlacher Research Ltd.* Retrieved May 20, 2001, from the World Wide Web: http://www.durlacher.com.

Eastlick, M.A., & Lotz, S. (1999). Profiling potential adopters and non-adopters of an interactive electronic shopping medium. *International Journal of Retail and Distribution Management, 27*(6), 209-223.

Fang, K. (1998). An analysis of electronic-mail usage. *Computer in Human Behavior, 14*(2), 349-374.

Golden, P. (2000). Wireless Security—Part I and II. *M for Mobile Devices Analysis, October Issue,* Retrieved November 21, 2000, from the World Wide Web: http://www.mformobile.com.

Green, P.E., Helsen, K., & Shandler, B. (1988). Conjoint internal validity under alternative profile presentations. *Journal of Consumer Research, 15*(3), 392-97.

Green, P.E., & Krieger, A.M. (1991). Segmenting markets with conjoint analysis. *Journal of Marketing, 55* (October), 20-31.

Green, P.E., & Srinivasan, V. (1990). Conjoint analysis in marketing: New developments with implications for research and practice. *Journal of Marketing, 4*,3-19.

Hoffman, D.L., Novak, T.P., & Peralta, M. (1998). *Building Consumer Trust in Online Environments: The Case for Information Privacy.* Working Paper, Vanderbilt University, 1-10.

Holak, S.L., & Lehman, D.R. (1990). Purchase intentions and the dimensions of innovation: An exploratory model. *Journal of Product Innovation Management, 7*, 59-73.

Kaasinen, E., Aaltone, M., Kolari, J., Melakoski, S., & Laakko, T. (2000). Two approaches to bringing Internet services to WAP devices. *Computer Networks, 33*, 231-246.

Karahanna, E., & Limayem, M. (2000). E-mail and v-mail usage: Generalizing across technologies. *Journal of Organizational Computing and Electronic Commerce, 10*(1), 49-66.

Karahanna, E. & Straub, D.W. (1999). The psychological origins of perceived usefulness and ease-of-use. *Information & Management, 35*, 237-250.

Koranteng (2000). Dial "m" for e-commerce. *Advertising Age International*, May, 1-24.

Labay, D.G., & Kinnear, T.C. (1981). Exploring the consumer decision process in the adoption of solar energy systems. *Journal of Consumer Research, 8* (December), 271-78.

Lederer, A.L., Maupin, D.J., Sena, M.P., & Zhuang, Y. (2000). The technology acceptance model and the World Wide Web. *Decision Support Systems, 29*, 269-282.

Lilien, G.L., & Rangaswamy, A. (1997). *Marketing Engineering: Computer-Assisted Marketing Analysis and Planning.* Reading, MA: Addison-Wesley-Longman.

Mahler, A., & Rogers, E.M. (1999). The diffusion of interactive communication innovations and the critical mass: The adoption of telecommunication services by German banks. *Telecommunications Policy, 23*, 719-740.

May, P. (2001). *Mobile Commerce - Breakthroughs in Application Development: Opportunities, Applications and Technologies of Wireless Business.* Cambridge, UK: Cambridge University Press.

Moon, J., & Kim, Y. (2001). Extending the TAM for a World Wide Web context. *Information & Management, 38*, 217-230.

Newell, F., & Newell-Lemon, K. (2001). *Wireless Rules.* New York, NY: McGraw-Hill.

Ortt, J.R. (1998). *Videotelephony in the Consumer Market.* Doctoral dissertation, Delft University of Technology.

Ostlund, L.E. (1973). Perceived innovation attributes as predictors of innovativeness. *Journal of Consumer Research, 1* (September), 23-29.

Plouffe, C.R., Vandenbosch, M., & Hulland, J. (2001). Intermediating technologies and multi-group adoption: A comparison of consumer and merchant adoption intentions toward a new electronic payment system. *Journal of Product Innovation Management*, 18(2), 65-81.

Rogers, E.M. (1962). *Diffusion of Innovations*. New York, NY: The Free Press.

Rogers, E.M. (1995). *Diffusion of Innovations* (4th Edition). New York, NY: The Free Press.

Ruyter, J.C. de, Wetzels, M.G.M., & Kleijnen, M.H.P. (2001). Customer adoption of e-services: An experimental study. *International Journal of Service Industry Management, 12* (2), 184-207.

Sawtooth. (1985-87). *ACA System Manual*. Sawtooth Software, Inc.

Stone, A. (2001). Mobile gaming update: Can doom rescue m-commerce. *M-Commerce Times*. Retrieved November 20, 2001, from the World Wide Web: http://mcommercetimes.com/Services/185.

Sutherland, E. (2001). Gaining m-trust. *M-Commerce Times, February Issue*. Retrieved March 19, 2001, from the World Wide Web: http://www.mcommercetimes.com/Solutions/86.

Tarasewich, P., & Warketin, M. (2000). Issues in wireless e-commerce. *ACM SIGecom Exchanges, 1*(1), 19-23.

UMTS Forum. (2000). *Enabling UMTS / Third Generation Services and Applications*. UMTS Forum Report 11. Retrieved June 20, 2001, from the World Wide Web: http://www.umts-forum.org.

UMTS Forum. (2001). *The UMTS Third Generation Market—Phase II: Structuring the Service Revenue Opportunities*. UMTS Forum Report 13. Retrieved June 20, 2001, from the World Wide Web: http://www.umts-forum.org.

Verhoef, P.C., & Langerak, F. (2001). Possible determinants of consumers' adoption of electronic grocery shopping in The Netherlands. *Journal of Retailing and Consumer Services*, 8, 275-285.

Chapter XII

Mobile Data Technologies and Small Business Adoption and Diffusion: An Empirical Study of Barriers and Facilitators

Jeanette Van Akkeren and Debra Harker
University of the Sunshine Coast, Australia

ABSTRACT

The technological environment in which contemporary small- and medium-sized enterprises (SMEs) operate can only be described as dynamic. The exponential rate of technological change, characterised by perceived increases in the benefits associated with various technologies, shortening product life cycles and changing standards, provides for the SME a complex and challenging operational context. The primary aim of this research was to identify the needs of SMEs in regional areas for mobile data technologies (MDT).

In this study a distinction was drawn between those respondents who were full-adopters of technology, those who were partial-adopters, and those who were non-adopters and these three segments articulated different needs and requirements for MDT. Overall, the needs of regional SMEs for MDT can be conceptualised into three areas where the technology will assist business practices; communication, e-commerce and security.

INTRODUCTION

This chapter presents findings from a two-phase study on the perceptions, needs and uses of mobile data technologies by Australian small business owners. In Phase I, focus groups were conducted, and rich information obtained on possible uses and applications of Mobile Data Technologies (MDTs) for three usage groups, that is, non-, partial-, and full-adopters of IT and Internet applications across many industry sectors. Based on findings from Phase I, the second phase of the study involved interviewing 500 small business owner/managers on mobile data technology adoption issues and perceptions of MDT usage.

The primary appeal of mobile data technologies, apart from mobility, is that associated 'services' are delivered on existing devices such as mobile phones, palm-tops, and personal digital assistants (PDAs). In the literature little empirical work exists on applications and services that would encourage the adoption of mobile data technologies by small businesses. This study provides empirical evidence on attitudes of small business owner/managers in a regional setting, Queensland Australia, to mobile data technologies, and identifies the most significant facilitators and inhibitors to adoption.

The development of gateway technologies for service providers supporting WAP are already available and on the market. Further, wireless applications have been developed that provide mobile devices with Internet content and e-business services. These mobile data technologies were expected to affect business in a similar fashion to the Internet and World Wide Web a few years ago (Semilof, 1999). The major appeal of mobile data technology is that it provides information to the mobile user such as reading news, getting stock quotes, sending e-mail, downloading data, locating other users, remote accessing of home and business sites, and making purchases on a device that consumers are comfortable with—the mobile telephone.

In Australia, a small business is defined as an organization employing less than 20 people, typically independently owned and financially controlled by the owner/manager; and [usually] locally based business operations (Annual Review of Small Business, 1998). The adoption of mobile telephones in the past five to 10 years for voice services and messaging, by both small and large organizations in Australia, has been high. Helping to accelerate the demand for the newer mobile data technologies in countries already embracing these technologies (such as Japan) is the explosive growth of the Internet and mobile computing (Clever, 1999). In Australia, this provides somewhat of a conundrum. Although adoption of e-commerce by large organizations in Australia has been relatively high, the same cannot be said for small businesses, where adoption has been slower than other countries, such as Singapore, the United Kingdom, the United States of America, and Japan (Forrester, 1997; Yellow Pages, 2000). In contrast, however, mobile telephone adoption and diffusion is relatively high by smaller organizations. The conundrum is therefore, will small businesses who have been reluctant to adopt e-commerce technologies in the past be more ready to go on-line with the merging of mobile voice and mobile data technologies? The main focus of this study is based on the question: "What are the needs of regional small businesses in relation to MDT, and do these needs differ depending on the level of IT adoption already in place?" Conducting a study on mobile data technology attributes will help to identify the important issues for small business owner/managers, whether they are early or late adopters of pre-existing technologies. Results from Phase I are presented followed by a discussion of the findings.

BACKGROUND
Adoption of IT/E-Commerce by Small Business

Empirical studies have identified a variety of factors thought to affect e-commerce/Internet technology adoption in small business (Julien and Raymond, 1994; Brooksbank, Kirby and Kane, 1992; Kirby and Turner, 1993; Icovou, Benbasat and Dexter, 1995; Thong and Yap, 1995; Harrison, Mykytyn and Rienenschneider, 1997). From the adoption factors identified in earlier studies, a framework (Figure 1) was developed based on the study on the adoption of e-

Figure 1: Framework of Small Business Adoption of IT Innovations (Source: Van Akkeren and Cavaye, 1999)

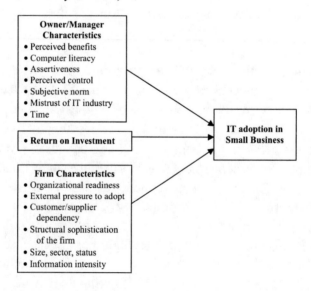

commerce technologies thought to facilitate or inhibit technology adoption by small business owner/managers (Van Akkeren and Cavaye, 1999). It was on this basis that attitudes and perceptions of small business owner/managers were assessed on the adoption of mobile data technologies for this current study.

Owner/Manager Characteristics

- *Perceived benefits* affect technology adoption in terms of the perceived ease of use and/or usefulness of the technology. If an owner/manager does not perceive the technology in a positive way, he/she will be reluctant to adopt.
- The *computer literacy of the business owner* can also influence technology adoption. If the owners are unaware or do not understand the technologies available, they are unlikely to adopt them into their own business.
- The *level of assertiveness, rationality, and interaction of business decision processes* can also impact on IT adoption. If owners of the firm are assertive in business decision processes, understand the benefits and applications for the technology in their organization, and are able to rationalize how that information can be useful, they will be more likely to adopt IT.
- *Perceived control* relates to the amount of requisite opportunities and resources (time, money, skills, cooperation of others) someone possesses to be able to carry out the course of action (technology adoption). For example, a small business owner may decide that connection to the mobile Internet is an important competitive use of IT. Yet if there is a possible budget shortfall, their decision to adopt will be influenced.
- *Subjective norm* is thought to affect technology adoption in terms of the strength of the person's normative beliefs that 'groups' think the behavior of interest (i.e., technology adoption) should or should not be performed. A person's motivation to comply with the group is also a factor affecting normative behavior.

Firm Characteristics

- *Organizational readiness* refers to the level of technology currently incorporated into business processes. If there is little technology incorporated, or outdated/inefficient technology being utilized, a firm is less likely to adopt new technologies.
- A small business will be reluctant to adopt innovative IT unless there is a specific request for it by their trading partners and/or customers. If this *external pressure to adopt IT* is not present in the industry sector, then the business owner may perceive the technology as a waste of resources.
- The *dependency of the small business customer on the supplier* is linked to the previous factor. Not only would the supplier need to have adopted the

technology to make it viable, the small business owner would need to recognise and understand the benefits to his or her firm in adopting the technology. In addition, an organization may perceive that their clientele was of a certain socio-economic level that would not readily benefit from the introduction of new technologies.

- The *structural sophistication of the firm* in terms of centralisation and complexity will also influence technology adoption in its ability to incorporate new technologies into its work practices. A particularly complex structure could either inhibit or facilitate technology adoption and would be dependent on whether the owner believed that IT could be easily incorporated and enhance operations or excessively disrupt operations.
- The *size, sector, and status of the organization* has been shown to influence technology adoption, particularly in relation to the sector and status. If competitors and trading partners within the sector have adopted IT, an owner may be more inclined to adopt as well. The size of the business can also influence technology adoption, as a very small business with only two or three employees may not have the time or expertise to devote to implementing and using new technologies.
- Finally, the *level of information intensity* within the organization may influence the owner to adopt or not adopt a technology. For example, if large amounts of data and information are part of the business processes, an owner may be more likely to adopt technologies that could streamline operations and lead to process efficiencies within the organization.

Other Factors
- The need by small business owners for an immediate *return on investment* is due to the necessity to be concerned with medium-term survival rather than the long-term attainment of market share. To make a substantial outlay of capital resources, the owner would need to see exactly where the return was going to be in the short term.

The three categories presented in the framework that impact on the adoption of IT innovations provided areas of discussion for the focus groups in Phase I of this research. Participants were encouraged to discuss their attitudes and perceptions of IT adoption in general and mobile technologies in particular.

Recent studies on reasons why small business owner/managers adopt or do not adopt information technology (IT) and e-commerce technologies have highlighted both inhibitors and facilitators to adoption and are similar in content to the factors described above (Van Akkeren and Cavaye, 2000; Fink, 1998; Chau and Pederson, 2000). Small business adoption is discussed as being determined by decision-maker characteristics, information system (IS) characteristics, organizational characteristics, and environmental characteristics (Thong 1999). In this study it was found that the need for IS to offer better alternatives to existing practices is critical to adoption by small businesses. Therefore, could the use of mobile data technologies provide the 'better service' that small business owner/managers seek?

Adoption and Diffusion of Mobile Data Technologies

Major innovations may have to 'prove themselves' in new markets before they can displace other technologies (Friar and Balachandra, 1999). It is the early adopters or innovators who will initially experiment with these technologies. In addition, the usefulness and ease-of-use will impact on owner/manager acceptance of the technologies (Agarwal and Prasad, 1997). Therefore the attributes of the technologies may improve their acceptance by small business owner/managers. The acceptance of Web-based technologies is also influenced by ease of use and perceived usefulness in terms of current IS sophistication, complexity of the new technologies, and perceived costs and benefits (Nambisan and Wang, 1999).

In Australia, the adoption of Internet/e-commerce technologies varies in different states, and further, between regional and city-based firms. However, the adoption of mobile phones is consistently high across states and regions within Australia. Mobile data technologies, which 'marry' mobile phones and e-commerce technologies, are seen as eliminating time and distance as barriers for regional businesses in their adoption of these technologies. Further, in Australia there is a strong "push" by government at all levels for small- to medium-sized firms to adopt innovative information systems, in particular, electronic commerce and associated technologies.

Estimates of mobile data technology usage vary: Greengard (2000) estimates that usage will be one billion worldwide by 2003; Thurston (2000) states that annual turnovers in the U.S. alone will be US$1.3 trillion. The International Data Corporation (2001) estimates use of handheld mobile devices at 10 million in the Asia-Pacific Region by 2003, and Datamonitor (2001) puts handheld device sales in the Asia region at 310 million by 2005. With such potential markets available to the vendors of mobile data services and devices, it is useful to understand reasons why the majority of potential end-users are so far resisting these new technologies.

Lack of speed is a barrier to adoption as mobile data technologies are slow and hence inefficient (Taylor, 1999; Saunders, Heywood, Doron, Bruno and Allen, 1999). Another barrier is the perception of a lack of standardized IT environment for developing mobile data applications as impeding the growth of the mobile data market (Harrison, 1999; Axby, 1998). Limited bandwidth, higher usage costs, increased latency, and a susceptibility to transmission noise and call dropouts are also possible barriers to adoption (Duffy, 1999; Johnson, 1999). It is possible therefore that adopters are 'sitting back' and waiting for at least some of these problems to be corrected before entering the mobile data market. Another area of concern for end-users is that the Wireless Application Protocol, the emerging technology used to send data to and from handheld devices, has no security mechanisms built into it (Riggs and Bachelor, 1999; Chan, 2000). This is of potential concern not only to the business user, but also to the customers of the business as well.

Capturing users requires 'transparency,' that is, users want information or communication access whenever and wherever they need it, using whatever device is most convenient at that moment (Osowski, 1999). Small business owner/managers do not buy technology; they buy business benefits (Duffy, 1999). Mobile data technology benefits include easy communication through e-mail, ready access to information (wherever/whenever), entertainment, and improved lifestyle through e-commerce and banking.

Clearly, the literature on mobile data technologies to date underlines the importance of highlighting the benefits of using the technologies, and the ease of use to potential users. Small business owner/managers are not interested in the architecture, standardisation issues, or the technologies themselves. Instead they require a device that provides efficient, effective access and communication applications personalised to their individual needs.

Most literature on the adoption and marketing of mobile data technologies is not empirically based and is limited to discussing the technologies in terms of their application to business, rather than adoption barriers. Mobile computing is seen as eliminating time and distance as barriers for regional businesses in their adoption of these technologies, particularly in relation to the design of work and for reaching potential markets. However, many regional areas of Australia have been less prepared to adopt Internet and e-commerce technologies compared to their city-based counterparts.

The importance of this study is that it addresses the gap in the literature by providing empirically based research on what small business owner/managers see as potential applications for MDTs. By segmenting the market by adoption status, readers are provided with greater insight into why owner/managers do or do not adopt innovative technologies. Specifically the study identifies the types of applications and attributes that would provide the most benefit to owner/managers in encouraging them to adopt MDTs.

RESEARCH DESIGN

Given the exploratory nature of this study, the focus of Phase I/II is a qualitative analysis of the factors that are influencing the adoption behavior of small business in the mobile technology arena. Data gathered from the focus groups and telephone surveys provide a set of mobile data technology attributes that can be used to study their adoption and diffusion by small business. In the initial part of this two-phase study, two focus groups were conducted with respondents in each usage group—non-adopters, partial-adopters, and full-adopters of information technology—each covering a variety of industry sectors. Market segmentation is a powerful tool in identifying different subsets of the population with similar needs. Therefore, the market was segmented this way as it has been found in previous studies that different adopter levels have different needs and attitudes to technology and innovations (Van Akkeren and Cavaye, 1999). By segmenting the participants in this way, this research was able to identify specific needs of those who are familiar with IT compared to those that are less computer literate. Descriptions of each group are:

- **Full-Adopter:**   Used a computer and the Internet for business, e-mail, e-commerce, website.
- **Partial-Adopter:**   Used a computer for business, some use of the Internet for business/home, but no website.
- **Non-Adopter:**   No use of the computer for business purposes.

As much as possible, a range of industry sectors was represented in each of the focus groups as different industry sectors have different needs in terms of technology adoption and usage (Van Akkeren and Cavaye, 1999). Every respondent was the owner/manager of the firm as they make all the management decisions relevant to the enterprise, including technology adoption and usage policies.

The six focus groups ran for approximately one-and-a-half hours each and were video taped with the permission of respondents. Questions were asked about current technology usage, and respondents were encouraged to discuss their experiences, attitudes, and perceptions of information technologies. After showing a brief video that demonstrated mobile data technologies, respondents were then encouraged to discuss their reactions to these new technologies and possible

Table 1: Focus Group Attendance

Focus Group	Category	Respondents
Focus Group 1	Non-adopters	10
Focus Group 2	Non-adopters	9
Focus Group 3	Partial-adopters	15
Focus Group 4	Partial-adopters	11
Focus Group 5	Full-adopters	14
Focus Group 6	Full-adopters	12

applications to their industry sector. The focus group topic guide was informed by the model developed from the literature review and in collaboration with Nortel Networks. The data analysis was conducted using the content analysis approach.

In Phase II, potential respondents were drawn from the Yellow Pages Online with different business types grouped into their industry sector. The database comprised approximately 5,500 potential respondents in total, and a sample of 500 respondents were interviewed. Interviewers worked through their respective database in a random manner to contact potential respondents, and a policy of three callbacks before disregarding the potential respondent was employed.

Following completion of the 500 telephone interviews, and in conjunction with the data entry process, questionnaires were screened to gauge their usability. Of the 500 survey response sheets submitted by interviewers as "completed," 18 were deemed 'not usable' due to substantial insufficient collection of data on certain variables, or due to the respondent falling outside certain sampling parameters. The final sample size was thus 482, derived from 1,251 telephone calls (not including call-backs), indicating an overall response rate of 39% which is above the accepted norm for this type of research. Data was coded and entered into an SPSS Data File.

Setting the p level at .05, as was done in this study, succeeds in filtering out weak correlations, thus we can be 95% confident that the results are actually true. Unless otherwise stated, and for the duration of this report, the p-level is significant at .05 or less, indicating that there is a 5% probability that the relation between the variables found in the results is a chance occurrence.

Descriptive output (including frequencies, means, modes, simple cross-tabs, etc.) was generated from the data file. This output was then assessed using various statistical techniques to identify significant differences between certain groups. T-tests were conducted to identify significant differences between groups of two (such as male versus female). One-way Analysis of Variance (ANOVA) was conducted to identify differences between more than two groups (for example, adoption levels and industry groups). These ANOVA tests were run with a 'Tukey's post hoc evaluation' to determine significant differences between groups at the 5% signifi-cance level. More advanced multivariate statistical techniques such as logistical regression, factor analysis, and discriminant analysis were also attempted however, the data did not meet the strict assumptions required to run such analyses.

The research was concerned with understanding the needs of small businesses for MDT. However, the literature suggests that it is foolish to treat all small businesses the same in respect of IT adoption and usage. Therefore, a measure of IT adoption was used again in the second phase of this study to ascertain whether a full-adopter of technology, for example, had the same needs for MDT as did a partial- or non-adopter. Thus, the rate of IT adoption is becoming increasingly important to business longevity in the 21st century; indeed, 62% of small businesses are in the process of becoming online businesses, with a further 29% recognizing the need to do so (Dearne, 2001). Further, more than nine in 10 (95%) medium-sized businesses are now connected to the Internet (Dearne, 2001). With this in mind, ANOVA was conducted to identify differences between more than two groups (for example, adoption levels). The ANOVA tests were run with a 'Tukey's post hoc evaluation' to determine significant differences between groups at the 5% significance level.

FINDINGS (PHASE I[1])

Problems with, and Praise for, IT

After a brief introductory session, full- and partial-adopter respondents were asked to discuss any problems with, and praise for, IT in general.

Problems with IT

While there was some overlap in the areas of problems with IT cited by both full- and partial-adopters, certain issues were mentioned by only one group. For example, 'pricing and costs' were areas of concern for full-adopters only, and they perceived a link between IT and these areas. Combining new ways of trading (for example on the Internet) with the move to relocate manufacturing plants to offshore locations, full-adopters felt that competition has risen to new heights:

"There's always a cheap copy available somewhere ... people appreciate the quality but they're not prepared to pay for it." (F, Clocks, 3)

"People will access the Internet, find the cuckoo clock they want ... press their button, and they've got it there!" (F, Printing, 25)

A second area of problems with IT, recognised by both full- and partial-adopters, related to the 'reliability and support' of IT products and services. There was criticism about the rate of change in the area of innovation and, in more practical terms, strong feelings of frustration about the usefulness of IT manuals:

"The gap between the promises and the deliverables is quite huge."
(F, Health Foods, 2)

"It's the pace of change—six months and it's out of date." (F, Retail, 2)

"I'm on the phone constantly every 2 or 3 days, 'How do I do this?'"
(P, Convenience Store, 2)

"I'd like to be able to read a manual." (F, Apartments, 2)

"The information on how the hell the damn thing works is a nightmare!"
(P, Travel Agent, 5)

However, there were useful suggestions to address the latter concern:
"...an instruction manual on a video disk ... instead of words there's a picture of someone." (F, Electronics, 2)

Similarly, both full- and partial-adopters cited 'compatibility' as an area of frustration with IT. This related to computers, consumables (such as printer cartridges), and attachments in e-mail:
"You get something that's sent by e-mail and you can't open it up and that's a real pain." (P, Signage, 3)

A key difference between these two groups of adopters was the issue of 'fear.' Full-adopters did not display any fear about working with technology, however partial-adopters realised they needed to use IT to demonstrate the currency, and therefore dynamism, of their business, in spite of their reservations:
"One good thing IT gives is the appearance you know what you're doing." (P, Hairdresser, 5)
"If I present something into one of my shops which is a little bit high tech, the girls will look at it and think 'I'm a hairdresser, not a computer whiz,' so they're a little scared of this technology." (P, Hairdresser, 5)

Being low users of technology, non-adopters in the groups were asked why they did not use it. The most important reason cited by them was 'cost,' together with:

• losing data	• maintenance
• obsolescence	• lack of suitable training
• no need/benefit	• viruses
• time consuming	• break downs
• power failures	• impersonal

Understandably, there was also the issue of 'fear' with non-adopters—fear of buying the wrong technology, or about their own ability to learn new technology:
"If you write things out, you don't have to worry if you press the wrong button and something gets lost." (N, Home Maint., 2)

All of these problems were couched in business terms, that is, all of the areas were felt to negatively affect business practices.

Innovations

Respondents in all groups were asked to suggest the main innovations in IT over the past five years. Table 2 displays the responses of all groups:

Table 2: Innovations

Innovation	Benefits	Typical Comments
Mobile Phone	Freedom, flexibility, convenience, availability.	*"If you miss out on a deal, you miss out on money." (N, Boutique, 2)*
Computer	Speed, control (inventory, accounting), data storage, letters, retrieving information.	*"We used to have two rooms of books, now we have one shelf of CDs." (F, Solicitor, 4)*
Internet	Information, world trends, check prices, website, competitor information, cost effective (e.g., sending samples of work via email), convenience, speed, current.	*"It's more convenient than getting the Yellow Pages out." (F, Resort, 9)*
EFTPOS	Speed, streamlined service, cashless society, enhanced security, keeping up with competitors, add-on sales, impulse buying, security, keep up with competitors.	*"We rarely take cheques ... generally credit cards, and that means big business." (F, TV Service, 14)* *"The money is immediately in your bank." (N, Shoes, 3)*
Desktop Banking	Immediacy, ease of moving money around, payroll, bill paying, remote operation.	*"We operate shops in Rockhampton and on payday ... just highlight the people we want to pay and then schunk!" (F, Printing, 25)*
Fax	Speed, time and money saving, direct communications, visual.	*"You come in and the floor's full of it, all the orders." (N, Boutique, 2)* *"We do all our work by fax." (P, Dry Clean, 9)*
Barcoding/ Scanning:	Currency of information (e.g., stock levels), competitive advantage.	
Databases	Tracking, reports, mail outs.	
Networking	Linking PCs together, reduces costs, increases speed.	
Operating Systems & Software	Moving from DOS to Windows and NT–better for business.	
Printers	Higher quality, cheaper now.	
Scanners	Time and cost of producing high-quality work reduced now.	
Video Cassette Recorder	Demonstrate business (if appropriate) (only mentioned by non-adopters).	

Product/Service Applications Video

The Nortel Networks video was shown in all groups and then responses elicited about product/service application needs and benefits. The video gave viewers an insight into the possible uses of MDTs for both business and personal use. Each application available on a device was explained and demonstrated in the video with participants shown how each application could enhance their business operations and personal lives. Adopters were visibly stimulated by the video, displaying knowledge of the area and being keen to discuss future applications:

> "We're not very far away from a lot of these things … I witnessed a digital camera take a photo and he hooked it up to his mobile phone and sent it to someone." (F, Printing, 25)

Partial-adopters were unenthusiastic about the technology demonstrated in the video, being very wary, cautious, and fearful:

> "I wouldn't want one of those because it'd be 'where are you?' and they would be onto me." (P, Dry Clean, 9)
>
> "I wouldn't like it." (P, Signage, 3)

Table 3: Services (Full-Adopters)

Service	Benefits	Typical Comments
1. 2-way communication: video/voice/voice recognition	Relationship marketing, the power of face-to-face persuasion, personalised and improved customer service, speed, time, planning.	*"International business is a very personal thing."* (F, Exports, 17)
2. Prioritizing & screening messages	Delegation of work to others, freedom, cost savings on labour, rent overheads.	*"No one would have to be in one particular office space."* (F, Fencing, 6)
3. Remote access	Ability to check home/business from another location, security, peace of mind, working smart.	*"If it wasn't okay I could push another button and get Security guys there straight away."* (F, Property Developer, 2) *"I could check the chlorine levels in my pool."* (F, Apartments, 2)
4. Online information	Voice searching on the Internet or specialist databases.	*"Like 'intestacy' – what sort of documents do I need?"* (F, Solicitor, 4)
5. Attachments, downloading	Ability to send information to others.	
6. Navigation	Directions, deviations, detours, speed, time, planning.	
7. Translation	Useful for international transactions.	

The reaction of the non-adopters was somewhere between the full- and partial-adopters; when asked, half of the non-adopters wanted one of the handsets, mainly to stay in touch with the youth and technology in general, while the other half did not see a need. However, they were overwhelmed, initially, by the technology demonstrated:

"We couldn't take it all in!" (N, Shoes, 3)

Table 4: Services (Partial-Adopters)

Service	Benefits	Typical Comments
1. Remote access & security	Control	*"You don't have to go into work ... just put that machine on and see what's happening." (P, Coffee House, 12)* *"I don't need to sit at the computer ... just walk around the store chatting to it!" (P, Conv. Store, 2)*
2. 2-way communication: video/voice/ voice recognition	Speed, communication, mobility, immediacy, ease of use, improved communication means improved customer relations, build loyalty.	*"You can do more than one thing at a time!" (P, Newsagent, 5)*
3. Navigation	Information–delays, detours, local knowledge, time and cost savings from better planning.	*"The courier guy ... hasn't had to think about it, he's done it in half the time ... made his run." (P, Festival Org, 1)*
4. Prioritizing & screening messages	Screen messages, screen ads out.	

Table 5 – Services (Non-adopters)

Service	Benefits	Typical Comments
1. 2 way communication: video/voice/ voice recognition	Sending samples, time saved on travel and answering calls, freight costs reduced, business networking opportunities	*"You can see who you're talking to." (N, Hairdresser, 2)*
2. Navigation	Save time.	*"The map business ... I loved that, that was excellent!" (N, Boutique, 2)*
3. Remote access & security	Both business and home needs.	
4. Shopping		

Respondents then participated in an exercise whereby the services demonstrated in the video were listed, benefits associated, and then the services ranked in order of importance. Tables 3, 4, and 5 display this information for each adopter level:

Full-adopters had the clearest thinking about how they would use this technology—immediately. Their discussions centred around how their top three ranked services would interact and the impact this would have on their business

Table 6: Top Three Rankings, All Groups

Rank	Full-Adopters	Partial-Adopters	Non-Adopters
First	2-way communication: video/voice/ voice recognition	Remote access & security	2-way communication: video/voice/ voice recognition
Second	Prioritizing & screening messages	2-way communication: video/voice/ voice recognition	Navigation
Third	Remote access & security	Navigation	Remote access & security

practices. For example, they were the only group of adopters who included 'prioritising messages' in their top three (Table 6), with the key benefit of delegation of tasks. There was also discussion in these groups about screening of junk messages and to prioritise messages, suggestions included screening by time of day, and by different types of callers.

The partial-adopters were not keen to have this technology at all but when asked to rank their services, they included the navigation service, as did the non-adopters.

All groups raised concerns about the services that were shown to them. Partial-adopters were concerned about the reliability of the system, confidentiality, their lack of free time, civil liberties, and concerns about 'big brother'. Non-adopters mentioned the issue of invasion of privacy. Security of such a system was a recurring theme amongst all groups:

"What happens if you lose it (the handset)?" (F, Property Developer, 2)
"Someone else could access your home security before they go in and rob you." (F, Windscreen, 4)

Remedies for the security concern included using thumbprint recognition, retina scan, or voice recognition.

FINDINGS PHASE II
Reaction to MDT by IT Adopter Level

The findings from Phase I identified one of the key issues that was addressed in Phase II: testing how the different adopter levels compared to each other in their

Table 7: Adoption Status—Sample Characteristics

Adoption Status	No.	% of Total Sample
Non-Adopters	70	15%
Partial-Adopters	137	28%
Full-Adopters	275	57%
Total	482	100%

attitudes, needs, and approach to technology. This is because it is suggested that the more IT-literate people are, the more aware they will be of their current and

Table 8: ANOVA Differences Between Adopter Levels

Statement	Full-Adopter	Partial-Adopter	Non-Adopter
Wait before investing (n=474)	2.09	1.74	*
Being able to email (n=460)	2.30	2.66	2.90
Owner/Manager feels excited (n=471)	2.54	2.87	3.04
Trade with customers (n=456)	2.61	3.02	*
Being able to navigate (n=437)	2.73	3.11	3.13
Being able to monitor or operate equipment (n=447)	2.83	*	3.42
Live 2-way video (n=445)	2.92	3.23	*
Staff would feel threatened (n=452)	3.79	3.54	3.05
Overall interest in acquiring (n=474)	**2.49**	*	**2.93**

* only statistically significant results are shown.

future new technology needs. Table 7 displays adoption status characteristics of the sample.

Table 8 displays the results of the ANOVA analysis, showing the key differences between the three adopter groups in terms of MDT needs.

The ANOVA results showed that there were clear differences in the attitudes and needs of respondents, depending on their different levels of adoption of technology. Full-adopters were more excited about the prospect of this technology compared to the partial- and non-adopters. This state of mind is carried through the data as, in terms of waiting before investing in the technology, the full-adopters indicated that they would not wait as long as the partial-adopters. Similarly, in the business environment, there are key differences between the three groups. The non-adopters, for example, are significantly different from the other two groups in relation to staff reaction. It would appear that the less familiar the owner/manager is with current technology, the higher their perception is that their staff would likewise be uncomfortable with new technology.

Clearly, the acceptance of technology will influence attitudes and approaches to it; it will also influence perceived needs of different market segments. A key bank of questions in the questionnaire focused respondents on the use of MDT in a business setting and asked them to state to what extent they agreed or disagreed with a set of statements. To complement this bank of questions, an important, overarching question was also posed but placed later in the survey to alleviate any bias. This later question gauged the level of interest in acquiring the MDT, if it was available and affordable tomorrow. Thus, by focusing on those respondents who answered either 'strongly agree' or 'agree' with each statement in the 'key benefits' bank, it can be shown which features of MDT are important to each adopter level.

Full-adopters were most comfortable with technology compared to the rest of the sample. Table 9 displays the rankings of key benefits for the full-adopter group, together with their responses regarding their 'overall interest in acquiring.'

Table 9: Full-Adopters who 'Strongly Agree' or 'Agree' with Statements

Rank	Statement	Full-Adopter Number (n=275)	Full-Adopter % (n=275)
1	Wait before investing	236	86
2	Being able to email	217	79
3	Owner/Manager feels excited	176	64
4	Trade with customers	171	62
5	Banking and other admin.	162	59
6	Trade with suppliers	162	59
7	Monitoring business premises	162	59
8	Being able to navigate	154	56
9	Being able to access the net	146	53
10	Being able to monitor or operate equipment	146	53
11	Competitive pressure to adopt	143	52
12	Live 2-way video	135	49
13	No need for this technology	96	35
14	Being able to shop	82	31
15	Staff would feel threatened	41	15
	Overall interest in acquiring	**165**	**60**

While full-adopters agree that they will wait awhile before investing in the mobile data technology (86%), the owner/manager was excited by the prospect of it (64%). In terms of addressing the needs of the full-adopter small business owner/manager, the most important benefits were found to be using MDT for mobile e-mail (79%), e-commerce via trading with customers (62%) and suppliers (59%), and to bank at their own convenience (59%) or monitor their premises (59%). Six in 10 (60%) full-adopters were very interested or interested in acquiring this technology.

Table 10 displays the rankings of key benefits for the partial-adopter group.

Nine out of 10 (93%) partial-adopters agreed, or strongly agreed, that they would wait a while before investing in this technology. However, when focused on how the technology could help them in their business, partial-adopters mainly felt that being able to deal with e-mail (64%), banking (60%), and monitoring their business premises (56%), all in a remote fashion, would be advantageous. Six in 10 (59%) partial-adopters were interested or very interested in acquiring this technology.

Non-adopters have little knowledge of IT in general, and use it least compared to the other two groups. Table 11 displays the rankings of key benefits for the non-adopter group.

While non-adopters indicated that they, like the other two groups, would wait a while before investing in MDT (87%), they did not feel as strongly on this as the partial-adopters of technology (93%). The features of MDT that the non-adopters valued most were e-commerce via trading with suppliers (62%) and customers (51%), remote banking (52%), and e-mail (51%).

Table 10: Partial-Adopters who 'Strongly Agree' or 'Agree' with Statements

Rank	Statement	Full-Adopter Number (n=137)	Partial-Adopter % (n=137)
1	Wait before investing	127	93
2	Being able to email	88	64
3	Banking and other admin.	82	60
4	Monitoring business premises	77	56
5	Trade with suppliers	73	53
6	Owner/Manager feels excited	66	48
7	Trade with customers	63	46
8	Competitive pressure to adopt	63	46
9	No need for this technology	63	46
10	Being able to monitor or operate equipment	60	44
11	Being able to navigate	58	42
12	Being able to shop	58	42
13	Being able to access the net	55	40
14	Live 2-way video	52	38
15	Staff would feel threatened	36	26
	Overall interest in acquiring	**81**	**59**

Table 11: Non-Adopters who 'Strongly Agree' or 'Agree' with Statements

Rank	Statement	Full-Adopter Number (n=70)	Non-Adopter % (n=70)
1	Wait before investing	61	87
2	Trade with suppliers	43	62
3	Banking and other admin.	36	52
4	Trade with customers	36	51
5	Being able to email	36	51
6	Being able to access the net	33	47
7	Live 2-way video	32	45
8	Being able to navigate	32	45
9	No need for this technology	31	44
10	Monitoring business premises	31	44
11	Staff would feel threatened	30	43
12	Competitive pressure to adopt	30	43
13	Owner/Manager feels excited	29	42
14	Being able to monitor or operate equipment	21	30
15	Being able to shop	20	29
	Overall interest in acquiring	**31**	**44**

In a marked difference to both the full- (60%) and partial-adopters (59%) of technology, less than half (44%) of the non-adopters were 'very interested' or 'interested' in acquiring this technology.

Industry Differences by Adopter Levels

This research project was concerned with understanding the differences between different adopter status groups in relation to MDT needs, including that the industry sector variable into this discussion at this point strengthens the analysis. By displaying a cross-tabulation of these two aspects of the data, greater insight is provided in terms of the profile of industry sector by adopter status (Table 12).

Table 12: Industry Sector by Adopter Status (%)

Industry Sector	Full -Ad.	Partial -Ad.	Non -Ad.	Total Industry
Property and Business Services (n=71)	85	15	0	100
Health and Community Services (n=45)	73	25	2	100
Agriculture, Forestry and Fishing (n=15)	60	13	27	100
Manufacturing (n=35)	60	17	23	100
Transport and Storage (n=18)	56	33	11	100
Wholesale Trade (n=28)	55	28	17	100
Retail Trade (n=106)	52	27	21	100
Construction (n=79)	47	36	17	100
Accommodation, Cafes and Restaurants (n=35)	45	33	22	100
Personal and Other Services (n=38)	32	55	13	100
Communication Services (n=4)	25	50	25	100
Total (n=482)	**57**	**28**	**15**	**100**

The Property and Business Services (85%) sector has the highest proportion of full-adopters of technology, followed by the Health and Community Services (73%) sector. Industries with the lowest proportions of full-adopters include Communication Services (25%); Personal and Other Services (32%); and Accommodation, Cafes, and Restaurants (45%).

DISCUSSION

It is possible to argue that there are many different factors that will impact on a small business owner/manager's decision to adopt or not adopt mobile data technologies. The focus groups found significant differences in the way different types of people view and use technology. The industry sector the firm belongs to, the current IT adoption status of the firm, the level of mistrust of the IT industry, and the cost of the technologies are highlighted in this study as possible barriers or facilitators to adoption. The features of the mobile device, including the applications on offer (which directly relates to perceived business benefits of the technologies) are also raised as having a possible impact on adoption.

Full-adopters of technology are very open to new ideas and innovations, compared to the partial-adopters, although they raise concerns about the cost of keeping up with the rate of technological change. Non-adopters, while being fearful of technology generally, have an open mind when presented with futuristic new technology, especially when compared to the partial-adopters. One reason for these differences could be that partial-adopters have been 'forced' to adopt technology before they were really prepared; this adoption may have been in response to customer, competitor, or supplier pressure. Thus, it could therefore be hypothesized that they are fearful of over-reliance on technology, of being hoodwinked by IT companies, of being made to feel inadequate, and/or of being overtaken by others in terms of business practices.

When asked about innovation in IT in the past five years, most discussion in the groups centered around the mobile phone, computers, the Internet, EFTPOS, desktop banking, and the fax machine. Benefits associated with these innovations included freedom, flexibility, speed, convenience, increased information, competitive advantage, ease of use, and direct communications. These reactions bode well for mobile data technology suppliers and the future of WAP, as the key features of WAP provide all of these benefits on a mobile device.

In terms of Internet use by the group members, full-adopters have no fear and use the technology for business and pleasure. Partial-adopters tend to have Internet technology forced upon them by competitors, customers, or suppliers, while the non-adopters commented on family use and pressure, and demonstrated a desire to 'keep up' with the youth of today, and tomorrow.

The Nortel Networks video that was used in the groups facilitated much discussion, excitement, and fear, but it also biased responses somewhat, especially in relation to the number of handsets or PDAs people would tolerate. Initial reactions to the video by respondents were mixed. Full-adopters were visibly excited by the prospect of the technology demonstrated, partial-adopters were unenthusiastic, cautious, and fearful, while non-adopters were evenly divided in their interest.

Full-adopters, many of whom could be considered early adopters and innovators, would use this technology tomorrow if it were available, as long as it was affordable. The service they would most like to 'buy' is the two-way communication with video/voice recognition. This group of adopters was the most focused in their intended use of the technology, being very business-focused and seeing a need for almost all of the services on display. The non-adopters also cited the two-way communication with video/voice recognition as their favourite service, and both groups felt that this service would give them improved customer relations with the face-to-face persuasion, increased speed, and better use of time. Partial-adopters were reticent about the technology but felt that if they had to choose, the ability to remotely access sites would be useful and the security aspect appealing. The greatest benefit for them would be control of the working environment. All three adopter groups were concerned with security of the system.

Benefits are very important to small business owner/managers who do not have the luxury of time to train, research, or upgrade their technology without losing business. Thus, Phase II of the study focused upon technology use and associated benefits across different industry sectors, gauging levels of concern/comfort with technology. Findings from Phase II identify that overall, Australian small businesses on the Sunshine Coast would wait before investing in this technology. However, this is not surprising given that Australian small businesses have generally been slower to adopt e-commerce technologies compared with other developed countries such as Japan, the U.S., and Singapore (Forrester, 1997; Lawrence et al., 1998; Van Akkeren and Cavaye, 1999). This study has shown that small businesses have an overriding need for **communication**, closely followed by a need for **e-commerce** capabilities and **security**; thus, specific applications that address these needs include:

- access to e-mail,
- trading with customers,
- trading with suppliers,
- banking and other administrative tasks, and
- monitoring business premises.

The ANOVA analysis showed that there are in fact key differences between the adopter level groups and, in marketing and adoption terms, this should be taken into account. There is a distinct difference, for example, between full- and partial-adopters when considering how long to wait before investing in MDT: the full-adopters would adopt quicker than the partial-adopters. Equally, the more comfortable the owner/manager is with current technology, the more excited they feel about the MDT, and this is further supported when dealing with staff, as the more IT-literate the owner/manager, the less threatened they say their staff would feel. Table 13 summarizes the top-ranking features of MDT needed by adopter group.

The industry sector to which respondents belonged also had some influence on attitudes towards MDT and this supports findings in previous studies that industry sector will influence adoption (Thong, 1999; Yellow Pages, 1999). Interest in acquiring the technologies was strongest in the Transport and Storage, Communication, and Personnel and Other Services sectors.

Overall the reaction to the new technology was very positive as, in every industry except one, more than half of the respondents were either 'very interested' or 'interested' in acquiring mobile data technologies. (The industry least interested in acquiring the technology was Construction at 46%.) Within these industry sectors, the top ranking applications that support the underlying principle of mobile data technologies (business anywhere, anytime) were consistently noted as e-mail (communication), trading with customers, trading with suppliers, banking and other administrative tasks (e-commerce), and monitoring business premises (security).

FUTURE TRENDS

In Australia, MDTs are slowly becoming more commonplace, with providers of these technologies focusing on wireless mobile phone applications. In particular, banking, receiving/sending e-mails, booking tickets to shows, and downloading Internet contents are the main thrust of marketing campaigns. Largely, small business uses and applications of MDTs have so far been very limited in the media.

Table 13: Top Ranking Features of Mobile Data Technology by Adopter Groups

Statement	Full-Adopter (n=275)	Partial-Adopter (n=137)	Non-Adopter (n=70)
Being able to email	1st	1st	3rd
Trade with customers	2nd	-	3rd
Banking and other admin.	3rd	2nd	2nd
Monitoring business premises	3rd	3rd	-
Trade with suppliers	3rd	-	1st

However, as this research indicates, there is an enthusiasm for these technologies with owner/managers identifying possible strategic business uses to their own organizations. It is important for managers to ascertain how mobile data technologies can enhance their business processes. This is particularly important for small business owner/managers whose bottom line is crucial to their survival. Capital expenditure on technology needs to be carefully explored so that benefits to the firm can be identified before purchasing innovative technologies.

Findings from this study have allowed refinement and development of the framework presented earlier (Figure 1). The refined framework, Figure 2, incorporates a range of factors impacting on regional small business adoption of MDT specifically, and it is hoped that this model can provide direction for future research on the adoption and diffusion of MDT for small businesses.

Small businesses, by definition, do not have the luxury of time and money that bigger firms do. If they are going to invest money in new technology, whether it be a mobile phone, computer, PDA, or MDT, they need to be able to reap immediate

Figure 2: Factors Impacting Small Business Adoption of Mobile Data Technologies

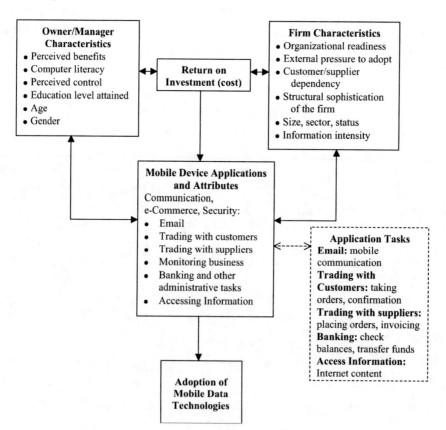

rewards. The most important aspects of small business owner/managers' daily business lives are concerned with communication and being in constant contact with the business in order to pursue contacts, orders, invoices, and so on. However, they are also aware of the way the world is changing and, thus, identified a need of being e-commerce capable. These managers recognise the inherent cost savings of conducting electronic business and banking, and this is important to them. Finally, security is a vital issue for most small businesses; their business premises are their livelihood and any untoward actions (such as staff pilfering, burglaries) have an immediate and dramatic effect on their bottom-line.

Future Research Opportunities

Given the exploratory nature of this study, there are many opportunities for further research into the area of mobile data technology adoption and diffusion. Phases I and II have identified the underlying issues for owner/managers in the adoption of these technologies for their businesses: communication, e-commerce, and security.

This study has provided insight into the needs of small business owner/managers for mobile data technologies. Findings from Phases I and II have identified that they are more likely to purchase these technologies if applications such as mobile e-mail, the ability to trade with customers and suppliers, and banking anywhere/anytime are made available on the device. Additionally, one could hypothesize that the following have been identified in this study:

- Firstly, the applications available on the device will impact on the adoption rate of MDTs.
  "You can do more than one thing at a time!" (P, Newsagent, 5).
- Secondly, cynicism by owner/managers in relation to the IT industry as a whole is a substantial barrier to the adoption of MDTs.
  "The gap between the promises and the deliverables is quite huge." (F, Health Foods, 2).
- Thirdly, if an owner/manager can readily identify the benefits to their firm by adopting MDTs, they are more likely to adopt.
  "The courier guy ... hasn't had to think about it, he's done it in half the time ... made his run." (P, Festival Org, 1).
- Finally, the current adoption status of the firm will impact on the decision by the owner/manager to adopt MDTs.
  "We couldn't take it all in!" (N, Shoes, 3).

Further research and testing of these hypotheses may elicit information that could be useful to the development of applications and adoption of MDT. Each area of further research is a significant project in its own right, and different research methodologies should be employed to uncover the truth in each case.

The findings from this study provide a major contribution to both the theory and practice in this area. Theoretically, it has contributed to knowledge and learning in the field of adoption of IT/innovations. Practically, managers of SMEs are now much more informed on the role of IT and innovation in enhancing business practices.

CONCLUSION

At the beginning of this study, MDTs were a very new concept to owner/managers of small businesses, particularly in regional areas of Australia. The interest and enthusiasm shown by the majority of participants in this research bodes well for MDT hardware and application on-sellers. However, there exists a certain degree of cynicism by owner/managers towards information technology in general and specifically to IT professionals. This was evident in many discussions by owner/managers about the "hype" used by salespeople in the IT profession to sell their "latest and greatest" products that in the long run provide little benefit or return on investment to their organization. Overcoming the bias held by many owner/managers that MDTs are simply another "gimmick" could well prove to be a daunting task. However, there is little doubt that there are benefits of MDTs to business owners, and the marketing of these benefits may go a long way towards overcoming current perceptions.

ENDNOTES

1   Where direct quotations are made from respondents, the reference is given as 'full-adopter' (F), 'partial-adopter' (P), 'non-adopter' (N), followed by an industry descriptor and number of employees.

REFERENCES

Agarwal, R., & Prasad, J. (1997). The role of innovation characteristics and perceived voluntariness in the acceptance of information technologies. *Decision Sciences, 28*(3), 557-582.

Annual Review of Small Business. (1998). Department of Workplace Relations and Small Business, Australian Government Publishing Service, Australia.

Axby, E. (1998). Creating a market for mobile data. *Telecommunications, 32*(9), 37-39.

Brooksbank, R., Kirby, D., & Kane, S. (1992). IT adoption and the independent retail business: The retail news agency. *International Small Business Journal, 10*, 53-61.

Chan, T. (2000). Cracks in the WAP. *America's Network, 104*, 26-27.

Chau, S. & Pederson, S. (2000). The emergence of new micro businesses utilising electronic commerce. In Cable, G. G. & Vitale, M. R. (Eds.). *Proceedings of the 11th Australian Conference of Information Systems (ACIS)*, Brisbane, [CD-ROM].

Clever, M. (1999). Mass market solutions for mobile data. *Telecommunications, 33*, 40-49.

Datamonitor. (2001). *Projected Handheld Device Sales in the Asia Region.* Retrieved December 2001 from http://www.datamonitor.com

Dearne, K. (2001). SMEs learning the e-ropes. *The Australian*, August 14, 41.

Duffy, R. (1999). Wireless set to take the lead. *Telecommunications, 33*, 24-26.

Forrester Research Incorporated. (1997). *Forrester Ranks World Economics for E-Commerce*, Retrieved August 1998 from http://www.forrester.com.

Friar, J. H., & Balachandra, R. (1999). Strategies for marketing new technologies. *Research Technology Management. 42*, 37-43.

Gefen, D., & Straub, D. W. (1997). Gender differences in the perception and use of e-mail: An extension to the technology acceptance model. *MIS Quarterly, 21*, 389-400.

Greengard, S. (2000). Going mobile. *Industry Week, 249*, 18-22.

Harrison, D. A., Mykytyn, P. P. Jr., & Rienenschneider, C. K. (1997). Executive decisions about IT adoption in small business: Theory and empirical tests. *Information Systems Research, A Journal of the Institute of Management Sciences, 8*, 171-195.

Harrison, H. (1999). WAP: The key to mobile data. *Telecommunications, 33*, 96-98.

Hom, D. (2000). E-business portals ease end-user service creation. *Telecommunications, 34*, 43-45.

Iacovou, C.L., Benbasat, I., & Dexter, A.A. (1995). Electronic data interchange and small organizations: Adoption and impact of technology. *MIS Quarterly, 19*, 465-485.

International Data Corporation. (2001). *Asia-Pacific Use of Handheld Devices 1998 to 2003*, Retrieved December 2001 from http://www.idc.com/en_US/home.jhtml.

Johnson, A.H. (1999). WAP. *ComputerWorld, 33*, 44-69.

Julien, P.A., & Raymond, L. (1994). Factors of new technology adoption in the retail sector. *Entrepreneurship: Theory and Practice, 18*, 79-90.

Kirby, D., & Turner, M. (1993). IT and the small retail business. *International Journal of Retail and Distribution Management, 21,* 20-27.

Lawrence, K.L. (1998). Factors inhibiting the utilisation of electronic commerce facilities in Tasmanian small- to medium-sized enterprises. *8ᵗʰ Australasian Conference on Information Systems.*

Mahajan, V., & Muller, E. (1999). When is it worthwhile targeting the majority instead of the innovators in a new product launch? *Journal of Market Research, 35,* 488-495.

Mobile Data Conference. (1999). Sydney, NSW.

Nambisan S., & Wang Y. (1999). Roadblocks to Web-based technology adoption. *Association for Computing Machinery, Communications of the ACM, 42,* 98-101.

Osowski, K. (1999). Unifying communications. *Communications News, 36,* 18-20.

Riggs B., & Bachelor B. (1999). Vendors address issue of security for wireless devices. *Informationweek, 765,* 25-26.

Saunders, S., Heywood, P., Dornon, A., Bruno, L., & Allen, L. (1999). Wireless IP: Ready or not, here it comes. *Data Communications, 28,* 42-68.

Semilof, M. (1999). Hitting the road? Take the Internet along. *Computer Reseller News, 864,* 1.

Taylor, M. (1999). The need for speed. *Communications International, 26,* 41-42.

Taylor, S., & Todd, P.A. (1995). Understanding information technology Usage: A test of competing models. *Information Systems Research, 6,* 144–176.

Thong, J., & Yap, C.S. (1995). CEO characteristics, organizational characteristics, and information technology adoption in small business. *Omega, 23,* 429-442.

Thurston, C. (2000). Economic waves wash onto all shores. *Global Finance, 14,* 121-123.

Van Akkeren J. K., & Cavaye A. L. M. (1999). Factors affecting entry-level internet adoption by SMEs: An empirical study. *Proceedings of the Australasian Conference in Information Systems, 2,* 1716-1728.

Weiers R. M. (1988). *Marketing Research.* (2nd ed.), NJ: Prentice Hall.

Wexler, J. (1999). Cut the cord. *Upside, 11,* 201.

Yellow Pages Australia (1999). *Small Business Index: Survey of Computer Technology and E-Commerce in Australian Small and Medium Business.* Melbourne, Pacific Access.

Chapter XIII

We Know Where You Are: The Ethics of LBS Advertising

Patricia J. O'Connor
Queens College – City University of New York, USA

Susan H. Godar
William Paterson University, USA

ABSTRACT

Privacy is the most significant and complex ethical issue facing LBS. While LBS is more than the combination of e-commerce and telemarketing, we use the ethical failures of those two media to show that consumers will seek legislative action to protect themselves from invasions of privacy using the new medium. The alternative is effective self-regulation by the industry; we conclude with a proposed model for such self-regulation, involving existing trade groups.

INTRODUCTION

Just five years ago, there was very little notice of e-commerce and little advertising on the Web. When Hoffman and Novak wrote their article in 1996, they were introducing this advertising tool to marketing researchers. Now, Web advertising is the subject of numerous academic articles. A great deal of research attention has focused on e-commerce and how it has changed, and is changing, both the marketing and the management of companies. We know, for example, that much of the allure of the World Wide Web lies in the exploitation of its "worldwide" capabilities: marketers have access to a large group of customers, regardless of their physical location. Service providers have made use of the technology to inform customers about their offerings, expedite responses to customer complaints, and facilitate reservations and purchases.

Little research attention has yet been paid, however, to the emergence of mobile commerce (m-commerce), which differs from e-commerce in providing a new way to market to local customer bases—and to consumers who are merely in transit through a particular location. With the m-commerce application of location-based services (LBS), the physical location of a moving customer is identified. Based upon that location, the customer is then directed to the nearest service provider. The "directing" takes place through portable, mobile devices designed to allow consumers to access the Internet whatever their location.

A primary reason why this new marketing tool has not yet been much studied is that, so far, its use is not well-advanced. Many companies worldwide are devoting much money and effort to making LBS m-commerce more ubiquitous, and once a standard protocol has been adopted it seems likely that this new technology soon will be at least as transformative of the marketing and management of companies as e-commerce has been.

In this chapter, we explain the three features that differentiate LBS m-commerce from e-commerce: mobile location identification, synchronous two-way communication, and provider power. Then, we argue that the most complex ethical issue confronting marketers as they begin to use this new tool is an escalated form of the ethical issue raised by telemarketing and by e-commerce: privacy. It is our belief that LBS will create for itself the same type of restrictive legislative environment that now constrains telemarketing if effective industry self-regulation does not take place. We conclude by articulating a model for self-regulation that we think will allow the industry to avoid otherwise inevitable legislative action.

We focus on the situation in the U.S. While privacy is an issue worldwide, three factors make it likely that the U.S. will be a testing ground for privacy in relation to LBS. First, the government's mandate that location information be available for all cellular phone calls will mean that the ability to track location will soon exist in the U.S. Second, regulations on the use of consumer data are not as restrictive in the U.S. as in other countries. This opens the door to the possibility of unethical behavior by companies. Finally, many Americans believe that privacy is a right, and are perhaps more likely than citizens of other nations to take action to defend that right.

LBS vs. E-Commerce

We have identified three features that differentiate LBS from e-commerce: mobile location identification, synchronous two-way communication initiated by the marketer, and the power which may be exercised by service providers. These features have a potential impact on efforts to use the new medium in marketing. All have the potential to raise ethical questions.

The first differentiating feature is that LBS seeks to exploit a "micro-environment," marketing to a small group of consumers located in (or passing through) that specific target area. This is in contrast to the approach of e-commerce: seeking to exploit the Internet's "worldwide" capabilities by marketing to a large group of customers, regardless of their physical location. Using the information made available by wireless service providers, m-commerce marketers will know their customers' physical location with a high degree of precision. Using cell phones or PDAs, they will be able to send consumers messages as they near, for example, a retail establishment. A customer driving down the street could be called on her cell phone and told of an oil change special at a nearby garage (Sonnen, 2001); or a PDA could beep and she would be able to see the web pages of all nearby establishments. If Bluetooth or a similar purchasing capability were added and service providers made those purchase records available to marketers, the latter would also be able to determine instantly whether their efforts were successful, and to refine further their advertising to a particular consumer.

LBS information and technology also open the possibility for true interaction with a local customer, which is the second feature differentiating LBS and e-commerce. In an e-commerce setting, marketers must simply hope that a customer will click through to a page or an ad. The new technology allows marketers to reach out to consumers, sending messages directly to them without any specific customer request. Thus, in LBS the marketer can initiate the interaction, and if a consumer chooses to respond, will have the opportunity to engage in synchronous communication. The possibility of real-time two-way communication will engender new opportunities and challenges for marketers.

Finally, providers may become much more significant with this technology than they have been in e-commerce. Rowley (2000) contends that there are two approaches to information seeking on the Web: browsing and directed search. The small size of wireless devices and the difficulty of using a telephone keypad as an input device may help to explain why NTT DoCoMo, the major cellular telephone service provider in Japan, has found that 85% of their customers do not stray from their homepage by more than two clicks. It therefore appears likely that a kind of limited browsing will be the norm in that future in which Shaffer (2000) contends that 25% of all Internet commerce will be wireless. Many items will probably be accessed only by clicking on ads that are on or very close to the homepage.

Marketers will not have an incentive to advertise their individual URLs as they currently do, but will instead probably find it more efficient to establish contracts with successful providers. This will consolidate personal information about consumers in the hands of a few large companies.

The features that differentiate LBS from e-commerce seem likely to raise ethical issues, foremost among which is invasion of privacy. We believe that the clearest way to begin thinking about the ethics of LBS advertising is to examine the ethical failures of existing marketing activities.

For the purposes of this discussion, we think of LBS as a combination of telemarketing and e-commerce. The consumer's phone number is known, and one can in principle allow the use of that number to send push advertising to someone at that phone number: telemarketing. The consumer's location, as well as many other facts about the consumer, are known, and one can in principle sell that information to someone else so that the consumer can be targeted for the sale of particular goods and services via the wireless device: e-commerce. LBS cannot, of course, be reduced simply to a combination of telemarketing and e-commerce. This framework, however, provides a way of using the past to understand and to predict what the ethical issues of this future marketing medium will be, and a way to learn from mistakes that business has made and is making. It also highlights the importance of the service providers' ethical choices: since they will be providing access to consumers, it is they who are able to function as ethical gatekeepers. Nonetheless, marketers also face ethical challenges in LBS advertising.

The Importance of Ethics

There are at least two reasons why business ought to attend carefully to ethical issues, both of which are clearly visible provided we focus steadily on the bottom line. Ironically, Milton Friedman's infamous remark about the sole responsibility of business being to increase profits, so often cited as a justification for business not troubling itself with ethical concerns (and, in fact, apparently intended by Friedman as just such a justification), is precisely the reason why business ought to attend to ethical issues very carefully indeed. To see why this is so, we need only look briefly at the goal of ethics and the goal of business.

Speaking extremely generally, ethics is a set of rules, or a decision procedure, or both, intended to provide the conditions under which the greatest number of human beings can succeed in flourishing, where "flourishing" is defined as living a fully human life. Speaking equally generally, business is the attempt to provide human beings with some of the things that contribute to their flourishing. These "things" take a great variety of forms, of course, and depending on an individual's relation (e.g., employee or consumer) to a particular business, what the business provides will be quite different. In the case of an employee, what is provided may be not merely the means to purchase some of the wherewithal of flourishing, but also satisfying work and a sense of participating in something larger than him- or herself, both of which human beings also need to flourish. In the case of the consumer, the business is supplying something that he or she has come to believe will help him or her to flourish. Clearly, at least at this level of generality, not only are ethics and business not opposed to one another, but both have precisely the same end in view: human flourishing.

The problem arises—and the oft-heard comment that the phrase "business ethics" is an oxymoron is one indication of the pervasiveness of the problem—from the business community's persistent failure to recognize, and to enact its recognition of, two obvious truths. One is that there are dimensions of human flourishing in addition to the possession of money. The other is that large numbers of human beings are very well aware of the first truth, and are increasingly willing to act on the basis of that awareness. Businesses focused on the bottom line—those that want to make a profit—should attend very carefully to ethical issues to avoid either direct or indirect stakeholder actions.

It is quite clear that stakeholders are willing to take direct action to punish companies that act in ways that damage people's ability to flourish. The phenomenon of stockholders attending annual meetings to insist, on ethical grounds, that companies modify their business practices is on the increase. Consumers also engage in product boycotts to protest unethical corporate actions and policies.

Consumers may take indirect action by pressing their elected representatives for new legislation to restrict what they regard as unethical business practices. Consumers don't believe the cliche that you can't legislate morality. They often seek, and frequently receive, legislative protection against unethical behavior by business; many "consumer protection laws" result from precisely this process. A particularly interesting example of successful indirect consumer action, directed against an activity similar to m-commerce, are the newly enacted laws restricting telemarketing.

During the period when their activities were virtually unregulated, telemarketers frequently invaded the privacy of consumers, stole their time, and refused to cease unwelcome contacts when told to do so—all of which are clear instances of behaving unethically. Consumers responded directly by installing the equivalent of "locks" on their telephones (caller ID boxes and various "anonymous call rejection" services provided by local telephone companies are two examples of these) to safeguard their privacy, time, and freedom from harassment. When these measures proved insufficient, consumers successfully lobbied their elected representatives for enhanced protection against the unethical practices of telemarketing (Ferguson, 2001). By July of 2001, twenty-three states had implemented a "no-call," "no-sales," or "black-dot" law allowing consumers to protect their privacy by prohibiting telemarketing calls (Murphy, 2001). In short, the consequence of telemarketers behaving in ways that consumers found unethical has been an explosion of laws restricting telemarketing.

Moreover, as Donna Gillin (2000) points out in her brief review of contemporary U.S. privacy legislation, some of the new laws may have consequences beyond the restriction of telemarketing. She observes that some legislation is not sufficiently nuanced to distinguish between research calls and sales calls. Hence researchers, who have long been sensitive to the privacy and other concerns of their subjects, are likely to be hampered in their ability to collect data by consumers' indirect action to restrict unethical practices in which the researchers did not themselves engage. Legislation can be a blunt instrument that unintentionally renders business practices that are not morally offensive just as illegal as the practices that caused the legislation.

In his aphoristic comment about the "sole" responsibility of business, Friedman elided the obvious truth that there are dimensions of human flourishing in addition to the possession of money. Perhaps even more culpably—but we must remember that he was writing nearly two generations ago—Friedman's aphorism fails to acknowledge what is, in the U.S. in 2002, certainly equally obvious: that large numbers of human beings are very well aware of the first truth, and are increasingly willing to act on the basis of that awareness. Businesspeople who, citing Friedman, excuse themselves from considering the ethical dimensions of their practices are, judged by the aphorism itself, being irresponsible. One dimension of our discussion in the next section is the suggestion that it is precisely because e-commerce has largely ignored consumer privacy concerns that the public is now extremely sensitized to privacy issues. LBS will suffer for this sensitivity unless its practitioners, utilizing a medium that inherently poses even greater threats to privacy, act more responsibly from the beginning by engaging in effective self-regulation based on explicitly ethical grounds.

The Complexities of Privacy

Marketers and managers who have been attending to the literature on e-commerce must already be aware of the explosion of consumer concern about privacy. Groups, organizations, and conferences advocating data privacy have proliferated online: the Electronic Privacy Information Center (www.epic.org); Junkbusters (www.junkbusters.com); and the Web sites devoted to the annual MIT-sponsored Computers, Freedom, and Privacy conferences. Academic journals have sponsored special issues, such as the Spring 2000 issue of the *Journal of Public Policy and Marketing*, on privacy.

E-marketers, and e-businesspeople more generally, vary widely in their responsiveness to consumer concerns about privacy. On the one hand, Scott

McNealy, CEO of Sun Microsystems, is quoted as saying, "You already have zero privacy. Get over it." (Milne, 2000, p. 240), and Microsoft has on occasion "deliberately designed features into its software that invade users' privacy" (Kronenberg, 1999, note 25). On the other, industry groups such as the Online Privacy Alliance, the Network Advertising Initiative, and the Internet Advertising Bureau are all attempting to articulate recommended privacy guidelines for their members. Given the phenomenon noted above (of morally offensive business practices leading in time to legislation rendering those practices illegal), the intent of the industry groups to self-regulate clearly is in the interests of business, while the belligerent arrogance of a McNealy or the sneakiness of a Gates clearly is not.

It is evident from the e-commerce literature that one difficulty confronting those who would like to be responsive to consumer privacy concerns is understanding exactly what constitutes a violation of privacy. Without such an understanding, it is not possible, even for companies that would like to do so, effectively to self-regulate. When it comes to pinning down exactly what is at stake for those concerned about privacy in regard to computers and the Internet, Deborah Johnson (2001) captures the sense of many when she writes:

> The term *privacy* seems to be used to refer to a wide range of social practices and domains, for example, what we do in the privacy of our own homes, domains of life in which the government should not interfere, things about ourselves that we tell only our closest friends. Privacy seems, also, to overlap other concepts such as freedom or liberty, seclusion, autonomy, secrecy, controlling information about ourselves. So, privacy is a complex, and, in many respects, elusive concept" (p. 120).

Simplifying and elucidating this elusive concept has proven difficult, in part because it has a very long intellectual history, even within the U.S., and in part because, as Johnson suggests, it overlaps to some degree with other important concepts. We suspect that because privacy is not a simple concept, e-businesses have been tempted to conclude either that it is not important, or that it is too "messy" for a concern with it to be operationalized in a company's practices, and that it can therefore safely be ignored.

To understand why these are not safe conclusions, it is helpful to recall that within the U.S. privacy is most often understand as a "right," akin to the right of free speech. William Brown, referring to the locus classicus of the concept of privacy as a "right," reminds us that Judges Warren and Brandeis "articulated a general right to privacy as a 'right to be let alone.' While they stated that no constitutional right to privacy was explicit, they argued that a right to privacy is implicit in a number of places in the Bill of Rights and is, therefore, a derivative right" (Brown, 1996, p. 3). He notes further that Warren and Brandeis argued that "the violation of privacy is an incursion on something hitherto inviolate, something primal and rooted deep within the person, part of their 'inviolate personality,'" and that this line of argument about what is at risk when privacy is violated has been expanded during the intervening century to include "concepts such as human dignity, individual uniqueness, integrity of self, and individual autonomy" (p. 6-7). Kronenberg (1999) adds that "privacy is based in part on the notion of each of us having one unique identity and the guarding of that identity, so as to both preserve it and not have it appropriated by others" (p. 16).

In short, at least within the United States, not only is privacy understood as a right, but violations of privacy are understood to threaten one's individuality, dignity, and freedom. It is probably not necessary to underscore the fact that Americans are extremely insistent on having their rights respected, or the fact that threats to individuality and freedom are taken very seriously in the U.S.

A second warning that business—whether e-, m-, or traditional—cannot simply take the short way with the complexities of privacy (by ignoring it altogether) comes from a line of argument that suggests information about consumers should be understood as property. This has the advantage of shifting the grounds of the discussion *from* the elusive right to privacy, *to* a right business understands better. It seems, in some ways, a natural move; as Richard T. DeGeorge (1999) notes, "The question of privacy quickly slips into the question of ownership. Ownership of information is a central issue laden with ethical implications. . . . Who owns information about individuals? Can such information rightly be owned?" (p. 5). It also seems a natural move because the practices of businesses clearly indicate the corporate perspective is that all data, including consumers' personally identifiable information (PII), is in fact property once it is in the hands (or on the servers) of business. Rent is charged, trades are engaged in, fees are paid—or else others are accused of "theft." Until now, the corporate perspective on DeGeorge's questions has been predicated on the unexamined assumption that PII has cash value only after having been collected by a business—as though data were lumps of ambergris found on the seashore.

As Edmund F. Byrne points out, however, mere parity of reasoning shows this is not the case: "If, as data collectors claim, data are property, then the collectors too should pay for whatever they take. Control of privacy... begins not with the value added but with the original taking; and on this it is up to the original owner to set a price or, if he or she so chooses, not to sell. This is the way it is with private property" (Byrne, 2001, p. 8). If we dispense with the right to privacy and treat PII as property, will the results be favorable for business? The answer "no" seems obvious, but the question is far from rhetorical. At least one consumer has already been sufficiently outraged by what is, alas, an all-too-common invasion of privacy by traditional businesses (the practice of renting/trading/selling mailing lists) to bring suit using a state law that prohibits the use of a person's PII without the individual's consent (Kirsh et al., 1996). Thus, we may soon have an opportunity to begin finding out whether this shift in perspective will in fact hurt business—if only by requiring a massive revision in how PII can be used after they are gathered.

Consumers understand privacy as a right. Threats to it are understood also to threaten one's individuality, dignity, and freedom. Attempts to avoid engaging the issue of the right to privacy seem to lead quite directly to instead subsuming personally identifiable information under property right—a path business surely does not want to take. Under the circumstances, we suggest that it would be far better for managers who wish to exploit the opportunities of LBS to get to grips with consumer privacy concerns than it would be to attempt to ignore them. In order to do so, however, some definition of privacy that is more concrete than the Warren/ Brandeis "right to be left alone," and less complex than Johnson's list of domains and concepts, must be given; neither of them can readily form the basis of self-regulatory standards and practices that would appear to be business's best hope of staving off consumer-driven demand for PII privacy legislation.

Sissela Bok provides a definition of privacy that appears to be precisely what is needed to provide the basis for such self-regulation. Privacy, she says, is "the condition of being protected from unwanted access by others—either physical access, personal information, or attention. Claims to privacy are claims to control access to what one takes... to be one's personal domain" (Bok, 1984, p. 10-11). The key terms in Bok's definition are these: protection, access; control. The areas to be protected are of three kinds: physical, personal information, and attention.

Extending an argument she attributes to James Rachals, Deborah Johnson (2001) explains why control of information is a particularly salient aspect of privacy when consumers interact with businesses. Johnson writes that "Rachals seems right about the way information affects relationships. We control relationships by controlling the information that others have about us. ... [L]oss of control of relationships comes with the loss of control of information" (pp. 121-122). The relationship between an individual and a corporation necessarily contains a power asymmetry, and especially under these circumstances, "what is important to the individual is that the individual have some power or control in establishing or shaping the relationship" with an organization (Johnson, 122). By maintaining control over his/her personal information, an individual could partially redress the inherent asymmetry of the relationship, but as things stand in the world of e-commerce, Johnson suggests, "it would seem that individuals have very little power in these relationships. One major factor making this possible is that these organizations can acquire, use, and exchange information about us, without our knowledge or consent" (p. 122-3).

As we show in the next section, what has resulted from the practices of e-commerce should be understood as a warning to those interested in entering LBS.

If the self-regulatory efforts of those engaged in LBS lead consumers to believe that they have true protection that provides them with control over their personal information, control over who has access to their persons, and control over who can focus attention upon them, we believe that they will not press for additional privacy legislation. An impetus to engage in effective self-regulation is provided by the pressure for privacy legislation that is mounting in the area of e-commerce.

The Failure of E-Commerce

As we suggested earlier, the best way to begin to see the magnitude of the privacy challenge facing LBS is by looking at the current state of e-commerce. The business practices in this arena that are of greatest concern to consumers are data mining and the processes used by corporations to ensure that there are data to mine. Kronenberg says bluntly: "The most direct challenges to privacy arise from the practice of data mining," and "the Internet has become the largest single source of data mining information on individuals" (1999, p. 7). That is: using Bok's definition the perception already is that business has violated privacy by not allowing consumers control over who attends to them, on the basis of what personal information.

It is reported that "a whopping 87% of Web users believe that they should have 'complete control' over the demographic information that Web sites capture" (Hoffman et al., 1999, p. 3). Typically, however, consumers are offered no control at all of these data. Instead, "while consumers clamor for full disclosure and informed consent, the few Web sites that do tell their visitors they are tracking them and recording their data, follow the traditional opt-out model. [These] policies place the entire information protection burden on the consumer, offer none of the control, and set up an environment of ipso facto mistrust between the Web provider and the consumer" (Hoffman et al., 1999, p. 4). The Web sites that do even as much as this are, indeed, "few"; Milne (2000), reporting on the dismal failure of the FTC reliance on "fair information principles to guide privacy regulation and industry practice in the United States," reveals that of 365 organizations belonging to the Direct Marketing Association that were surveyed, "less than half practice the fair information principles of notice and choice" (p. 2).

Consumer responses to these failures to provide control over personal information have been predictable. On the basis of the 1997 GUV 7th WWW User Survey, Hoffman et al. (1999) report that because Internet users are not offered the ability to control how information will be used, "fully 94% of Web users have declined to provide personal information to Web sites at one time or another when asked and 40% who have provided demographic data have gone to the trouble of fabricating it" (pp. 3-4). They also report that "over 71% [of Web users] believe there should be new laws to protect their privacy online" (p. 3). Kronenberg assures us that "Internet privacy pressure groups are fierce, strong, and persistent; and over time their impact is likely to be increasingly successful" (1999, pp. 11-12).

We concur. We believe that, even in the legislative environment prevailing since the September 11, 2001, terrorist attacks in the U.S., Web users will eventually obtain data privacy laws. We also believe that, as in the case of the legislation restricting telemarketing, in their effort to safeguard consumers against ethically offensive business practices, the laws will be blunt instruments that also render illegal business practices that are not offensive. In fact, given that many elected representatives are presumably familiar with the telephone but are anything but au fait with the Internet, there is good reason to suppose that the Internet legislation will be even worse in this regard than the telephone legislation has been. To avoid this outcome for m-commerce, managers must erect an ethical structure for the medium that will give consumers an unprecedented amount of control over business access to their personal information, and over business attention to their physical location.

Challenges for LBS

Location-based services raise the specter of consumers being kept under constant, detailed surveillance. Surveillance is a particularly egregious invasion of privacy, since it constitutes a trespass in all three areas marked in Bok's definition for control by an individual: attention, physical access, and access to personal information. As Beth Schultz notes with good reason: "When it comes to privacy, location-aware applications, which use knowledge of a user's exact location, bother people most. ... A commonly described scenario for location awareness is of a retailer zapping an ad to someone walking by a brick-and-mortar outlet. While some shoppers may be ecstatic to learn about a sweater sale 10 feet away, others would surely find this a creepy invasion of privacy" (2001, p. 2).

It will not work for service providers to attempt to allay this fear by pointing out that they are already required by the FCC's "E-911" regulation to be able to report the location of 95% of their customers within 150 yards. Commercial use of location information will nevertheless be seen as ethically problematic. Given the reason for the FCC regulation, few consumers could reasonably object to providers making this information available, upon demand, to emergency services. Using the terms laid out by Bok (1984), when the situation warrants attention to one's person, provision of the personal information necessary for that (possibly life-saving) attention to arrive expeditiously can hardly be understood as an invasion of privacy.

It is a short technological step from "able to report" to the government, for the purposes of swift provision of emergency service, to "actually reporting" to marketers, for the purposes of targeted provision of advertising. But it is a giant leap across the privacy line. As already noted, Bok's definition makes clear that all three areas people are most likely to wish to control for themselves would be breached through mobile location identification. One need not have recently read *1984* to find extremely objectionable the idea that Big Brother has one under constant surveillance. Many, if not most, U.S. consumers will likely thus have serious objections to marketers' use of location information.

The counter-argument that can be anticipated, especially from service providers, will be an attempt to blur the distinction between location information being provided on demand to governmental agencies charged with the protection of citizens, on the one hand, and location information being constantly provided to other citizens (whether corporate or individual) so that they can attempt to profit by it, on the other. Cellular providers are already maintaining that, since they have paid the infrastructure costs of collecting the information, they should be able to use it as they like; and what they would like is to "exploit the ability of mobile networks to pinpoint the whereabouts of each phone," to the extent of "book[ing] $2.5 billion from selling data on users' whereabouts" by 2003 (Reinhardt, p. 27).

What service providers must keep in mind is that the two uses—emergency assistance and profit-making—of location information data are completely different from one another. They are different in time ("on demand" as contrasted with "continually"). They are different in agent ("to the government" as contrasted with "to the entrepreneur"). And they are different in purpose ("to protect health and safety" as contrasted with "to make a profit"). The self-evidently ethical first use of the data cannot possibly provide an ethical justification for a second use that is so different in purpose. Specious and self-serving arguments to the contrary will only add insult to what consumers will correctly perceive as the injury of having their privacy invaded.

It may be tempting to respond, in regard to this issue, that it is moot, since current law provides that consumers must "opt-in"—that is, give explicit permission for a provider to send them messages via their cell phones. As already noted, however, the industry appears quite willing to advocate on behalf of changes to the statutory status quo. The business press has pointed out, in regard to the "opt-out" provision of the Financial Services Modernization Act of 1999, that service providers have gone to some length to make certain that consumers will *not* understand the choices with which they are being presented. France writes, "The [privacy] notices are about as easy to digest as car warranties. They're packed with legalese, written in small print, and violate almost every known rule about how to make complex ideas comprehensible to the average consumer" (2001, p. 83).

Taken together, these factors hardly inspire confidence that consumers will truly be able to exercise informed consent. If the industry is unsuccessful in changing the statutory environment, it may, based on the evidence not only from the implementation of the Financial Services Act, but also from current practices of Web sites (which frequently hide "opt-out" notices well below the "submit" button), be expected to take the low road of obfuscating the meaning of "opting in." There is a clear ethical issue here: it seems extremely likely that without some form of industry self-regulation, people will find themselves receiving marketing messages they are convinced they did not agree to receive via their cell phones or PDAs. Wireless access providers should implement LBS in such a way that only those consumers who genuinely agree to having their location information made available to marketers will receive the advertising contacts.

Model for Self-Regulation

It would be wise for companies interested in LBS to remember both the price that telemarketers and others have recently paid, and what seems poised to happen in e-commerce. Business has engaged in unethical behavior; consumers have strenuously objected to practices that included invasions of their privacy; increased statutory restrictions are already in place in one of these arenas, and appear to be on the consumer agenda in the other. It is easy to predict that m-commerce applications such as LBS will go through the same process unless business people bind themselves to behave ethically. Timely action is essential, since only the current lack of a standard transfer protocol stands in the way of an explosion of wireless technology and the proliferation of devices that will make LBS an extremely powerful marketing tool. Once that tool is in hand, it will be too late for a thoughtful discussion of the best rules for using it, and the industry will have chosen, by default, a course that we predict will parallel the developments in the other two media.

We advocate that companies (such as advertisers and service providers) that are planning to enter LBS form groups, organized by their intended role within the industry, to formulate ethics standards and articulate sanctions for violating the standards. The standards themselves should also be incorporated, in an appropriate form, in the corporate ethics policies or "credos" devised by each group member to govern its own business practices. (Many corporations already have such policies, and while many of them suffer from serious flaws—such as conflating "unethical" with "illegal"—they are arguably better than nothing, in that they make clear that the corporation recognizes some limits on the means by which it is willing to pursue profit.) This step will help to move the understanding that there are industry-wide ethics standards all the way down to the level of the individuals who are designing software, selling and designing advertising, using location information, and so on. Such "deep dissemination" of the standards is necessary if they are to have a chance of success.

We also advocate that the industry form a separate, coordinating group, which would serve at least two roles: representing the industry to the public, including the

FTC and other government agencies; and training and retaining a cohort of individuals charged with the responsibility of enforcing the standards and applying the sanctions. These monitors should be empowered to verify that participating businesses adhere to the standards, train employees in those companies in the standards, and investigate consumer complaints. The mere existence of a trade organization that allegedly requires adherence to certain standards as a condition of membership is not sufficient to ensure that members will comply with the standards in practice; recall Milne found that of 365 organizations belonging to the Direct Marketing Association that were surveyed, "less than half practice the fair information principles of notice and choice" on their Web sites (2000, p.2).

To managers who are not familiar with what has, belatedly, begun to take shape in e-commerce, ours may sound like an impossible series of suggestions. It is not. Some of the structure to implement this approach already exists. As noted earlier, a number of industry groups have formed in response to consumers' e-commerce data privacy concerns. Among the best of these groups are the Network Advertising Initiative and the Wireless Advertising Association (http://www.waaglobal.org), each of which has connections with the Online Privacy Alliance (http://www.privacyalliance.org). We evaluate these groups as "among the best" because the current statements of their standards are compatible with, although they do not (yet) make explicit reference to, Bok's definition of privacy. As we argue below, it would be extremely useful if they modified their standards to incorporate this definition, as well as made some other changes.

The published standards of the first two groups are substantive, specifically directed toward their members' activities. The third group has chosen to articulate "meta-standards." OPA has specified that companies wishing to be members must have a privacy policy, and four areas must be addressed: notice and disclosure, choice/consent, data security, and data quality and access. This approach would suit the Online Privacy Alliance to be the "coordinating group" envisaged above. Its meta-standards are brief and readily understandable to the public, and it already applies a sanction, though a mild one: companies without privacy policies that meet the criteria cannot be members of OPA.

If what these groups have already done can be taken, for the sake of discussion, as a starting point, it is evident that it must be developed further in order to conform with the model we have suggested. With the exception of the WAA, the organizations were formed to respond to privacy concerns only, and that in the context of e-commerce. While this is a good first step, the reach of their attention must be lengthened to include LBS and other m-commerce applications, since—as we have shown—this new marketing tool brings with it additional concerns about privacy. Stronger sanctions will have to be put in place. Personnel must be trained, and empowered, to verify that participating businesses adhere to the standards, to train employees in those companies in the standards, and to investigate consumer complaints.

CONCLUSIONS

This chapter has argued the point that all organizations must make it very clear that standards should be promulgated and that these standards must be *ethical* standards. There are two reasons to insist upon this. First, it is true. It is evident that the best of the extant standards and draft standards referred to above have one or more of the following four antecedents: 1) the 1973 "Code of Fair Information Practices," which provided the model for 2) the "Privacy Act" of 1974 (Johnson, p. 130); 3) the 1980 OECD "Guidelines on the Protection of Data Privacy and Transborder Flows of Personal Data"; or 4) the European Community "Directive on Data Protection," which came into effect in 1998. It is no accident that all of these documents refer occasionally, and some of them frequently, to what is needed for people's "welfare." Recall Bok's definition of privacy: "the condition of being protected from unwanted access by others—either physical access, personal information, or attention. Claims to privacy are claims to control access to what one takes... to be one's personal domain" (1984, p. 11). It is universally the case that individuals need some degree of control over personal information in order to flourish, and the source documents for the standards currently being devised recognize this explicitly. The standards themselves should include a similar recognition; we urge the incorporation of some form of Bok's definition, together with statements about deception and theft. In addition, an account of how adherence to standards is monitored, contact information for personnel empowered to investigate consumer complaints, and a list of sanctions that will be imposed for non-compliance should all be included.

The second reason to insist upon explicitly grounding the standards in ethics is a practical consideration. We argued earlier that ethically offensive business practices lead to pressure for consumer protection legislation that, among other things, may be insufficiently nuanced, resulting in practices that are *not* unethical also being rendered illegal. Regarding privacy in particular, we demonstrated that in the U.S. it is regarded as a right, and is bundled with other highly emotive ideas, including individuality, human dignity, and freedom. Standards that are not articulated—both for the use of the companies themselves, and for dissemination to the public—in concepts and language that *match how consumers think about these issues* will, we believe, not be effective in obviating the progression to restrictive legislation. To put the matter briefly: if the effort at self-regulation is not seen to be firmly grounded in ethical considerations, consumers (who have good reason to be wary) are almost certain to see it merely as a cynical and self-serving attempt to stave off legislative action, rather than as an attempt by the industry to safeguard human flourishing from irresponsible managers.

As marketers begin to use LBS in advertising, they will be faced with a substantial ethical issue: respecting consumer privacy. We have argued, from the parallel cases of successful pressure to limit access by telemarketers and growing consumer pressure for restrictions on data use in e-commerce, that consumer backlash is likely to prompt legislative action severely limiting the potential of this new medium. To minimize the risk of unethical behavior and its impact on profits, businesses should actively engage in the creation of an effective mechanism for industry self-regulation.

REFERENCES

Bok, S. (1984). *Secrets: On the Ethics of Concealment and Revelation.* New York: Vintage Books.

Brown, W. S. (1996). Technology, workplace privacy, and personhood. *Journal of Business Ethics,* 15(11), 1237-1248.

Byrne, E. F. (2001). The two-tiered ethics of electronic data processing. *Techné,* 2(1). Retrieved October 1, 2001, from http://scholar.lib.vt.edu/SPT/v2n1/byrne.html.

DeGeorge, R.T. (1999). *Business Ethics* (5th edition). New York: Prentice Hall.

Ferguson, K.G. (2001). Caller ID—Whose privacy is it, anyway? *Journal of Business Ethics,* 29(1), 227-237.

France, M. (2001). Why privacy notices are a sham. *Business Week,* (June 18), 82-83.

Gillin, D. (2000). How privacy affects us all: Friction between the researcher's need for information and a respondent's privacy grows. *Marketing Research,* 12(2), 40-1.

Hoffman, D., & Novak, T.P. (1996). Marketing in hypermedia computer-mediated environments: Conceptual foundations. *Journal of Marketing,* 60(7), 50-68.

Hoffman, D.L., Novak, T.P., & Peralta, M. (1999). Building consumer trust online. *Association for Computing Machinery: Communications of the ACM,* 42(4), 80-85.

Johnson, D. (2001). *Computer Ethics* (3rd ed.). Upper Saddle River, NJ: Prentice Hall.

Kirsh, E., Phillips, D., & McIntyre, D. (1996). Recommendations for the evolution of cyberlaw. *Journal of Computer-Mediated Communication,* 2(2). Retrieved September 15, 2001 from http://www.ascusc.org/jcmc/vol2/issue2/kirsh.html.

Kronenberg, V. (1999). Beware of geeks bearing gifts: Ethical implications of current market models of the Internet. *Business and Professional Ethics Journal,* 18(3&4), 125-152.

Milne, G. R. (2000). Privacy and ethical issues in database/interactive marketing and public policy. *Journal of Public Policy and Marketing,* 19(1), 1-6.

Murphy, B.P. (2001). Telemarket backlash: Giving cold calls the cold shoulder. *Business Week,* (July 2), 2.

OECD. (1980). *Guidelines on the Protection of Privacy and Transborder Flows of Personal Data.* Retrieved October 3, 2001, from http://www1.oecd.org/dsti/sti/it/secur/prod/PRIV-en.HTM.

Online Privacy Alliance. *Guidelines for Online Privacy Policies.* Retrieved October 3, 2001, from http://www.privacyalliance.org/.

Reinhardt, A. (2001). Wireless Web woes. *Business Week e.biz,* (June 4), 24-7.

Rowley, J. (2000). Product search in e-shopping: A review and research propositions. *Journal of Consumer Marketing* 17(1), 20-35.

Schultz, B. (2001). Have wireless Internet device, will buy. *Network World Fusion.* Retrieved October 2, 2001, from http://www.nwfusion.com/ecomm2001/mcom/mcom.html.

Shaffer, R.A. (2000). M-commerce: Online selling's wireless future. *Fortune,* (July 10), 262.

Sonnen, D. (2001). What does the future hold for mobile location services? *Business Geographics,* 9 (January), 14-17.

Wireless Advertising Association. (2001). *WAA Advertising Standards Initiative Draft Standards.* Retrieved October 3, 2001, from http://www.waaglobal.org/press/standards_press.html.

Chapter XIV

A Perspective
on M-Commerce

Mark S. Lee
Coca-Cola North America, USA

ABSTRACT

Several statistics from several industry sources have forecast staggering growth for m-commerce over the next five years. But assuming we believe the statistics, marketers need to understand the dynamics of mobile usage and position themselves to take advantage of this substantial opportunity. While most marketers understand that wireless consumers have different application needs and usage patterns than standard online users, many may be perplexed in finding a logical starting point for developing a marketing approach to m-commerce.

This chapter outlines some of the key differences in online consumer behavior and provides a perspective on how marketers might use mobile commerce to stimulate consideration and purchase of their products and services. The chapter shares an approach and an existing application used by The Coca-Cola Company to provide a reference point and to help other marketers understand and leverage mobile commerce as another viable tool in their marketing arsenal.

INTRODUCTION

We've all heard the buzz and the hype about how mobile commerce, or m-commerce, is poised to change the world. Recent statistics listed in *The Industry Standard* and other trade publications forecast m-commerce to account for $332 million in the U.S. alone by 2003 (up from $22 million in 2001), reaching $1 billion by 2004 and \$3.7 billion by 2006 (Anonymous, 2001). This type of staggering growth seems daunting to say the least. But assuming we believe the statistics, how will marketers in general and The Coca-Cola Company in particular position themselves to take advantage of this substantial opportunity?

At The Coca-Cola Company, we believe that context is everything with regard to understanding and meeting consumer needs. Typically, we look at consumer needs based on the particular *occasion* that the consumer is in and try to understand what is most relevant and important to the consumer in that occasion. Looking at it through this filter, you can easily imagine that consumer needs and expectations on the wired web are vastly different from their needs and expectations on the wireless web. In fact, we believe that there is a significant paradigm shift in the way that consumers use these two mediums that must be deeply internalized by marketers.

Wireless consumers do act quite differently than standard online users. A study by the Boston Consulting Group released in November 2000 found that most wireless users spend less than five minutes using m-commerce applications and only 8% use m-commerce services for more than an hour a week. By contrast, the average U.S. consumer surfs the Internet for 31 minutes per session. Usage patterns also vary by the user's location. The study found that American consumers prefer mobile devices for e-mail; surfing; and getting news, travel, and regional information, while some entertainment services that are popular in other countries, like downloadable ring tones, aren't interesting. American users are also not as concerned about sending credit card information over the wireless networks as their Japanese and Swedish counterparts. About 59%, however, fear that location-based services, which can pinpoint a consumer's whereabouts at any given time, will compromise their privacy (Sirkin & Dean, 2000). In light of these dynamics, where is a logical starting point for a marketer like Coca-Cola with respect to m-commerce?

To begin to answer this question, we would first want to understand those occasions that might be most relevant to the mobile consumer. Mobile purchases tend to be more of an impulse buy than a planned expenditure. While most consumers would hardly think about buying a car on a mobile device, they may likely be open (and, in fact may find it very convenient) to receiving an alert on their mobile device when it is time for an oil change and schedule an appointment and pay for the services in advance to save time and perhaps money. For Coca-Cola, fast food occasions are more impulse-driven than the weekly grocery trip to Kroger or Wal-Mart. In fact, we know from our own fast food consumer segmentation research that 70% of fast food consumers haven't decided what kind of food they want prior to getting in the car. The significance of this for marketers is that these consumers are already *predisposed to a purchase occasion* (they know they're going to buy fast food in this example). And since they haven't decided exactly which fast food item they want, we have the opportunity to influence their decision and have a direct impact on their purchase behavior. In addition to fast food, some of the other impulse-driven occasions where Coca-Cola can play a role might include trips to convenience stores, movie theatres, and casual dining restaurants.

Given this, you may be thinking that m-commerce might make logical sense for The Coca-Cola Company, except for the fact that we don't sell our products over the web. That may be true, but we don't sell our products on the TV, radio, or through mail-order subscription in your favorite magazine either. But all these mediums provide us with an opportunity to get our brands into the consideration set in the minds of the consumer prior to the purchase occasion. And typically, the closer we get to that purchase location, the more impact our messages have on that purchase decision. With wireless and m-commerce capability, we not only have the opportunity to just get into the consideration set, but *we can actually "cement" the deal by giving the consumer the ability to complete the transaction on the spot*. Instead of just viewing a list of options for an item of interest, m-commerce allows the consumer to take action at the moment that their purchase intent is at its peak. Often times, this may be of tremendous convenience to the consumer to be able to complete their transaction(s) without having to engage in a separate activity or wait in line at the store.

Once we've determined the most relevant purchase occasions for the mobile consumer, we would then want to understand the likes and preferences of mobile consumers and prioritize our communication or messaging to that audience based on our understanding of their needs. At the highest level, Coca-Cola (like many other marketers) could use the mobile medium to do at least four things:

1.   Build awareness of our products and services
2.   Facilitate a transaction
3.   Develop a relationship with the consumer
4.   Monitor our progress and/or results

Building awareness may be as simple as sponsoring theatre listings with a tag or ad from Diet Coke. It's about getting into the mindset or consideration set of the consumer at a time when it is relevant for them. If you're thinking about a recreational activity like going to the movies, you might also be open to ad-sponsored tags—especially if the advertiser provides valuable assistance to your activity (e.g., Diet Coke might sponsor movie synopses or movie trivia to give the consumer more context around the movie listing options).

Facilitating a transaction is about making it easier for the consumer to make a purchase. If you can view movie listings on your mobile phone, chances are that you would also be interested in purchasing your tickets and, perhaps even a bag of popcorn and a Diet Coke in advance if it's easy to do so and it saves you time and/ or money.

Developing a relationship with the consumer is all about creating a dialogue with the consumer on *their* terms. You develop that relationship as you would a friendship…you want to learn about their preferences (likes and dislikes), understand when they want your support and guidance, and know when they need some space. It's about being helpful without being intrusive and it's built on a platform of trust and common interest.

Monitoring our progress or results is about translating information into insight. For a marketer, it's crucial to understand how the actual results varied from the expected results of each program. With this discipline, we can maintain a cycle of continuous improvement and ensure that we are offering more and more value to the consumer. This type of effort helps build differentiation and preference for your brands and, ultimately, leads to competitive advantage.

As mentioned previously, part of the attraction to the wireless arena for marketers is that, in many cases, mobile consumers may already be predisposed to a purchase decision. They've left the house, the office, or some other location for some predetermined destination with the intent to take action (e.g., attend a meeting, shop, eat, etc.). To the extent that we can give them relevant and valued information to complete their mission or simply make their lives a little easier, we have the opportunity to greatly influence their purchase decision. Naturally, we would only want to do this on their terms to respect their privacy and to build a relationship that they value. Spamming the consumer with unwanted messages would only do more harm than good. Therefore, it's important to let the consumer dictate the terms of the relationship while the marketer adheres in a helpful, non-invasive manner.

To help bring these points home, let's use a real example of how Coca-Cola North America is currently leveraging the wireless medium in general and mobile commerce in particular.

In November 2000, Coca-Cola North America formed a strategic partnership with go2 Systems, a privately held company based in Irvine, California. Go2 is essentially a consumer directory and locator service—the "Yellow Pages" of the wireless web. Go2's key point of difference is that they drive traffic to *physical store locations* (our customers' outlets) versus to other websites. When our customers (retailers) sign up with go2, their outlets are listed on both the wired and wireless web.

On the traditional web, their outlets are listed on go2's website (http:// www.go2online.com) as well as in a dedicated website that go2 establishes for each customer. On the wireless web, go2 has partnerships with most of the major carriers, so consumer coverage is very broad at 80+% of wireless web users.

Consumers register for go2's free service on the go2 website (or on a participating retailer's website) and set up "start" locations for their most common geographic reference points (e.g., home, work, Boston airport, etc.). Once consumers have established their start locations, they can use go2 from any web-enabled device to find virtually anything they're searching for in the physical world by category (restaurants, theatres, golf courses, bookstores, etc.) or by unique customer URL (go2BurgerKing.com, go2JiffyLube.com, etc.).

Each customer is listed in two different categories as well as by their own unique address (e.g., for AMC theatres, listings are under go2movies, go2theatres, or go2amc.com). When consumers initiate a search on the go2 system, all locations for their search topic are listed within a 10-mile radius of their start location. Search results initially yield all the various outlet names and their distance from the start location. Once a particular outlet is chosen, information includes outlet name, street address, distance from start location, outlet phone number, and turn-by-turn directions on how to get there.

As you might imagine, this type of basic information alone could be very helpful and valuable to consumers who are on the go. Since The Coca-Cola Company's beverages are usually more of an incidental item than a "destination" purchase (consumers are primarily visiting these outlets to buy food, watch a movie, buy gas, etc.), we want to "remind" consumers of our beverage products' availability, appropriateness, or specials. Where possible, we also want to give the go2 consumer the ability to conduct a mobile commerce transaction with our participating retailers.

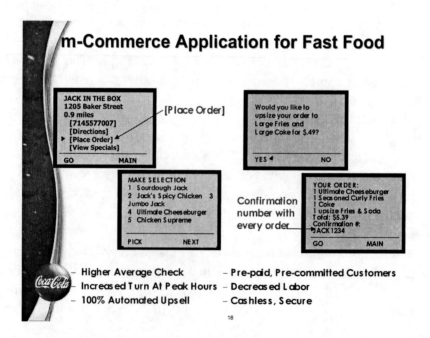

18

By providing this type of valuable service to our retailers and our consumers, we hope to build a better relationship with them so they continue to interact with the service and our brands again and again. An important benefit for wireless marketers is that transactions are easily traceable on an aggregate basis (individual privacy is not compromised) so we can monitor our progress and results and improve upon our usefulness and service to the consumer.

In other parts of the world (e.g., Japan, The Netherlands, Australia), Coca-Cola is leveraging m-commerce in other arenas, including vending. Wireless vending or "Dial-A-Coke" transactions can make a tremendous amount of sense from a convenience factor if consumers trust wireless as a safe means to transmit currency digitally.

Coca-Cola is just beginning to scratch the surface of m-commerce opportunities. In the final analysis, we'll evaluate the feasibility and acceptability of m-commerce applications just as we would for any other medium. But if the growth rate and interest in m-commerce applications continues at even a fraction of the rate of the past few years, our opportunities are considerable.

REFERENCES

Anonymous. (2001). Numbers, market download. *The Industry Standard Magazine*, (July 23).

Dean, D. & Sirkin, H. (2000). *Mobile Commerce: Winning the On-Air Consumer*. White Paper, Boston Consulting Group, November.

Chapter XV

Location-Based Services: Criteria for Adoption and Solution Deployment[1]

Joe Astroth

Autodesk Location Services, USA

ABSTRACT

This chapter provides an overview of location-based services and insight into the pivotal importance of location-sensitivity to the success of wireless data services. This chapter argues that mass-market adoption of wireless data services will only occur if these services enhance productivity and/or convenience for end-users; transforming novelty into a "must-have." It is the author's view that m-commerce is inextricably linked to location and that the incorporation of location-sensitivity will transform these transactions into a relevant, personalized and actionable experience for the user, thereby encouraging the kind of uptake required to fulfill market potential and bring revenue to carriers. The author will provide examples of location-sensitive wireless data servces in consumer and enterprise environments. A specific case study showcasing a next generation solution jointly developed by TargaSys, a division of Fiat Auto and Autodesk Location Services will describe key elements of a successful model for location-based services. Future directions, revenue models, and key technology enablers for successful deployment will also be discussed.

OVERVIEW OF LOCATION-BASED SERVICES

The intersection of two powerful technologies—Internet connectivity and wireless communication—is driving the proliferation of mass-market wireless data services in the first decade of the 21st century. The Internet provides a ubiquitous means for the delivery of information and services from a wide variety of heterogeneous sources wherever a network connection exists. Wireless communication networks enable those connection points to float free of the geographical constraints of the wired telephone or cable infrastructure.

Internet connectivity and wireless communication are augmented by a third key technology, location determination technology (LDT), wireless data services that are customized for a specific place, time and individual can be economically delivered to a mass market. LDT, which identifies the current position of a free-floating network user and automatically reports that position to a service application, is the key component behind the most promising and profitable segment of the wireless data market: location-based services.

What are location-based services? A location-based service is any applications that offer information, communication, or a transaction that satisfies the specific needs of a user in a particular place. Traffic information for the highway a user is currently driving on, or a discount (that expires in 15 minutes) for a coffee shop around the corner from where a user is walking, are both classic examples of location-based service. Mobile commerce (m-commerce) represents the transaction component of location-based services, but the universe of location-based services broadly incorporates many other applications where money may not change hands.

Do users want location-based services? In addition to enterprise applications such as fleet management, horizontal consumer offerings have begun to emerge that appear to have strong user demand among the adopters. In this chapter we will discuss the experience of one of these seminal consumer offerings, auto concierge services, which has demonstrated many of the benefits of location-based services for users. We will also identify the key criteria for successful location-based service adoption, the primary architectural issues involved in deploying a comprehensive location-based services solution, potential revenue models for this new form of business and some expectations for the continuing evolution of location-based services.

THE MOBILE NETWORK OPERATOR OPPORTUNITY

There are many actors who are seeking to exploit location-based services: content and application providers, network infrastructure vendors and location determination vendors among them. But the commercial deployment of location-based services cannot proliferate without significant financial investment by a central market player: the mobile network system operator. Government mandates, which require the addition of LDT to mobile telephone handsets to enhance public safety, are an important prerequisite that enable the deployment of location-based services. But government action alone is not driving system operators to invest in the additional infrastructure necessary to make location-based services a commercial reality.

Mobile network service operators enjoyed rapid growth throughout the 1990s. New subscribers were acquired at high double-digit growth rates, and network infrastructure was upgraded from first-generation (1G) analog phones to second-generation (2G) digital phones to meet the increasing demand for voice traffic and the initial demand for data traffic over the mobile network. Network operators enjoyed the double benefit of rapid subscriber growth and increasing voice-traffic-per-subscriber, even as per-minute tariffs were declining.

As the era of initial mass adoption of mobile phones ended in the industrialized countries, new subscriber growth rates and average revenue per user (ARPU) began to moderate. High subscriber turnover, "churn," churn rates, which operators could accept in periods of rapid subscriber growth, began to affect profitability. Despite network operator efforts to differentiate and brand themselves based on voice quality, voice traffic became commoditized and pricing highly competitive.

Location-based services offer mobile network operators a means to truly differentiate their product offerings. The degree to which operators are successful in localizing services for users, responding to user preferences, as well as pricing and ease of integration with existing services will determine competitive advantage. Service and content differentiation is enabling network operators to increase their ARPU, and subscriber acquisition and retention rates. As users invest more time and effort in customizing their many location-based service options through their network operator, they will be less inclined to switch providers to seek the latest and cheapest 'bucket of minutes' incentive plan.

Location-based services also enable operators to dramatically increase their value to corporate accounts. By integrating location-based services into critical enterprise applications, a network operator moves beyond commodity voice traffic to offer a range of value-added services to the enterprise. A network operator that, for instance, uses location information to automatically transmit customer records or repair histories to a sales person in the field can have a real and measurable impact on the enterprise's customer service expenses and profitability. As enterprises reduce the number of suppliers, they increasingly select network operators based on the location-based services they offer executives, and the effectiveness of the platform they offer the organization for integrating location-sensitive enterprise applications that increase corporate efficiency.

For both the consumer and corporate markets, location-based services enable network operators to strengthen their relationship, and their pricing power, with subscribers. It is the revenue opportunity of location-based services—realized through higher fees, lower churn and market share growth—that justifies network operator investment in 2.5G and 3G transmission technologies, technologies that increase bandwidth and enable new wireless data service offerings.

CRITERIA FOR ADOPTION OF WIRELESS DATA SERVICES

What are the criteria for market adoption of wireless data services? The constraints of technology and human behavior compel wireless data services to exhibit three key attributes to achieve market adoption: wireless data services must be personalized, localized and actionable. Wireless data services are most successful when they incorporate the critical attribute of locality that defines location-based services.

The mobile handset is the inevitable vehicle for mass-market wireless data services. The display, interface, and bandwidth constraints of the mobile handset, however, create a unique set of challenges. Unlike a Web surfer at a desktop computer, who is connected through a 56K dial-up or wired broadband Internet link, the mobile handset user will not click through multiple pages to indicate preferences, locate information or complete a transaction. With a small screen and a limited keyboard, a mobile handset user requires a far more customized and limited set of interactions.

Reducing the complexity of user interactions (compared to the complexity of user interactions on the traditional, wired Internet) is therefore a key challenge for vendors of wireless data services. To minimize complexity, wireless data services must be personalized and localized. Minimizing complexity, however, is not sufficient to drive consumer use of wireless data services. A third attribute, actionability, is essential for consumers to demand access to wireless data services whenever, and wherever they are. Wireless data services that are personalized, localized and actionable have been shown to attract users from mere interest into daily reliance. The effective combination of these three attributes into a single product transforms a wireless data service into a location-based service.

Personalized

To simplify interactions, location-based services must be highly personalized. This is a goal that many conventional Internet services share, but the importance of personalization for location-based services is much greater. We are all creatures of habit. Our preferences for types of services, membership in awards clubs, credit card numbers, banking relationships, commuting routes and thousands of other daily choices can be preset through the rich, highly interactive computer keyboard and display interface. These preferences can then be recalled with a single click on the more concise handset interface exactly when needed.

If a business person's itinerary changes while on a road trip, for instance, it would be highly convenient if location-sensitive wireless data services made it possible to book a new flight reservation using the mobile handset. It would be even more convenient, however, if the service took into consideration the traveler's most convenient options for accommodation, dining and provided directions based on the new itinerary, thereby providing significant value in terms of convenience and productivity for the subscriber and encouraging regular use. Further value would also be derived from personalization of the service such that the traveler's airline preference; frequent flyer number; credit card number; and class of service, seating and in-flight meal preferences would all be pre-loaded into the system.

With location-based services like these, subscribers are encouraged to invest time and effort in specifying their preferences in order to save considerable time in the future. This user investment then creates a significant switching cost that cements subscriber loyalty to network operators.

Localized

A second attribute that reduces the complexity of user interaction is localization. 'Search' is one of the most heavily utilized online Internet functions, but it is a terribly cumbersome operation on the minimal interface of a mobile handset. With knowledge of the user's position, applications can filter-out vast amounts of irrelevant information and transaction choices. Localization enables an application to present only those ATMs that are within walking (or driving) distance, for instance, only those customer records for the address where a sales rep is visiting, or the location of only those parking lots that are located near to the location where the user is driving.

Combined with personalization, localization further limits the presentation of irrelevant information: displaying only ATMs that belong to the user's banking network, or only parking lots that accept the user's credit card. Location-based services that combine these two attributes can quickly refine users' options to those few that completely meet their needs and preferences.

Localization is also how advertisers and commerce vendors can segment, and derive value from, the undifferentiated universe of mobile handset users. Just as billboards are the highly effective (if sometimes under-respected) workhorses of the advertising market, localized advertising on subscriber handsets has the potential to drive significant foot and auto traffic to local merchants. Through location-sensitive advertising, merchants can effectively address a dynamically changing audience of geographically accessible consumers, and thus attract a high volume of opportunistic m-commerce transactions.

Actionable

The third attribute that contributes to a successful location-based service is actionability. Information requests or m-commerce transactions that cannot wait, and whose response options have been filtered through localization and personalization constraints, are the basis for location-based services that see enduring consumer demand. When users know what movies are starting in the next 20 minutes, know which parking lots are already full at their destination or know current traffic conditions on the road they are already travelling on, they possess timely, actionable information that can be used to make immediate decisions and transactions.

Early deployments, such as the Autodesk case study of the Fiat Targa Connect,™ application verify the success of wireless data services that incorporate the three critical attributes of personalization, localization, and actionability.

CASE STUDY: FIAT TARGA CONNECT

Auto-based telematics are the first mass-market location-based services to gain broad acceptance,[2] and the experience of telematics is an important and early case study for future vendors of location-based services. Auto-based telematics differ in some important respects from mobile handsets—auto telematics hardware enjoys less rigorous constraints on battery life, display size and local data storage—yet telematics market adoption and usage characteristics should provide useful insights into the kinds of location-based services that will be successful, particularly as handheld devices evolve.

Autodesk Location Services, with which the author of this study is affiliated, provides the core technology for one of Europe's seminal auto telematics services, Targa Connect, which is operated by TargaSys, a division of auto-maker Fiat. Targa Connect offers an onboard navigation and personal assistance system as an optional add-on to owners of Fiat automobiles including, at time of print, the Alfa Romeo 147, the Fiat Stilo and Lancia, and other Fiat models in the near future. Targa Connect combines an in-car Global System for Mobile Communication (GSM) mobile phone and a Global Positioning System (GPS) as the enabling hardware. Targa Connect subscribers pay an installation fee (at the time of sale) of Euro 1500 (US\$1,350), and an annual service fee of Euro 200 (US\$180). Users pay no additional per-minute, per-transaction or premium fees for using the service.

Using the Autodesk Location Logic technology platform, Targa Connect integrates a wide range of location-based services, including real-time weather and traffic information, roadside and emergency assistance, point-of-interest (POI) information (ATMs, service stations, pharmacies, etc.), personalized news, and online medical assistance. Targa Connect also offers concierge services, like hotel, restaurant and entertainment booking, which are delivered by voice contact with multilingual human operators. Targa Connect location-based services are available to all subscribers in the eight largest European countries.

Targa Connect has partnered and integrated content and applications from dozens of third-party vendors, including hotel chains and the highly respected Michelin Guide. TargaSys maintains user security, system performance and a consistent user experience by integrating and hosting third-party applications on the integrated platform provided by Autodesk. The Autodesk Location Logic platform, which includes reusable application frameworks and flexible application programming interfaces (APIs), enables third-party developers to create and deploy applications for Targa Connect with minimal development effort. Third-party application deployment for Targa Connect is accelerated by the existence of an extensive Autodesk development community.

The User Experience

Market research by TargetSys indicates that potential adoption of location-based services would be inhibited by the limited usability of mobile phone data entry, the user's perceived long learning curve for new technology and user preference for human interaction when engaging with new technology. To address these issues, Targa Connect was designed for maximum ease-of-use, with data entry limited to selection of menu choices, exceptional customer service for new users and the option of interacting with human operators to complete tasks.

Users invoke the Targa Connect system by selecting a menu option. The in-car GSM phone initiates a short message service (SMS) transmission to the Targa Connect operations center. The SMS includes the subscriber's identity and the GPS-determined location. The requested information is retrieved and personalized according to a pre-set subscriber profile, then transmitted by SMS or voice command back to the user's in-car hardware. The user can speak with a human operator. The human operator accesses the subscriber profile, location and requests on the operator console.

The Targa Connect "Follow Me" navigation system is an example of a location-based service that embodies the benefits of personalization, localization and actionability. When subscribers request directions to a selected POI, Follow Me calculates the best route, taking into account the latest local weather and traffic conditions. Directions are delivered to the subscriber by their preferred channel

(SMS or voice command), and repeatedly updated mid-journey as conditions change.

TargaSys has experienced considerable success with its Targa Connect location-based service. Based on independent market studies, an adoption rate of 2-4% was expected prior to launch. Initial adoption rates exceed expectations by 3 to 1.

Location-Based Services Deployment

As can be seen by the Targa Connect case study, effective deployment of location-based services requires the coordinated efforts of multiple actors, each of which provides specific components of the total solution. These multiple actors include the mobile network operators that provide telephone services to the user, handset suppliers who design and sell mobile phones (either to network operators or directly to users), network infrastructure providers who design and sell the equipment that enables wireless data and voice communication, LDT vendors who design and sell positioning technology (which can be integrated into the handset or the network infrastructure equipment), content providers, application providers, solution integrators, platform providers, aggregators, portals and others. Of special interest are content, application, and platform vendors, as these actors are essential new entrants into the established commercial environment of the mobile telecom industry.

Location content vendors provide maps and geo-coded data, such as the location of ATMs, restaurants or pharmacies, and their operating hours, services and other information. This data can be static, in the case of street maps or POI locations, or dynamic, in the case of Yellow Pages listings or traffic conditions. These content databases typically represent a significant investment in the collection and collation of relevant data, and these resources can be prohibitively expensive to replicate. In many cases, the location-based services market represents a new opportunity for established content providers to leverage their existing intellectual property and brand, as in the case of the Michelin Guide in the Targa Connect case study.

Application developers and providers bring the broad range and utility of location-based services to the user. The network operator relies on these nimble, market-sensitive players to discover what services subscribers want (and will pay for). The application provider designs and delivers these services through the network operators' infrastructure. While operators may undertake to develop or offer some of the more frequently required services themselves, they usually turn to outsourced application vendors to offer the widest range of mass-market and niche services to their subscriber base.

Integrated Platform Model

Location platform providers are among the least visible actors—at least to the end user. Platform vendors provide a controlled and coordinated environment where content providers and application providers can perform their individual tasks on behalf of the network operator. The platform vendors' software performs the critical middle-ware functions of new subscriber provisioning, subscriber profile management, content management, billing and other housekeeping functions for all of the applications that the network operator supports. The platform provider must offer a range of appropriate interfaces for the content, application, and LDT provider communities. All of these administrative and interface functions that individual applications require are packaged by a platform vendor into a single software product and sold directly to the network operator.

The network operator who invests in an integrated platform retains their critical role as the sole interface to the subscriber. This role allows the network operator to maintain a consistent user experience, to monitor usage patterns and preferences for product and relationship development, and to generate a single-user bill for all location-based services. With this approach, the network operator preserves their full investment in new subscriber acquisition, and can differentiate the user experience to build brand recognition for the operator. An integrated platform also provides improved performance and security.

Autodesk Location Services has developed and deployed an example of an integrated platform that offers network operators full control of their subscriber base, as in the Targa Connect case study. The Autodesk platform integrates core content and applications, as well as middle-tier administrative services for billing, provisioning, map rendering, profile management and content management. The Autodesk integrated platform also incorporates a comprehensive set of APIs and templates that provide access to the platform and accelerated development for outsourced application providers.

It is essential for platform vendors to offer a rich set of interface options (XML, SOAP, Java, UDDI, etc.) so that a wide portfolio of independent application providers can choose the most effective and efficient integration strategies. This interface flexibility will result in the widest possible array of services for the subscriber.

LOCATION-BASED SERVICE REVENUE MODELS

Developing applications and deploying systems to deliver them are the stepping stones to revenue. But how will operators extract revenues from location-based services? There are ultimately three sources of revenue for any location-based service: the subscriber, the operator and third parties.

The first source of revenue for location-based services is the actual subscriber. Subscribers have proved willing to pay for location-based services in early system deployments, such as the Targa Connect case study. A number of billing models exist, including flat monthly fees; per-access, per-transaction or per-minute fees; and free basic plus premium service fees. Location-based services succeed when they offer utility to the subscriber, and subscribers are accustomed and willing to pay for mobile network services.

Network operators have also considered absorbing the cost of location-based services. New services attract and retain new subscribers, and network operators have been willing to subsidize location-based services to build market share and brand recognition.

Third parties who offer goods or commerce through location-based services are another source of revenue for network operators. Advertising, sponsorships and m-commerce transactions that result in a sale are all billable events. Content providers who want to build wider recognition of their offline offerings may also be a source of third-party revenue or subsidy. A restaurant guide, for instance, may subsidize its content in order to attract users to its offline publications.

It is clear that subscriber and third-party revenue models require detailed transaction and billing data. The ability to collect highly granulated data should be a critical requirement for every network operator as they evaluate location-based service platform vendors. Solving the micro-payments problem will allow network operators to extract revenue from their subscribers, something the online Internet word is still struggling to accomplish.

FUTURE TRENDS IN LOCATION-BASED SERVICES

A clear trend that will affect all wireless data services is the inevitable migration from 2G to 2.5G and 3G technologies. Like other wireless data services, location-based services will be able to leverage the increase in wireless bandwidths from 9.6 kbps to 56 kbps to 384 kbps and beyond to offer richer applications and improved user interaction. Location-aware advertising will migrate from text to graphics to video; localized movie listings can be accompanied by miniaturized movie clips. More bandwidth means faster response times and richer interaction through multiple menus and Web pages.

Increased bandwidth will also improve location-based services for the enterprise market. Schematics and diagrams can be instantly available as a supplier drives up to a job site; detailed invoices can be presented and accepted at the time and point of delivery. The demands of location-based enterprise applications may require the larger display capabilities of mobile, networked PDAs.

As LDT is incorporated into more handsets, peer-to-peer location-based services will emerge along side server-to-handset applications. The location-based service equivalent of instant messaging, where users are alerted when members of their buddy list are nearby, is likely to be as popular as traditional online chat applications. Happy accidents, like running into an old college roommate at O'Hare Airport, will occur more frequently with such ubiquitous location-based services.

In fact, the most exciting applications cannot be imagined. They will emerge from the development labs of thousands of application and content providers, many of which do not yet exist. Network operators will likely remain at center stage as the location-based service revolution evolves. They must continue to employ flexible and easily accessible platforms so that the most novel and useful location-based services can be seamlessly integrated into their network offerings for the benefit of subscribers.

ENDNOTES

1   Edited by Spencer Horowitz.
2   A North American telematics operator announced its one millionth subscriber in April 2001.

Chapter XVI

M-Commerce in the Automotive Industry: Making a Case for Strategic Partnerships

Mark Schrauben and Rick Solak
EDS, USA

Mohan Tanniru
Oakland University, USA

ABSTRACT

The telematics technology, intended to streamline the information processing requirements of consumers driving a vehicle, has brought to surface the need for integrating the information technology architectures of various service providers with the manufacturing technologies of various automotive firms. While system integration has always been an issue when multiple vendors are involved in providing enterprise-wide solutions in business, this issue takes on greater prominence when it can impact the privacy and security of the driving public as a whole. This article briefly looks at various opportunities telematics can provide to satisfy the "mobile" society, and discusses the organizational behavior required to operationalize the technological capability and inter-company behavior to enable flexible business models. It will also discuss the role multiple business and government leaders have to play to ensure that these opportunities do not come at a significant social cost.

Mobile commerce (M-commerce) in the automotive industry could be characterized in parallel and in conjunction with another term framed within the industry—*telematics*. The "telematics" industry is an emerging business area that allows car manufacturers and aftermarket producers to provide innovative solutions for information services. These information services include automatic and manual emergency calls, roadside assistance services, GPS, traffic and dynamic route guidance, Internet communications and personal concierge services.

In the highly competitive automotive business, both product quality and competitive pricing no longer provide sufficient differentiation to capture and retain a consumer. A manufacturer must provide each consumer with an attractive and desirable design, customer care, user experience and an overall *vehicle service package,* which includes telematics. The extent to which automobile manufacturers are successful in providing such a package is becoming a differentiating factor in the consumer's buying decision (Hogan, 2001). In fact, a firm's ability to provide information technology capabilities in an automobile may become as important as *cargo capacity and mileage* of an automobile for both retail customers and business service providers, i.e., leasing agencies, transportation and distribution companies, etc. While there are already some mobile applications such as fleet management, other applications such as in-vehicle computing, navigation and location-based services will start to take shape to support both customer groups.

In this chapter we will provide an *analysis of the market* that is moving the automotive industry in this direction, as well as the *business opportunities* that lay over the horizon. We will then discuss a few major issues that will, when resolved, ultimately dictate the potential growth and success of the M-commerce market in the automotive sector.

MARKET ANALYSIS

The telematics marketplace is experiencing explosive growth in unit sales and user-acceptance in the domestic market and abroad. The in-vehicle information systems market for personal and commercial vehicles will rise from $300 million in 1999 to $5.1 billion in 2003 (Greengard, 2000). In-vehicle navigation has already seen acceptance overseas, and will account for a global market of $16 billion by the end of 2004. Several other studies also point to the anticipated explosive growth in this market (Kalakota, 2002; Thurston, 2000). This anticipated growth has created an extraordinary opportunity for the automotive industry and consumer electronics marketplace. It has the potential to generate significant automotive, marine, and heavy truck OEM and aftermarket product sales, as well as alternative and additional revenue streams.

According to Larry Swasey, VP of Communications Research for Advanced Business Intelligence, a number of phenomena are creating huge opportunities in this marketplace. Rising traffic levels, an increase in the amount of time spent in a vehicle and the technology-savvy consumers behind the wheel are all driving the need for the latest communications technology to be available in the vehicle (Swasey, 1999). Those who actually need to be in a vehicle for extended periods of time will be looking to take advantage of a plethora of services ported from the Internet such as weather, traffic and location-based information. This information will become available to the masses as GPS units fall in price. There is also a cultural shift and changing expectations of what is the norm for today, in terms of services offered in a vehicle, and in the future.

Recognizing these changes, one can already see a steady shift among automobile firms from the manufacturing of goods to providing service in order to realize additional revenue and profit streams. Many firms will start providing value-added technical and consumer support services, and the demand for such services is hastened by several other shifts in the North American and global economies in the recent decade, as evidenced by the following:[1]

- Robust niche/sub-economies within larger economy
- A technologically astute society; Gen-x'ers are more receptive to the use of technology
- Consumer thirst for wireless communications, computing and entertainment conveniences
- Consumers' demand for convenience services accelerating technology and consumer services
- Personal safety
- Federal initiatives
- Federal investment in creating a wired society for transportation effectiveness and security
- Automotive technology integrated with high-tech industry
- Hardware/software and service providers demonstrating a willingness to partner/develop creative business models
- Drive to develop effective consumer marketing programs to retain customers
- Business desire for capital and cost reductions
- Demand for efficiencies in the supply chain
- Reduction in human capital to execute transaction processing
- Immersive brand management
- "I have to invest now because my competitor has the technology"

This evolving marketplace is *converging the interests of many industries* such as technology, communications, consumer electronics and Internet content providers, as they work with the automotive industry in providing the vehicle service package. For example, the convergence of voice and data technologies and services is reducing the barrier to designers and marketers within the automotive industry. Mobile communications equipment, cell phones and palm computers are becoming cellular computing devices capable of GPS, navigation, cellular telephony and computing functions, such as office applications and e-mail. These are all connected via wireless Internet and will enable consumers to operate these while in the vehicle.

This evolution is also *creating cross-industry dependencies*. For example, wireless Internet communications software, features and functions are becoming standard for consumer handheld, desktop and mobile devices. Such dependencies are causing entire business communities to cooperate in ways never seen before. Standard-setting organizations, consumer electronic organizations, communications, technology and software vendors are all looking to develop standards locally, regionally and globally. This will inevitably cause tremendous unrest within the automotive community due to its impact on the automotive product development and release cycle. The auto industry might want to let the leaders in both the high-tech consumer and service sectors manage the dynamic rate of change, but work with them in developing a technologically independent, open standard for automotive connectivity that can be leveraged across its business community.[1] While this may help the auto industry focus on its product development cycle, the business opportunities afforded by this mobile technology on the product development itself can't be ignored.

BUSINESS OPPORTUNITIES

The telematics marketplace will start to have a significant impact on many activities on the automotive value chain such as:

- Platform development and management
- Embedded systems engineering software and development
- Marketing and brand management
- Consumer services
- Repair and service parts operations
- Customer care
- Warranty services

Many of these impacts can be categorized as B2B [Business-to-Business], B2C [Business-to-Consumer] and B2Me [Business-to-Me]. Note that B2Me is an individual-centric portfolio of content and services controlled by the consumer, as opposed to B2C, where the focus is on a mass/targeted message pushed out to targeted demographic groups. Stated differently, while B2C views services as being pushed by the firm to the consumer, B2Me views services as being pulled by the consumer from the firm.

Business-to-Consumer (B2C)

[Example: Model Navigation Services]

In the case of "navigation services," the consumer is the recipient of this service on an "always on" basis or a "one-click request" basis. The service provider will promote this service through wireless carrier providers or the automotive OEM. The consumer will have the ability to, based on service parameters, use the service on a per transaction fee for a contracted service package, or as an offered feature of an automotive lease package. Other examples include enhancing user experience by providing movies on demand, serving coupons in real time for families on vacation, and providing convenience services for business people.

Business-to-Business (B2B)

[Example: Construction/Heavy Equipment/Leasing Services]

In a B2B model, a manufacturer has the ability to leverage telematics technology to integrate the supply chain including assembly line and order-to-cash processes. In the heavy/construction equipment sector, for example, an OEM manufacturer behaves as an automotive leasing agent. The OEM leases the vehicles and provides business services via the telematics network. The OEM provides telematics location-based services such as scheduled maintenance, equipment location, equipment use, vehicle diagnostics, troubleshooting, warranty analysis and emergency services for weather- or injury-related inquiries. All of these have a business impact in the areas of loss/theft management, equipment utilization and longevity, and utilization and product life cycle management, while providing useful information to the engineering communities within the product supply chain.

Business-to-Me (B2Me)

[Example: Digital Music Service Portal]

In a B2Me model, at the consumer's request/approval, a manufacturer or service provider brings an amalgamation of selected products such as insurance and automotive care to the consumer all day based on the consumer's request for such information. For example, in the case of digital music service, the consumers give their basic and customized preferences for styles and types of music to the business, and are able to obtain this service at any given moment, time of day, day of week, etc. This eliminates unwanted distraction from business messages intended for mass audience, or constraints associated with various regions (i.e., language spoken, music choices available, etc.). For obtaining this type of customized service, the consumers agree to hear/listen to specific advertising content. The service provider aggregates content and provides these services to the consumer under an agreed-upon contract.

Each of these business environments provides some unique benefits, and the automotive firm needs to look at these from multiple vantage points. As a service provider, a firm may work within a business model to generate additional revenues. As a manufacturer of "point" or a specialized solution to enhance customer experiences, a firm may provide safety and convenience to the driver/consumer. As a business, a firm may use telematics to leverage operational efficiencies within its distribution and supply chain, where transactions are executed automatically through mobile wireless applications and database technologies. Initially, the profitability achieved through efficiencies in order-to-cash processes, Just-In-Time commerce transactions and mobile transactions embedded in the supply chain may make the B2B model more attractive. The B2C and B2Me models will come later as the consumer base grows in number and across regions.

In summary, with telematics, the consumer wins with convenience, while the OEM or service provider gets closer to the consumer's preferences, lifestyle buying behaviors and consumption trends. However, a truly viable telematics market needs a *flexible platform independent of proprietary technologies and a regulatory environment that supports the industry, while protecting the privacy and security of the public at large.* These issues are discussed next.

ISSUES @ LARGE

The telematics and M-Commerce environment will have a profound impact on companies, consumers, business processes and automotive technologies. Enhanced revenue streams using online consumer transactions is one of the crucial outcomes, but a service provider's ability to develop, implement and execute an end-to-end solution and develop niche products and services is critical to realizing this revenue stream.[1] As we will see in this section, several institutional entities play a key role in supporting a firm in generating these revenue streams, but not before they address several key challenges that lie ahead.

AUTOMOTIVE INDUSTRY ENVIRONMENT

The following trends demonstrate the dynamics of the automotive industry, as it plans to make decisions on the level of engagement in the telematics marketplace.

- The automotive/transportation platform becoming technologically current/ Internet-ready
- Wireless computing becoming pervasive in society
- Product and service prices dropping
- Rapid increase in unit sales
- Burgeoning demand for call centers and convenience services integration
- Rapid rate of technological innovation
- Potential supplier consolidation

The automotive manufacturer has to seek partnerships and industry relationships with various technology providers to develop products and services, and deliver these to the customers in order to be successful in this marketplace. This will become a challenge given the rate at which the technical environment is changing.

TECHNOLOGY ENVIRONMENT

The rapid rate of change in the technology industry, as well as in the consumer and communications environment, has a significant impact on the success of the mobile environment. Automotive and other technology providers must work together in establishing a robust, stable, ongoing technology and communications environment that can incorporate both emerging as well as existing technologies. Some of the technology challenges include developing:

- On-board systems, displacing many off-board systems
- Navigational systems as standard equipment
- Global open-standards in product development
- Easy-to-use and flexible interfaces
- Strategies to ensure data and personal information security

The technology environment supported by both the automotive and other technology and service providers should be transparent enough to allow the consumer to configure the services they need with relative ease, if it is to generate significant acceptance among the consuming public and make the business model profitable for all involved. However, customer expectations on access to information are changing rapidly with each new technology, and are creating a volatile customer and business environment.

CONSUMER & BUSINESS ENVIRONMENT

There are several forces that are driving the use of mobile technologies in the automotive sector, and these will ultimately shape the industry's future applications. Some of these forces include:

- Increasing consumer demand to have both computer and technology services at their finger tips
- Growing desire by consumers to access multi-media and communications while in their vehicle
- Growing business demand for the application of technology to gain operational excellence
- Growing need for storing and analyzing customer preference data (using data warehousing technologies) in order develop strong customer loyalty
- Growing concerns related to privacy and consumer protection by various public and private organizations

Ultimately, a consumer's willingness to adopt this new environment and the business community's openness to leveraging these new technologies into it's new business model, will both be dictated by how effectively the marketplace is regulated and supported.

REGULATORY ENVIRONMENT

Global standards and regional pressures will have a significant impact on any successful telematics initiative and service in the automotive space. Recently, federal agencies have determined that dialing a cell phone is more distracting than talking to a voice recognition unit with the driver's hands on the wheel. A regulation to reduce this distraction will have a negative impact on the handheld communication devices (e.g., cellular phones), and will be a boon to the voice recognition software and hardware developers. This illustrates how a study/research that deals with the safety of the driving public could deal a blow to certain technologies while promoting others in the blink of an eye.

Various regulatory bodies of the federal government have established policies impacting each component of the mobile commence and vehicular telematics communities. Many of these agencies (listed below) are engaged in providing development support, leadership, and establishing policy and procedural standards.

- NHTSA [National Highway & Traffic Safety Administration]
- FCC [Federal Communications Commission]
- NTSB [National Traffic & Safety Board]

- Department of Transportation
- Department of Defense
- CPSC [Consumer Product Safety Commission]
- NIST [National Institute of Standards & Technologies]

Given that many of these agencies operate under different charters, reaching a consensus is quite complex. Also, these regulatory groups have to work with various business/industry interest groups (see below), if this mobile industry is to support the development of products and services in the telematics marketplace.

- AMIC [Automotive MultiMedia Interface Consortium]
- ANSI [American National Standard Institute]
- ITS [Intelligent Transportation Society]
- CEMA [Consumer Electronics Manufacturing Association]
- Bluetooth Consortium
- American Automobile Manufacturers Association

These groups have to play a vital role, as a "watchdog" for the automotive industry, not only to monitor and influence legislation that affect the product development cycle, but also support the collaborative effort among the various constituents of this telematics marketplace. For example, Figure 1 illustrates the many individual players that have to work together in support of providing a telematics solution to the consumer. Besides the established players in the

Figure 1: Illustration of individual players working together in support of providinga telematics solution to the consumer

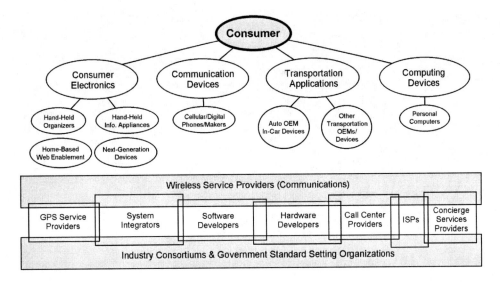

automotive tiers, this marketplace includes many new players that continue to bring new and innovative products and services to the market at a rapid pace. Given the volatility of this technology market, creating standards and seeking consensus is obviously quite a significant challenge for many of these industry groups.

THE BOTTOM LINE: PARTNERSHIP AND MULTI-THREADED STRATEGY

Industries must work together at the macro-level to create a technical environment that supports the amalgamation of business processes and services. Some of these industries include automotive, transportation, telecommunications, distribution, entertainment, technology, publishing and travel. Many of these industries have rarely or never worked together in the past. Each of these industries needs to understand that its business drivers will impact the decisions of others, and *partnership* is critical if telematics in the automotive environment is to flourish.

Also, each participant has to, with a varying degree, allow his/her technology and service to be supported in an open environment. For example, the E-portal model shown in Figure 2 describes how several satellite, electronic and automotive component suppliers work with various Internet, content and other service

Figure 2: Illustration of how several satellite, electronic and automotive component suppliers work to satisfy a consumer-related content request

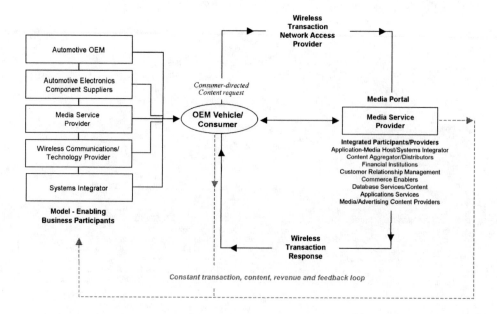

providers to satisfy a consumer-directed content request, while the consumer is in his/her automobile. To make this happen, several standards have to be agreed upon so that information can move from one technology provider to another, while ensuring the privacy and security of customer information. In other words, a technologically "open" standard has to be created to allow various business models to co-exist.

Given the diverse players involved in the telematics marketplace, partnering plays a major role for business growth. In this multifaceted and multilayered marketplace, partnerships and industry presence play the linchpin role for successful firms and these firms have to develop strategies that allow for multi-threaded competition. By offering hardware *and* services at various levels to OEMs, consumers and other services providers, a firm can take advantage of transactional revenue streams incrementally as opposed to the single-thread "if we build it, they will come" model. A firm can provide a "single, specialized service" in one target market and act as a "me too" provider in a different market segment within the same industry. The bottom line is that in this emerging marketplace, the consumer will define what is of value and what line of business will thrive. The participating firms must then determine which service lines to divest, invest, partner or purchase.

CONCLUSIONS

With its various safety and convenience-related service programs, telematics has moved from the novelty phase to becoming an integral part of an overall consumer package. The dynamic technology environment and its influence on the private sector; federal, state and local government agencies; and the consumer community has created enormous complexity. This complexity calls for leveraged "business and technology" processes never seen before in the automotive sector. New paradigms are being developed, business models are being created and consumers are thirsting for new services.[1] It has yet to be determined if the three legs of the stool—*industry, consumer and government*—can get it right and work together to make telematics applications flourish, or if they let the inherent complexities create an environment that makes the amalgamation of services, technology and M-commerce transactions difficult to manage in the short-run and not successful in the long-run.

Ultimately, it is difficult to say who will lead, how they will lead or if they will be allowed to lead. However, given the history of the automotive industry and its fortitude towards developing innovative and reliable consumer products, our contention is that this industry will play a key role in creating standards in support of the telematics marketplace. These standards will then make the incorporation of such products within an automobile as transparent to the driving public as many of the other successfully embedded technologies that came before.

ENDNOTES

1   Many of the observations put forth in this chapter are based on a collection of experience, insight and ongoing participation in the automotive, technology and consumer services industries. Working in, providing services and conducting innovative research in association with industry groups, trade organizations, clients and experts within the business community has afforded us access to innovative insight and business practices. This experience and research has enabled us to develop a broad, yet specialized view of technologies, trends and issues in this marketplace.

REFERENCES

Greengard, S. (2000). Going mobile, *Industry Week*, 249.

Hogan, M. (2001). Keynote, 2nd *Digital Detroit Conference*, November.

Kalakota, R. (2002). *M-Business: Race to Mobility*. McGraw-Hill

Swasey, L. (1999). Intelligent transportation systems: In-vehicle navigation and communications technologies. *Global Markets and Forecasts*.

Thurston, C. (2000). Economic waves wash onto all shores. *Global Finance*, 14.

Chapter XVII

Case Study: The Role of Mobile Advertising in Building a Brand

Minna Pura

Eera Finland Oy, Finland

ABSTRACT

Building a brand in the fragmenting media environment is a challenging task. Advertising should be integrated and personalized, it should utilize different channels, and reach the customer at the right place, at the right time, through the right channel, and in the appropriate context. Mobile advertising should be used as a means of creating value to the customers and serving the customers better. This paper gives an insight into the practical possibilities and pitfalls of mobile advertising as a brand building and customer relationship management tool. The case study describes how mobile advertising can be used to get the youth target group to give information about themselves to the company, and how this information can be utilized for future customer relationship management. The effectiveness of mobile advertising in a cross media context is analyzed through conversion and loyalty measures.

INTRODUCTION

According to research by Ovum (Pastore, 2001), in the next five years, 40% of the global market for Internet services will be attributable to multi-access services, i.e., the delivery of content and services to multiple devices over multiple networks. In the fragmenting media environment, advertisers have to meet new challenges in building a brand. Advertising needs to be integrated between different channels, taking into account the limitations and possibilities of each one. Wireless ads have the advantage of immediacy, reaching consumers closer to when and where they actually make purchase decisions. However, according to Jupiter analysts (Jupiter, 2001), the growth of marketing on post-PC platforms will remain marginal because of the lack of standards, audience fragmentation, and unclear return on investments. A Jupiter Consumer Survey (2000) found out that consumers willing to accept advertising on their mobile phone or personal digital assistant (PDA) said they preferred subsidized content and access (36%), followed closely by subsidized devices (35%). Nearly half (46%) of all users, however, said that no form of compensation would persuade them to receive advertising on their mobile phones or PDAs.

Jupiter (2001) projects online ad revenues to reach $16 billion by 2005, but post-PC advertising revenues will climb slowly and trail behind. iTV will reach only $4 billion and wireless $700 million by 2005. But the crucial issue is what the consumers consider as advertising. If the brand can be built by sponsoring a service that is valuable to the customer, attitudes towards mobile advertising can change. Trials across the world have shown that subscribers are willing to opt-in for value-added services and are highly likely to respond to multiplayer contests and branded promotions (Kotch, 2001).

Mobile as a marketing medium is in its pilot phase in Finland. Mobile operators and traditional media houses have implemented the first mobile media solutions (Heimo, 2001). So far, mobile advertising campaigns in Finland have primarily been based on SMS (short messaging service) text messages because of consumer familiarity with it. However, other forms, such as banners, logos, interstitials,[1] and even voice would be possible forms of mobile advertising, and new forms continue to emerge as technology enables new solutions.

The Internet and mobile are often considered personal channels that enable effective one-to-one marketing. Mobile advertising is often used as an integral part of Internet marketing. Currently, mobile advertising campaigns primarily include advertising messages sent via short messaging service (SMS) to registered users of a web portal. Banners on the other hand can be compared to mass media advertising if they are not targeted based on demographic or psychographic information. If the message is not relevant to customers, the acceptance of mobile advertising declines quickly. Therefore, mobile advertisers today should thoroughly consider how they use the mobile channel for advertising purposes. Customers should be given the possibility to choose where, when, and by whom they are contacted. The right message should reach the right audience at the right place and context, and at the right time through the right channel (e.g., mobile, Internet, digi-tv, e-mail, print, direct marketing, outdoor, point of sale).

The aim of this chapter is to discuss how mobile advertising is used today and to present a case study where mobile advertising was used as an integral part of the cross channel media mix. The effectiveness of mobile advertising is analyzed in a cross media context with the help of the case. There is very little academic research on mobile advertising and therefore this chapter contributes to research by bringing insight into the practical possibilities and pitfalls of mobile advertising as a brand building and customer relationship management tool. The case describes how mobile advertising was used to activate the youth target group into offering information about themselves to the company and how this information can be utilized for future customer relationship management.

STRUCTURE

First, the managerial aspects of mobile advertising as a branding tool are briefly discussed. The mobile market in Finland is described in order to understand customer behavior and attitudes to mobile advertising in the target market. The role of mobile advertising in building a brand is discussed and different forms of mobile advertising are presented. The possibilities of mobile advertising in customer relationship management are discussed. Before the case study's campaign analysis, the principles of mobile advertising effectiveness measurement are presented. The campaign is analyzed and the results give some managerial implications about how mobile advertising can be used.

BACKGROUND FOR MOBILE BRANDING

Aaker (1996) set forth the perspective that brand building is a strategic asset and that brand equity should be tracked through awareness, perceived quality, brand loyalty, and brand associations. The limitation of satisfaction and loyalty measures is that they do not apply to non-customers. On the Internet it is easy to measure whether the visitors are loyal to the brand website and visit the website frequently. It can be assumed that a high frequency of visits denotes a higher involvement in the brand and possibly a higher loyalty towards the brand. Nevertheless, loyalty towards a website might not correspond to buying behavior. This chapter describes how customer brand awareness can be built through mobile advertising in a cross media context and how its effects on website loyalty can be assessed.

Mobile Advertising as a Branding Tool

New technology affects brands and challenges the traditional ways of building a brand. As Micah Kotch, Marketing Consultant at Wireless Advertising Association (Kotch, 2001), states: "Trademarks are increasingly co-opted by a subculture of youth, which internalizes brands and rejuvenates them in the process. The emotional connection between a product and a customer will take center stage in the mobile channel. It will be a springboard for the distinctive names and symbols of today's companies—the trust marks that transcend brands and bind icons to the hopes and dreams of tomorrow's customers."

The primary aim of wireless advertising campaigns in the USA last year was brand building. Fifty-six percent of respondents reported brand building as the main goal of their wireless advertising campaign. Other objectives included lead generation and direct sales (WAA, 2000).

Mobile advertising has been compared to direct marketing because of its personal nature, but it can also be used for branding purposes. Mobile advertising should always be based on the customers' needs and it should be communicated in the right context. Customers benefit from giving permission to mobile marketing, which also makes mobile advertising more like a service than just marketing communication. Mobile is only one of the channels in the cross media context that enables the company to serve the customer better. According to Keskinen (2001), using mobile advertising for branding purposes requires specific targeting and communication of the core elements of the brand identity. Managing a superior customer experience and relationship is a long-term goal of marketing and measurable results are critical (Kotch, 2001). Advertising is an important step along the way, but if it is done without permission or is irrelevant, it makes meaningful dialogue with customers virtually impossible.

Mobile marketing must cater to each subscriber's individual preferences and lifestyles, thereby delivering value for the coveted one-to-one relationship (Kotch, 2001). Segmenting the customer base according to situational needs and lifestyle instead of psychographic or demographic factors helps the advertiser plan the messages that motivate the customer to action. One example of a service that cannot be targeted according to age or gender is a mobile service based on the Bridget Jones Diaries created by a Finnish wireless entertainment publisher, Riot-E (WOW Wireless, 2001). The service offers an "Ask Bridget" service, Bridget Jones' Guide To Life, and discussion of dieting, dating, self-help, and thigh circumference, plus how to handle men and manage friendships. People who would like to receive tips to single persons' daily problems, or humorous spam comments from the Bridget Jones mobile service, should be identified according to a certain single lifestyle. Moreover, the messages should be delivered in the right situation, when the customer desperately needs help and tips. Nevertheless, targeting according to lifestyle and situation is difficult and therefore many campaigns are still targeted traditionally according to demographic information.

In small mobile markets segmenting is often forgotten, because segmenting usually requires a critical mass. That is why it is easier to find different segments in Japan than in Europe (Järvelä, 2001). In rare cases a segment can also consist of one customer, as long as the customer is a loyal and important one to the company. If the target group is not segmented thoroughly according to customers' needs, desires, and motives, the services might not meet the needs of any segment properly. Targeting according to needs and desires might be risky, because current research does not reveal their future needs. In research on mobile services, Eriksson et al. (2001) identify problems with anticipating future needs: "Respondents often consider planned services good, but are not ready to pay for them. Respondents have difficulties in defining which mobile services or products they would like to use and would be ready to pay for." Future needs are therefore difficult to anticipate and new services require ongoing tracking and analysis of customer reactions such as sales, loyalty, and customers' opinions of the services. In this instance, the successful products can be recognized and developed further.

The strengths of mobile advertising include mobility, identifiability, measurability, and speed (Heimo, 2001). A mobile response channel is particularly useful in cross media campaigns. Traditional media effectiveness can be measured through, for example, conversion by mobile media, and the communication can be personified through information sent through the mobile response channel.

Advertisers must always consider the opinion and acceptance of the advertising forms in the target group. For example, spam mobile advertising is considered utterly annoying in Germany according to a survey conducted by a German Internet business association (Laine, 2001). Three-quarters of respondents were even prepared to switch service providers if they were not able to prevent spam mobile advertising. The European Community did not adopt a pan-European anti-spam law and therefore individual countries create their own policies for mobile marketing (Rohde, 2001). The Finnish legislation (Suomen Suoramarkkinointiliittory, 2001) requires natural persons' opt-in[2] permission to SMS or e-mail marketing messages. Therefore, in Finland messages can be sent only to customers or persons that have given permission to mobile marketing, which makes spam messages illegal. In Europe the current approach to spam advertising is fragmented. Five European countries—Finland, Denmark, Germany, Austria, and Italy—have passed opt-in regulation on spam.

The Finnish Mobile Market

Mobile penetration is high in Finland. In September 2001 (Åkermarck, 2001), 75% of Finnish people had a mobile phone in their own use. Possessing a personal mobile phone is probably as common today among 15- to 74-year-old Finns as having a wristwatch. In the last couple of years, mobile phones have also rapidly grown common among the under-15 age groups, as well as among pensioners. According to the latest Mobinet study by A.T. Kearney (Association for Interactive Media, 2001), adults of 35 years and over are increasingly using SMS. SMS usage grew by 10% since January 2001 in the United States, United Kingdom, France, Germany, Finland, and Japan. The sending and receiving of text messages has become very popular especially among young women in Finland. A total of one billion text messages, or close to 300 messages per one mobile phone user, were sent in Finland during the year 2000 (Nurmela, 2001). Nine out of 10 of the people in the 15- to 19-year-old age group use SMS daily and people read the messages almost immediately after receiving them (Heimo, 2001). Therefore, in the youth target group the mobile channel is an important part of the media mix.

Attitudes Towards Mobile Advertising

Cultural differences must be taken into consideration when assessing the possibilities of mobile advertising. The mobile device has a different role in people's lives in different countries and for different subgroups. In Finland the most popular mobile service target group has been children and youngsters. They are active mobile phone users and early adopters of mobile services. Thus, companies can strive for long-term customer relationships by attracting youngsters as customers (Järvelä, 2001). Similarly in the U.S., youth seems to be the pioneers of mobile service use. According to Teenage Research Unlimited (Kotch, 2001), 33% of U.S. teens already own a wireless phone and 87% say wireless is "in."

NetSurvey conducted a survey of members of its Internet panel in February 2001, and found that 34% of the Finnish respondents were "interested" to "very interested" in getting messages delivered to their mobile phones (Hausen, 2001). Finnish respondents were a bit more interested than other nationalities. For example, 31% of the Swedish respondents and 24% of the British respondents were interested to very interested in receiving mobile messages. The information respondents were interested in receiving was mainly connected to buying and consuming, especially in the age group of 15–19 years.

Another PC-based consumer panel survey conducted in Finland by Heimo in September 2001 (Heimo, 2001) revealed that the most appealing mobile advertising form was location-based advertising. The second most appealing form was sponsored content while the third was push advertising. Audio advertising, such as voice advertisements that interrupt conversations on the mobile phone, was the least preferred form of advertisement. In general, people who had received SMS advertising were more positive towards mobile advertising, regardless of the form. Another interesting result was that none of the people who had received SMS advertising had forwarded any advertising to their friends. Although the power of viral marketing enhancing the effect of e-mail marketing has been considered important, viral impact might not be so important in a mobile context because the mobile phone is considered too personal for forwarding spam messages.

MOBILE ADVERTISING

Mobile advertising includes push and pull messages as well as promotional sponsorships. Furthermore, it can be used flexibly as part of a cross-media concept, including media such as print, TV, radio, and point-of-sale material.

Push messaging is equivalent to spam e-mail. Typical push campaigns include offers sent to existing customers, mobile alerts, and information sent via a mobile device. Sending push messages illegally without permission is commonly called spam (Heimo, 2001). Context and value of the message have to be considered carefully before applying this strategy to approach customers. If the message is not relevant to the receiver, it can easily turn against the advertiser (Keskinen, 2001).

Pull campaigns attract the customer to order further information and other content through the mobile device. Customers can receive discount coupons or samples by sending their contact information via their mobile device. If SMS messages or picture logos are used as coupons, the advertiser should consider if the coupon should be personal or if spreading the message to friends should be allowed in order to create viral impact (Keskinen, 2001). Pull campaigns often include a cross-media approach (Heimo, 2001). Customers react to an advertisement in other media via the mobile channel. For example, consumers can order a catalog or additional information about the advertised product by sending a text message that includes a code from a poster or a magazine and his or her postal address or e-mail address to a mobile service number. The effectiveness of different creative solutions in a variety of media can be distinguished with the help of unique contact codes that are printed on the advertisements.

Sponsoring mobile service content is a viable option for mobile branding. As the customer of a mobile service orders information, he or she gets a marketing message at the end of the ordered information (Keskinen, 2001). Services can therefore be partly or fully financed by marketers. Nevertheless, the message might not be considered relevant because the advertisement is targeted according to a service provider's customer information instead of being based on customer preferences. Therefore, advertising that is presented in a context that supports the brand image is recommended. The customer's ability to respond directly to the message is vital for the success of the campaign and enables the advertiser to collect information about potential customers for an ongoing dialogue. By measuring response rate and tracking the customer's behavior, the company can evaluate the effectiveness of the campaign.

Reactions to Mobile Advertising

According to a survey conducted by the Wireless Advertising Association in the U.S. (WAA, 2001), the most popular way to respond to mobile advertising is by providing one's e-mail address, but clicking the advertisement directly or visiting a website are also popular approaches. Calling toll-free numbers is commonplace in the U.S., but not as common in Finland. Wireless advertising seems to be effective in attracting customers to register as well. Some respondents reacted to mobile advertising by visiting the brick-and-mortar store. These results indicate that even though the advertising message comes through the mobile channel, people react to the message in multiple ways, using all channels available, not just the mobile device. The mobile device might play an important role in integrating the advertising communication between different channels. Forrester Analysts (Johnson, 2000) conclude that integrating functions across channels will be rewarded by increased lifetime value and greater customer satisfaction.

Mobile Advertising and Customer Relationship Management

It is considered 5 to 10 times more expensive to acquire new customers than to keep existing ones (Gummeson, 1995). To encourage repeat purchases and build customer loyalty, companies must shift the focus of e-business from transactions to e-service—all the encounters that occur before, during, and after the transaction (Zeithaml et al., 2001). Companies can't survive in today's super-fast, hyper-competitive market without serving their customers with a better value proposition than their competitors. Thus, the mobile channel can also be used as a customer relationship management tool. For example, after the customer has contacted the company for the first time, the phone number can be used as an identification number and data on the behavior of the customer can be tracked (Keskinen, 2001). Combining this data with registration data on the Internet offers new possibilities to personalize the communication and to enhance customer relationships. The cross-media approach should enable customers to use the preferred channel combination where and when he or she wishes to. By collecting information from different channels, companies can serve their customers better. Customers are willing to give information about themselves if they are motivated by some perceived benefit, if they perceive that they are in control of the information that is gathered, and if they can trust the company and perceive the service to be satisfactory (Keskinen, 2001). After creating customer trust and gathering integrated information from all the channels, the information should be updated continuously with the help of an ongoing dialogue with the customer. Anticipating customer problems and needs help to keep them loyal (Johnson, 2000).

Measuring the Effectiveness of Mobile Advertising Campaigns

Mobile advertising effectiveness measures are not yet as refined as online advertising measurement techniques. It is not yet possible to determine what was actually seen by the target group. One can only measure the impressions sent out to the wireless devices. Paging and one-way SMS cannot confirm delivery and viewing, whereas WAP[3] allows tracking, similar to Internet campaigns. Similarly, viral forwarding of SMS ads cannot be tracked without a web component (WAA, 2000).

Traditionally, online advertising has been measured by ad impressions and click-throughs. These measures are not comparable with offline measures. Online measurement tools allow measurement of conversion and sales. In order to estimate the effectiveness of interactive advertising in a cross-channel campaign, it is necessary to concentrate on conversion measurement and try to estimate the behavior of the target group after seeing the ad.

Since wireless measuring methods are still evolving, the use of the Internet as one part of the campaign concept can be recommended. In the online environment, the number of unique visitors to the campaign, as well as the number of registrations on the website, can be measured. The behavior of the registered visitors can also be tracked more specifically. Web traffic measurement is based on log-files or cookies that cannot be personified to a natural person (only to a cookie or an IP-address). In order to identify how a person behaves, it is necessary to motivate him or her to offer some information. Contests that require some kind of registration information or profiling, therefore, better support customer relationship goals. Knowing the target group helps the company serve it better. Peer groups can be used to test new products, and registered visitors can be targeted with personified advertising.

SUMMARY

Building a brand in the fragmenting media environment is challenging. Advertising should be integrated and personalized through different channels reaching the customer at the right place, at the right time, through the right channel, and in the appropriate context. Integrating functions across channels will be rewarded by increased lifetime value and greater customer satisfaction. Mobile advertising should be used as a means to create value to the customers and serve the customers better. Mobile advertising includes push and pull messages as well as promotional sponsorships. It can be used flexibly as a part of a cross-media concept including media such as print, TV, radio, and point-of-sale material. The strengths of mobile advertising are mobility, identifiability, measurability, and speed. In the youth target group, the mobile channel is an important part of the media mix. Currently, mobile advertising includes primarily SMS-messages to registered users of a web portal. Mobile advertising is restricted by legislation and requires the customer's permission in Finland. Mobile marketing must cater to a subscriber's individual preferences and lifestyles, thereby delivering value for the coveted one-to-one relationship. An ongoing dialog with customers helps to anticipate future needs and keep customers loyal. Since wireless measuring methods are still evolving, the use of the Internet as one part of the campaign concept can be recommended.

CASE STUDY: TUPLA -CHOCOLATE BAR BRAND SPONSORING THE "TOMB RAIDER" MOVIE

Eera Finland Oy conducts strategic business consultancy. In this case, Eera planned the creative campaign concept for mobile and Internet channels, produced the online campaign material, and tracked the web page during the campaign. The case describes a mobile advertising campaign and brings insight into how the mobile channel can be used in a cross-media context. The unique concept of matching TUPLA -chocolate bar consumers to meet each other at a sponsored movie is interesting and shows the power of a creative idea. As very little research exists an mobile advertising, the case plays an important role in helping brand owners understand the possibilities of mobile advertising and how the effectiveness of mobile marketing can be evaluated. Conclusions of the study and the managerial implications suggested for the use of mobile advertising are based mainly on website traffic measurement results. The study summarizes the role of mobile technologies in supporting product branding, interactive marketing, relationship marketing, and marketing metrics.

TUPLA -Brand

Leaf Oy belongs to the Leaf group, which is a part of CSM's confectionery division. CSM is an international Dutch food and ingredients company. Leaf's brand, TUPLA, is the market leader in chocolate bars in Finland. The brand's website (http://www.tupla.com) supports the brand and appeals especially to the youth target group. Information about visitors is gathered through registration on the website. The contests that are held at tupla.com are aimed at collecting information about TUPLA consumers. Registered visitors receive invitations to various contests.

Sponsoring the "Tomb Raider" movie

TUPLA sponsored the "Tomb Raider" movie in order to increase awareness about the TUPLA-brand in the youth target group. A contest concept was built around the movie theme. The contest included three phases. It was first promoted in point-of-sale material, radio-, and print advertising, as well as in online advertising and on the back of movie tickets. The registered users of tupla.com got a contest code via an SMS-message, or e-mail in June. The aim was to attract visitors to the website to input the contest code. The contest codes were created in pairs and the competitors were obliged to check the website as many times as possible during 48 hours to see if their pair had been registered on the site. Only the 250 first pairs registered got two free movie tickets to the "Tomb Raider" movie premiere in Finland on July 15, 2001.

In the second phase in August, another 100 contest codes were sent out. The codes included 15 winners and the winners could choose between receiving an Xbox-game console and taking part in a lottery for the main prize: a trip to the location of the filming of the movie "Tomb Raider." The probability of winning the trip was 20%. Ten competitors chose an Xbox-console and five took part in the lottery. At the end of August, in the third phase, the remaining five competitors received a contest code and *the third person* who registered the code on the website won a trip to London and Iceland to visit locations where "Tomb Raider" had been filmed.

Campaign Analysis

The campaign analysis is based on information gathered by a web tracking tool, that measured the traffic on the web page, as well as the click-through from the online media. Visitors were identified with cookie-information and therefore it was possible to track returning visitors during the campaign period.

In this case, neither the registrations of the website visitors nor the e-mail messages were tracked with the tracking tool and therefore it is not possible to identify which advertisement the registered visitors reacted to. Nevertheless, the registration database allows us to analyze the advertising effectiveness by comparing the amount of registrations with the advertising efforts and dates. In other words: Did cross media advertising create conversion? Which channel was successful in activating the target group to register? Did advertising affect website traffic? How often did people visit the web page during the campaign?

Online and Mobile Advertising Created
Traffic to the www.tupla.com website

Online ads were shown between May 28 and August 5 in youth-targeted online media. All in all, the online advertisements were shown about 1.5 million times (one ad was shown a maximum of three times to one visitor). Banner advertisements of five different sizes and styles were used in different online media and attracted the interest of the target group very well. Over 30,000 click-throughs were generated. The average click-through rate was 2.1%, which is very high compared to the average click-through rate for campaigns done by Eera Finland Oy (0.4%). In some online media the click-through rate even exceeded 8%.

SMS-messages and e-mail messages were sent on June 13 and August 9. Figure 1 illustrates clear peaks in the page impressions on tupla.com's front page right after sending out SMS-messages and e-mail messages. There were 12,447 page impressions on June 13 and 11,079 page impressions on June 14. The second peak was generated with the second SMS-message on August 9 generating 14,283 page impressions. Figure 1 clearly shows the effectiveness of the SMS-and e-mail messages in attracting the target group to the website to check if his or her pair had already typed in the code. One could win only if both people who had received the same code registered on the website.

The results indicate that the SMS- and e-mail messages were very effective in attracting the target group to visit the website. Both the prizes and the movie premiere tickets generated interest in the contest. According to the media-agency, recall of the TUPLA-advertisement rose about 25% during the campaign. The

Figure 1: Daily Page Impressions on www.tupla.com's Front Page

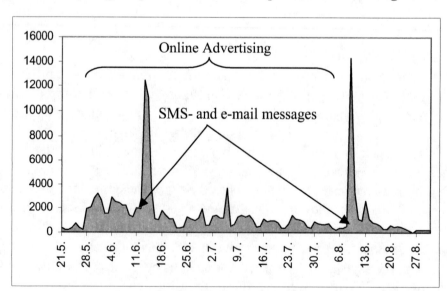

campaign was favorably received by the participating customers and was described as an exciting campaign.

Campaign Attracted New Visitors to Register on the Website

Results show that 40% of existing registered TUPLA customers entered the contest code on the campaign website. Most registrations came right after the campaign start launched on May 28. Figure 2 shows that registrations rose radically after the campaign began. But online media click-through peaks also illustrated in Figure 2 do not directly correspond to dates when online registrations were at their highest.

Therefore, it can be assumed that the cross-media approach as a whole contributed to the increased amount of registrations during the campaign, and that it was not only affected by online media. The campaign exceeded all expectations because registrations rose 28% during the cross media campaign.

Loyalty of Website Visitors was Relatively Low Due to Long Campaign Period

Only 0.9% of unique visitors returned to the www.tupla.com page during the whole four-month campaign period May 28 through August 28. During these four months the front page had 64,666 unique visitors, and 64,111 of them were new

Figure 2: Weekly Registrations on www.tupla.com

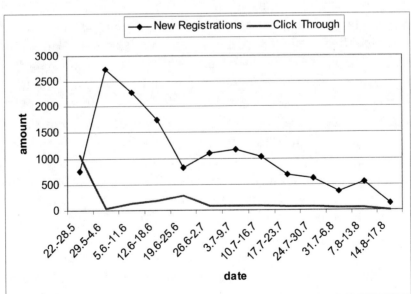

visitors who had not visited the website before, according to cookie files. In other words, only 555 unique visitors returned to the website. Based on these facts, the target group had typed in their contest code on the website, but after noticing that the probability of winning was small, they did not visit the web page again. Only a few lucky ones continued to visit the campaign page frequently. Therefore, the contest did not succeed in creating ongoing interest in the tupla.com website. Contests seldom have the kind of content that visitors want to follow on a regular basis. For that purpose, other appealing content had been created on the website. Nevertheless, the high amount of page impressions on tupla.com indicates that the campaign was very successful in attracting the interest of the target group.

Recommendations and Critiques for Advertisers Based on the Campaign Analysis

The TUPLA case illustrates two important conclusions: 1) the power of mobile SMS messaging in generating traffic to websites, and 2) cross-media advertising is a prerequisite for creating interest in new potential customers. Measuring the conversion helps in more effectively planning future integrated campaigns. If advertisers only measure click-through and ad views, the results may be biased. In this case, the loyalty figures and registrations gave new insight into the website traffic analysis. Even though the campaign was very effective in bringing traffic to the website, visitors were not very loyal in terms of visiting the brand's website. Contests of the type described here are an effective way of attracting people's interest and creating awareness, but are not effective in creating an ongoing dialogue or a lasting brand experience. In order to give visitors an enlightening experience with the brand, the website must include interesting content, entertainment, or games that attract the visitors to spend time with the brand. On the other hand, it would be interesting to know the existing loyalty to TUPLA-chocolate bars as well as usage patterns. Are they the same people that visit the website often and take part in the contests? Contests tend to attract people that are interested in winning, not necessarily people that usually are loyal to the brand. Existing chocolate bar consumers might be reached more effectively through point-of-sale communication. If the primary aim is to create brand loyalty, mobile advertising might not be the optimal way of promoting the brand. On the other hand, sponsoring a movie and building a campaign around a movie theme creates brand awareness and helps to create brand associations, which are an essential part of the brand equity in addition to brand loyalty and perceived quality.

Collecting customer data will be rewarded by better customer knowledge. Mobile advertising can be seen as an information channel as well as a tool for customer care and relationship management. Researching the media habits, values, needs, and desires of the target group help to build successful, integrated advertising campaigns. Customer behavior can be studied with the helps of peer groups through the web community and reached via a mobile channel. New products or sub brands can be tested in peer groups as well. Ongoing dialog is necessary in order to keep registered users interested and to promote active viral marketers.

FUTURE TRENDS

New techniques enable more effective customer relationship management by providing new possibilities for segmenting customers according to customer value, brand loyalty, etc., as well as targeting and personifying marketing messages according to behavior. For example, an e-mail or an SMS-invitation to an event promoting a new product could be sent to potential customers who recently read information about the product on the website. Rewarding the loyal customers by mobile discount coupons might be a good idea to maintain brand loyalty.

The key to success will always be relevant, interesting, and entertaining content that exceeds customer expectations and awakens positive feelings. However, in the future the content and product or service offering might not be enough to induce customers to act. Therefore, customer relationship management should be focused on the context. The content has to be relevant to the customer *and* delivered in the right context (Future Lab, 2001). The service or product must be valuable to the customer in a specific place and moment. In the future, technology-driven service development must give way to development that is based on problem situations in consumers' daily lives. Studies about customers' needs and desires need to be situation-specific, anticipating future needs and motives to use the service in a certain context (Eriksson, 2001). Advertisers realize that they must utilize an integrated "holistic" media mix in order to reach their target audiences. New technologies will leverage the opt-in data available from prospects and existing customers to deliver more intriguing commercial messages targeted to specific audiences. Rich media solutions (including text, moving picture, or voice) in online advertising increase the response compared to traditional text-based e-mail or traditional static banners (Aberdeen Group, 2000). Therefore, rich media will surely also become more popular in mobile advertising communication especially when bandwidth grows. New mobile ad forms will emerge as technology evolves. But for now, consumer attitudes towards the mobile device as an utterly personal device prohibits radical advertising approaches. Even though location-based services and customer-identification would allow marketers unique opportunities for novel ways of targeting customers and offering personal services, mobile advertising will remain marginal until the legislation and consumer attitudes have been adapted to the new possibilities of mobile marketing.

CONCLUSIONS

The context in which mobile advertising is used must be considered carefully. The consumer is delighted if he or she gets information that anticipates future needs or desires and creates added value. However, measuring the integrated effectiveness of campaigns will be essential in determining which forms of mobile advertising will be most effective and least annoying to consumers in the future. Mobile advertising can play an important role in customer relationship management and create value to customers by personal service. The sooner measurement techniques evolve, the sooner mobile advertising will find its way to the marketer's cross-media mix. However, customers themselves will ultimately decide how popular mobile advertising is going to be in the future. If mobile advertising is considered valuable, the mobile channel might become an important customer service channel and could be used actively for permission marketing purposes. The results of this case study indicate that contests are an effective way of attracting peoples' interest but not so effective in creating an ongoing dialogue or a lasting brand experience. Therefore, companies should create more personalized advertising and services that motivate people to engage a relationship with the company and the brand. However, since consumers cannot be expected to form deep, meaningful relationships with every brand, companies need to customize their relationship communication according to customer's wishes. So far, very little is known about what kind of messages are perceived as valuable and relationship building. There is a need for more in-depth studies of what creates value to the customer through the mobile channel and how ad messages affect customer image of the brand and loyalty towards it.

ENDNOTES

1   Interstitial is a whole screen advertisement that disappears after a couple of seconds. According to the Wireless Advertising Association, all standard sizes can be run as interstitials; an 80 x 31 pixel interstitial will run full screen on a four line high display, should time out at five seconds and the user must have the option to skip the ad.

2   Opt-in means asking the natural person's permission for marketing communication with automated telecommunication devices by allowing him or her to tick the option, "I am willing to accept advertising communication. Opt-out means that the alternative is ticked automatically and the person can remove the tick.

3   Wireless Application Protocol (WAP)

REFERENCES

Aaker, D. (1996). *Building Strong Brands*. New York: The Free Press.

Aberdeen Group. (2000). *Streaming Content and e-Commerce: A Radical Way to Know the Customer – and Improve Response*. An Executive White Paper. November.

Åkermarck, M. (2001). *Enää joka neljäs suomalainen ilman kännykkää.* (Only every forth Finn does not own a mobile phone). Report 17.9.2001. Retrieved October 1, 2001, from the World Wide Web: http://www.liikenneministerio.fi.

Association for Interactive Media. (2001, October 3). Mobile Internet/SMS Information. *AIM's Research Update Service-newsletter*. Retrieved October 9, 2001, from the World Wide Web: http://www.atkearney.com/pdf/eng/Mobinet_3_S.pdf.

Channel Seven Wireless AdWatch–Newsletter (2001, September 10). It ain't all about the money: The mobile marketing opportunity. Part 2. Retrieved October 7, 2001, from the World Wide Web: http://www.wirelessadwatch.com/insight/2001/insight20010910.shtml.

Channel Seven Wireless AdWatch–Newsletter (2001, September 24). It ain't all about the money: The mobile marketing opportunity. Part III. Retrieved October 7, 2001, from the World Wide Web: http://www.wirelessadwatch.com/insight/2001/insight20010924.shtml.

Eriksson, P., Hyvönen, K., Raijas, A., & Tinnilä, M. (2001). *Mobiilipalvelujen käyttö 2001 – asiantuntijoille työtä ja miehille leikkiä?* (Use of mobile services 2001—work for experts and fun for men?)·Kuluttajatutkimuskeskus, työselosteita ja esitelmiä 63, 2001.

Future Lab. (2001). *Barometeret: E-Marketing Marts 2001.* (E-Marketing Barometer, March 2001). Future Lab Business Consulting.

Gummeson, E. (1995). *Relationsmarknadsföring: Från 4P till 30 R.* (Relationship Marketing: From the 4Ps to the 30Rs). Malmö: Liber-Hermods AB.

Hausen, K., & Westerberg U. (2001). *Permission Marketing Report*. Netsurvey.

Heimo Oy Marketing Communication Agency. (2001). *Mobile Media Landscape – Mobile Marketing Tracking*. September. Finland – Consumer Report.

Johnson, C., Allen, L., & Hamel, K. (2000). Customer heuristics. *Forrester Report*. December.

Jupiter Research. (2001). Interactive advertising on post-PC platforms. *Emphasizing Modal Marketing*. 7 (January 22).

Järvelä, P., Lähteenmäki, M., & Raijas, A. (2001). *Mobiilipalveluiden kaupallisen kehityksen haasteet ja mahdollisuudet.* (The Challenges and Possibilities of Commercial Development of Mobile Services). Edita. Liikenne ja viestintäministeriö.

Keskinen, T. (2001). *Mobiilimarkkinoinnin käsikirja.* (Mobile marketing Guide). Vaasa: Mainostajien liitto.

Kotch, M. (2001). The mobile marketing opportunity. Part I: It ain't all about the money. *Channel Seven Wireless AdWatch–Newsletter.* Retrieved September 29, 2001, from the World Wide Web: http://www.wirelessadwatch.com/insight/2001/insight20010827.shtml.

Laine, J. (2001, October 2). 75 % ei edes avaa—Tutkimus: saksalaiset kuluttajat inhoavat spam-mainontaa.; (75 % won't even open—Research: German consumers hate spam-advertising). *Digitoday–Newsletter,* www.digitoday.fi.

Nurmela, J. (2001). The Finns and the information society. *Statistics Finland.* Retrieved October 7, 2001, from the World Wide Web: http://www.stat.fi/tk/yr/tietoyhteiskunta/matkapuhelin_en.html.

Pastore, M. (2001, February 26). Study: Advertising in the post-PC market. *AllnetDevices Newsletter.*

Rohde, L. (2001, July 13). European Parliament doesn't want to ban spam. *IDG News Service*\London Bureau. Retrieved on October 9, 2001, from http://www.e-businessworld.com/english/crd_european_649039.html.

Suomen Suoramarkkinointiliitto ry. (2001). *Kuluttajien henkilötietojen käsittely ja tietosuoja markkinoinnissa,* Televiestimien käyttö markkinoinnissa. (Handling of personal data and security in marketing, using telecommunication in marketing). Retrieved October 1, 2001, from the World Wide Web: http://www.ssml-fdma.fi/ohjeistuksia_kaytannot.html#7.

WAA. (2000). *Wireless Advertising Trials Research.* Wireless Advertising Association, September. Retrieved October 7, 2001, from the World Wide Web: http://www.waaglobal.org/research_trials.pdf.

WOW Wireless. (2001). Finnish Riot-E brings Bridget Jones to SMS. *Nordic WirelessWatch–Newsletter,* April 5.

Zeithaml, V., Parasuraman, A., & Malhotra, A. (2001). *A Conceptual Framework for Understanding e-Service Quality: Implications for Future Research and Managerial Practice.* Marketing Science Institute. Report Summary No. 00-115.

Chapter XVIII

Wireless in the Classroom and Beyond

Jay Dominick
Wake Forest University, USA

ABSTRACT

The deployment of wireless data networks at American institutions of Higher Education has increased dramatically since the establishment of the 802.11 standards by the IEEE. These networks are generally deployed as extensions of the campus network to provide additional functionality to an increasingly mobile student population. As these networks become ubiquitous, they will increasingly be used for both classroom management and business m-commerce applications. Institutions with or considering wireless networks are advised to establish policies for the cooperative management of bandwidth, spectrum and security. Over time, the deployment of ubiquitous, standards-based networks will enable the deployment of an entirely new set of applications that will be useful throughout the university.

INTRODUCTION

With the advent of open wireless networking standards, a new age of untethered computing has dawned across the broad vista of American Higher Education. Much of the focus has been on providing access to mobile computers in the classroom. Our experience at Wake Forest has been that there is a significant role for wireless and mobile computing not just in support of the classroom but throughout the Academic Enterprise.

Just as wired networks quickly spread around the campus and throughout our lives, so too will standards-based wireless local area networks. The transition is already beginning. It is happening on campuses throughout North America already. As these networks are deployed to support initiatives in teaching and learning, they will also increasingly support non-teaching activities. Beyond facilitating e-learning, the new wireless technologies when combined with developing classes of super-mobile and handheld computers will enable an entirely new class of mobile commerce and mobile workforce activities. Far from being a solution that only has applicability in the classroom, wireless and mobile technologies will have broad impact throughout the educational enterprise.

BACKGROUND

During the late 1980s and throughout the 1990s, there was a significant investment in campus networking in American Higher Education. Even as late as 1998, surveys indicated that networking and access to the Internet were top technology projects for both public and private higher educational institutions in the United States (Green, 1999). Many of these networks started out as departmental local area networks (LANs) that supported file sharing, print sharing and local e-mail. With the rapid expansion and commercialization of the Internet and its supporting standard (TCP/IP), stand-alone LANs were quickly assembled into campus-wide networks. The result, in many cases, was combined networks of different technologies, standards and support cultures. For many colleges and universities, control over campus LANs still remains a thorny issue. Providing networking on the college campus, in fact, became so omnipresent that it gained notice in the popular press. One popular technology magazine even devotes an entire issue each year to listing America's "Most Wired Colleges and Universities."

It seems rather counter-intuitive that there would be such interest in wireless networking after a period of intense investment in wired networking. Yet, during 2000 and 2001 wireless networking was a very popular topic on university campuses. Schools as diverse as Wake Forest University, Carnegie Mellon, the University of Oklahoma, the University of Kentucky, the University of North Carolina, the University of Oregon, Seton Hall University, Sacred Heart University and Buena Vista College have made significant investments in providing wireless access on their campuses while at the same time providing wired access as well (Oh, 2000; Young, 1999). Part of the reason that wireless networks have become so popular so quickly is that they seemingly fit quite well with the mobile lifestyle of today's laptop-toting student. Mobile access to the Internet for communications, research and fun are natural extensions to mobile computing.

Other reasons for investment in wireless include providing access in spaces that have traditionally been too difficult or costly to wire. From the library to the cafeteria to the quad, access to the electronic resources of the network are becoming available everywhere. At Wake Forest, we found that providing Wireless data access in three of our overseas houses was the only way to support the communications needs of the students without expensive and difficult wiring projects.[1] While five years ago, access to the Internet for our overseas students was not even considered desirable, today it is an absolute requirement. Increasingly the business of the university is done online and around the clock. Without access to our online systems, such as registration and student records, students at overseas learning programs are at an almost impossible disadvantage.

The relatively rapid deployment of wireless networks in higher education has also been spurred on by a creative approach to distributing the cost of the capital investment. Currently, several companies have developed programs that provide the wireless infrastructure for the university in exchange for the exclusive rights to lease or sell the cards to the students. These novel tactics make the decision to implement a wireless network much simpler and reduces the time to implementation. By outsourcing the deployment and management of these networks, schools can quickly deploy these networks without having to increase staff or obtain increasingly scarce capital resources.

Wireless networking is not just a phenomenon in American higher education. The ability to connect to the Internet through mobile devices seems to have caught on around the world very quickly. However, outside of the United States, wireless tends to mean a completely different set of technologies. In Europe, particularly, wireless networking generally involves use of the very robust cellular telephone network rather than deployment of campus-based wireless local area networks (WLANs).

For instance, in Germany students at higher educational institutions would access the Internet and university student records systems through their cellular telephones using protocols developed for the telecommunications industry (Brookman, 2001). In England the University and Colleges Admissions Services is considering providing wireless cellular telephones using the WAP protocol to sixth-formers so that they might keep track of their college admissions applications (Dean, 2000). The access methods throughout the world may differ, but increasingly there is the recognition that mobile access to information resources is going to be an important feature of the educational experience of the future.

WIRELESS COMPUTING IN THE EDUCATIONAL ENTERPRISE

For the purpose of this chapter, we will consider wireless to mean IEEE 802.11-based wireless local area networks using the unlicensed bands of the radio spectrum. There are several different variants of the IEEE 802.11 standards that involve different signaling methods, protocols and frequencies. The most popular standard today is the 11 Megabit per Second 802.11b standard that operates in the 2.4 GigaHertz band of the radio spectrum. The IEEE 802.11 protocols are certainly not the only wireless protocols, but they do seem to be the most prevalent. The IEEE 802.11 protocols seem to have taken such hold in higher education partly due to the focus on open standards. It is no coincidence that the rapid rise in the deployment of wireless networks coincides with the final ratification of the 802.11b standard by the IEEE in 1999.

Ubiquity and Wireless Networks

Unlike their wired counterparts, wireless local area networks are pervasive by their very nature. A radio transmitter placed inside a large room will indiscriminately cover an area of several hundred feet in diameter. Anyone with a standards-compatible wireless network card and the right set of permissions and passwords can access the network. There may be policy limitations as to how many people can be using that wireless network, but theoretically the limitations are not significant. With a wired network, there can be no more users than there are physical ports available. Once all the outlets are used, even though there may be more bandwidth available, there can be no more users. In a static environment such as an office building or plant floor, a physically controlled network is ideal. For a mobile population such as is typically found at a college or university, static networks represent a constriction on use.

Without the physical presence of network outlets, the user has a much more difficult challenge in trying to determine where the network actually is. Hence, a wired network user's first reaction upon entering a room is generally to look up at the ceiling for the telltale traces of radio antennas. Because it has only a very small physical presence, determining where a wireless network is located requires training and communications. Unless there is ubiquitous coverage for the network, users quickly tire of hunting down access points and the value of mobile access is diminished. We have found that unless coverage is ubiquitous and seamless, students find only minimal value in wireless access. Parallels with the cellular telephone network are very valid. When coverage is weak or unpredictable, it is much easier to avoid using the phone at all.

The need for ubiquity is further enhanced when the student is expected to pay some portion of the cost of access. Purchasing a network service that can only be used in a limited number of locations on campus presents a lot of frustration to most students. Switching between wired and wireless connections is not intellectually challenging, but it can present some unfortunate technical challenges. Students must understand where they are and in many cases must be able to switch network cards or reconfigure their network settings to obtain the proper network addresses. This process only has to fail a few times before the typical student becomes frustrated. Having a mix of wired and wireless networks throughout campus simply presents the typical student with too many variables.

The value of the wireless network increases as its coverage expands. The less that students (and faculty) have to think about whether or not they will have access, the more they are willing to use that access. The ultimate value is obtained when access to the wireless network is omnipresent. Our experience at Wake Forest has shown that a comprehensive presence is even more important than the absolute speed of the network.

Covering a campus is indeed a daunting challenge. This challenge extends even beyond covering the areas that students normally frequent. Providing a completely ubiquitous network further increases the value of the network because the financial and administrative sides of the university can begin to make use of some of the new mobile commerce (m-commerce) technologies under development.

Wireless Networking in Support of Classroom Activities

One of the main areas of interest for wireless access in the education area has been in facilitating classroom connectivity. Chan et al. describe a classroom of the future where wireless and new mobile computing, "school bag computers" along with other tools will "transform a static classroom into a highly interactive learning environment." At Wake Forest we have been experimenting with several mobile technologies for classroom use that present interesting possibilities (Chan, Hue, Chou, & Tzeng, 2001).

In order to help facilitate what Chan et al. describe as "complex problem learning," Wake Forest University has experimented with the concept of using these super mobile devices as classroom facilitators. Using devices like the Compaq iPAQ and software developed in-house, Wake Forest has created a portable classroom management and communication tool. The purpose is to experiment with how these new computing paradigms can begin to shape the "future classroom."

One such application is the portable server. The portable server is a handheld device with a wireless network card that is used in conjunction with either mobile devices or laptop computers owned by the students to facilitate feedback and group decision making in the classroom. In this case, the professor would carry the server (Palm Pilot or Windows CE device) from which she might distribute her notes, deploy quick quizzes or even record her lecture in real time. In addition to the standard capabilities of any server—downloading files during the class, taking online quizzes, etc.—the portable server can permit the faculty member to get immediate and private feedback from the students. By using the portable server as a fully functional web server to accept messaging (through a self-generated web page or instant message software), the faculty member can privately view feedback from the class as he or she wishes during the class period. The portable server has the capability to collect, calculate and display the results of the questionnaire for the instructor. For large lecture classes, this may provide an excellent way to obtain questions and comments during the lecture itself.

In addition to facilitating communications during the class, wireless computing devices can also be used to help manage the actual mechanics of the class. We have found that using the portable server as a remote control device for a laptop connected to a projector allows the instructor to move away from the podium and regain the freedom of movement. When that is combined with software to allow the instructor to privately view student feedback, a faculty member can quickly adapt the lecture to enhance student comprehension.

By using the mobile handheld as a server, the instructor does not need to worry about the potential security issues associated with running a webserver on her personal computer or with the potential administrative hassles of obtaining web space on a departmental server. At the end of the class or if there is a problem, the instructor can simply turn the device off. If more space is needed for files, memory upgrades are available through higher capacity Flash memory.

The current generation of wireless, handheld computing devices have several significant advantages over traditional laptop computers. First, they are significantly less expensive than the typical laptop computer. More importantly for the classroom experience, these new devices have instant-on features. Press the button and the device is up and running. In the classroom setting where time is the most precious commodity, having a device that can become functional instantly is very important.

There are, of course, significant disadvantages to devices like the Palm V or the Compaq iPAQ. They do not have the screen size or resolution necessary for extended or detailed use. There are also limited input capabilities—generally limited to tapping on the screen or using a stylus with a special character set. Clearly, these devices and perhaps the first generations of truly super mobile devices yet to come, will not be able to replace the laptop computer as a primary computing device. Finally, there is relatively little software available for these devices. The advent of operating systems like Windows CE that can support a wide variety of popular software will ameliorate that problem over time.

Using Wireless and Super Mobile Computing Outside of the Classroom

In addition to the classroom, disciplines that require data collection and field work can also benefit from the use of wireless and super mobile computing. With most wireless technologies, networks can be quickly deployed by designating one wireless card as the master—so called "ad hoc" networks. Data collection can be done through single applications or databases in real time rather than through multiple collection points that have to be aggregated and the data hand entered. Using wireless networking with handheld devices with barcode readers or cameras presents very interesting opportunities for field-based data collection.

One discipline that already makes significant use of Personal Digital Assistant (PDA) technology is medical education. A significant number of medical schools and hospitals provide or support PDAs such as those made by Palm, Inc. These devices are excellent for storing and retrieving information such as treatment tracking, drug lists, medical reference material, and even billing and coding protocols. In the typical medical teaching environment, both faculty and students are highly mobile and very busy. They need to have their hands free to attend to the patient. The low weight, small size and instant-on capability of the modern PDA makes them a much more practical support tool than are laptop computers. The presence of a wireless network provides a natural extension and enhancement to the utility of these super mobile devices (Shipman & Morton, 2001). While there is some concern that there may be interference issues with wireless technology in medical settings, a number of hospitals have installed wireless local area networks. Managing the electromagnetic spectrum is increasingly an issue for medical facilities in general (Gilfor, 2001).

Surprisingly, wireless PDAs are not generally being used to access clinical patient records systems. Recent federal regulations like the Health Insurance Portability and Accountability Act (HIPAA) make security and privacy of patient records absolutely mandatory. For these m-commerce technologies to be employed in this area, solutions will have to be developed that address electronic security issues both with the device and the wireless networks to which they are attached (Shipman & Morton, 2001).

Wireless Access for Mobile Classrooms

Wireless networking is an excellent solution for providing network access to transitional work areas. In the K-12 arena, many school districts handle the uncertainties of school census by providing portable classrooms. These "learning cottages" are often provided at the last minute and are rarely connected to the existing school network. For a school with a comprehensive wireless infrastructure, providing access to portable classrooms becomes a matter of proper site location. Locating the receiving antenna in the portable classroom (either through a wireless hub or a network card in a laptop) provides instant connectivity to the school network. Fairness issues related to portable trailers can be somewhat ameliorated by providing Internet access.

In Winston-Salem, North Carolina, USA, Wake Forest University has partnered with an elementary school in the Winston-Salem/Forsyth County School system to provide Internet connectivity to six portable classrooms at one school location. With only two 802.11 access points, we were able to provide up to 2 MB Internet access into the trailers. Combined with wireless laptops and training for the teachers, these portable classrooms became as well connected as the best classrooms in a matter of weeks. The 802.11 network was able to overcome the difficulties of penetrating through the aluminum trailers without any special antennas or configuration.

Wireless Networking in Support of the Business Mission.

Virtually all of the focus on campus WLANs recently has been on how they are used to support the teaching and learning process. This is entirely appropriate given the nature of the educational organization. What has been overlooked is that ubiquitous wireless networking will provide a significant opportunity for better managing the business of the university.

Most educational organizations—from K-12 right through the largest Carnegie Classification Schools—have substantial facilities and business functions. From police/security to warehousing to banking to food service to hotelling, the modern American educational enterprise is a pretty comprehensive business. Over the past decade the integration of wireless, hand held devices and barcode scanning has revolutionized the parallel for-profit industries. World-leading companies like UPS, Goodyear Chemical, Ricoh Norge and Herman Miller, not to mention Fedex and WalMart, use mobile, automated systems to manage their inventories and processes. A number of companies are using mobile and wireless technologies to tie into their Enterprise Resource Planning Systems (ERPs) from vendors like SAP, Inc. (Nesdore, 2001). Many of these same sorts of inventory management and tracking activities are performed at American educational organizations on a daily basis. Providing access to modern, connected applications for better managing those resources is a worthwhile undertaking.

The deployment of a standards-based ubiquitous network is essential to making the most of these back room applications. With the scarcity of spectrum available, it makes no sense to deploy incompatible or competing wireless networks around campus. While this may have been the model for wired networks, it becomes destructive when dealing with a shared resource like the Radio Frequency Spectrum.

Having a ubiquitous wireless network also permits the rapid deployment of new applications into that infrastructure. An example at Wake Forest involves a wireless application written for the Palm Operating System that is in the process of being deployed for checking students into fraternity and sorority functions. In the new environment of extreme sensitivity surrounding alcohol at campus functions, there is a tremendous interest in ensuring that only students of legal age consume alcohol on campus. In addition to the concern of the University, the Greek organizations also face a potential liability if they serve underage students. Consequently, a group of students in conjunction with our Residence Life and Housing Department, approached our Information Systems Department in the Spring of 2000 about developing a system for electronically verifying student identification and determining age. Over the next six months, we worked with a partner to develop a system that used a wireless PalmOS-based device (Symbol 1740) equipped with a barcode scanner to automatically determine the age of a student entering a party. The ruggedized (drop resistant and beer-proof) Symbol 1740 connects to the already existing wireless network at the party location to query our student records database. When the student arrives at the event, a barcode on their student identification card is scanned by the Symbol 1740. As soon as the identification is scanned, the application queries the student record database through the wireless network. If the student is over 21, that information is displayed on the Symbol 1740, and the student is logged into the database as having entered the party. No personal information is transmitted over the network, and all records of entry are time-stamped should there ever be a need to review that information. For the student the solution provides for quick entry to the event, substantially reducing lines. It also provides a much simpler way for Residence Life to track party attendance. For the Greek organizations, it reduces their potential liability, removes the question of fake identifications, and eliminates paperwork and record-keeping tasks. All of these benefits are available because of the combination of ubiquitous wireless networking and a new class of highly mobile computing devices.

This model of ubiquitous mobile access to the central records systems of the university has widespread potential benefit. This combination can simplify the process of conducting a number of onerous and time-consuming tasks at the university. One example is Residence Hall check-in and checkout. Another is a library application for book inventory. Any application that requires the entry of data away from a workstation can benefit from mobile access through a ubiquitous wireless network. None of these applications can necessarily support the deployment of a wireless infrastructure on their own, but they make financial sense when they can take advantage of an already deployed network.

For most university settings, the security of wireless networks is important, but not critical. Despite the recent disclosures about the relative insecurity of Wired Equivalency Protocol (WEP) in 802.11-based wireless networks and their coverage in the popular press, there seems to be little public concern expressed by university administrators. Part of this, no doubt, has to do with the general sense that student-related communication does not need a high level of security. As these applications begin to move to supporting the business applications of the educational enterprise, addressing the security issues with wireless will be critical (Borisov, Goldberg, & Wagner, 2001; Gomes, 2001).

MANAGING THE RESOURCE

As wireless networks become more widespread and critical to the educational enterprise, there will arise the need to carefully manage the electromagnetic spectrum on campus.   Most of the major standards (802.11, 802.11b) now supporting wireless networks on campuses, and those that are proposed for the near future (802.11a), operate in unlicensed parts of the radio spectrum. The convenience of not having to apply for spectrum licenses from the FCC makes the deployment of these networks very simple and easy. The downside to the unlicensed bands is that they quickly become overrun with a variety of different type of devices—from cordless telephones to microwave ovens to karaoke machines. Some of the existing standards are potentially incompatible in the same band. Without proper thought and management on the campus, the very openness of the spectrum can limit the effectiveness of the technology (Latimer, 2001).

Campus-wide planning for use of the radio spectrum becomes very important. Given the experience that most campuses had in deploying their wired LANs initially, the deployment of these new wireless networks provides an excellent opportunity to avoid problems in the future. Establishing a spectrum management team with organization-wide representation is a very important step in ensuring that the limited amount of spectrum available is used wisely.

CONCLUSIONS

As with the new class of applications evolving for the business applications, mobile and ubiquitous access to databases and other resources are opening up a whole world of new possibilities for computing. While wireless today may be seen mostly as a wired-line replacement program focused on providing connectivity in the classroom, its true potential lies in facilitating an entirely new set of applications. It may take time to evolve an understanding among faculty and administrators about how to make use of the new combination of technologies. In the interim, planning for a standards-based ubiquitous network is essential.

Security remains a significant stumbling block for developing significant m-commerce applications in some medical- and business-related portions of the educational enterprise. Nonetheless, the potential for wireless and super mobile computing in support of the educational mission is great. From students with portable devices for e-mail and chat, to educators enhancing the teaching process through feedback processes to researchers collecting data in far away lands, the potential for super mobile computing and m-commerce in education is high. More complete adoption requires some improvement in the input and output capabilities of these devices along with the maturing of the independent software market. In the long run, though, low cost, high mobility and new functionality are drivers that will make a compelling case for the adoption of super mobile computing and m-commerce in higher education sooner rather than later.

ENDNOTES

1   Wake Forest University has residential learning programs in Venice, London and Vienna.

REFERENCES

Borisov, N., Goldberg, I., & Wagner, D. *Security of the WEP Algorithm* [Web Page]. URL http://www.isaac.cs.berkeley.edu/isaac/wep-faq.html [2001, October 4].

Brookman, J. (2001). WAP lets students ring for results. *The Times Higher Education Supplement*, (January).

Chan, T.-W., Hue, C.-W., Chou, C.-Y., & Tzeng, O. J. L. (2001). Four spaces of network learning tools. *Computers & Education, 37*, 141-161.

Dean, C. (2000). Sixth-formers to get free mobiles. *Times Education Supplement, 4384*(July), 1.

Gilfor, J. *Wireless Devices and Electromagnetic Interference in Hospitals, Urban Myth?* [Web Page]. URL http://www.pdamd.com/features/interference.xml [2001, October 4].

Gomes, L. (2001). Silicon Valley's open secret—Wireless computer networks, often unguarded. Can be eavesdroppers' gold mine. *Wall Street Journal*, (April), 1.

Green, K. C. (1999). Encino, CA: The Campus Computing Project.

Latimer, D. (2001). Taking Control of Your Campus Wireless Spectrum, Educause Review, March/April, 60.

Nesdore, P. (2001). The Wireless Wave. 24. Cahners Business Information.

Oh, J. (2000). The Untethered Campus. *The Industry Standard*, (September).

Shipman, J. P., & Morton, A. C. (2001). The new black bag: PDAs, health care and library services. *Reference Services Review, 29*(3), 229-238.

Young, J. R. (1999). Are wireless networks the wave of the future? *Chronicle of Higher Education*, 45(22), A25.

Glossary[1]

2G or 2nd Generation—The currently available digital communication networks for voice and data communication (e.g., GSM, CDMA, PDC).

2.5G or 2.5th Generation—Represents an upgrade from the currently available communication networks. This protocol provides more bandwidth and enables packet-switch networks. In Europe this is associated with GPRS.

2.75G or 2.75th Generation—A network that includes a number of upgrades to the 2.5G networks and that allows for greater bandwidth. In Europe this is usually associated with a standard referred to as "EDGE."

3G or 3rd Generation—Mobile technology standard that corresponds to the IMT-2000 standard (e.g., UMTS in Europe). This standard will provide higher bandwidth and use packet-switch networks.

4G or 4th Generation—An anticipated future mobile technology standard that is based on new modulation schemes (e.g., OFDM) and the concept of a separate uplink and downlink channel. This will offer higher bandwidth than 3G networks.

ARPU or Average Revenue per User—A method of measuring revenue associated with the delivery of mobile commerce services by MNOs.

B2B or Business to Business—This is a commonly used term in e- and m-commerce to describe transactions carried out between businesses.

B2C or Business to Consumer—This is a commonly used term in e- and m-commerce to describe transactions carried out between a business and it customers.

Bluetooth—A wireless personal area network (PAN) technology protocol that was promulgated by the Bluetooth Special Interest Group (www.bluetooth.com). The Bluetooth Special Interest Group was founded in 1998 by a consortium of Ericsson, IBM, Intel, Nokia and Toshiba. Bluetooth is an open standard for short-range transmission of digital voice and data between mobile devices (laptops, PDAs, phones) and desktop devices.

CDMA - Code Division Multiple Access networks—A method of frequency re-use whereby many handheld phones use a shared portion of the frequency spectrum. CDMA uses spread-spectrum techniques to assign a unique code to each conversation.

CDMA2000—The North American version of the IMT-2000 3G technology.

CHTML or Compact HTML—An HTML-compatible markup language for handheld devices that was developed by NTT DoCoMo for use in the i-mode service.

D-AMPS—Digital Advanced Mobile Phone System.

EDGE or Enhanced Data for GSM Evolution—An enhanced version of the GSM and TDMA networks that increases bandwidth. EDGE is often called the 2.75G network standard.

GPRS or General Packed Radio Service—An enhancement to the GSM mobile communications system that supports the transfer of data packets. GPRS enables the continuous flow of IP data packets over the network to enable applications such as Web browsing.

GPS or Global Positioning System—A satellite-based navigation system that triangulates a user's signal via three or more satellites. The system was originally developed by the U.S. military but is now also available for commercial and private applications.

GSM or Global System for Mobile Communications—A digital cellular phone technology based on TDMA. This is the predominant network in Europe but is also used in the U.S. (e.g., VoiceStream) and around the world.

Handheld—Handheld computing device such as a mobile phone or PDA that serves as an organizer and/or communicator.

HDML or Handheld Device Mark-up Language—An XML-based mark-up language originally proposed by Phone.Com before the WAP specification was standardized. It includes a subset of WAP capabilities but also has some features that were not included in WAP.

i-mode—A packet-based information service from NTT DoCoMo (Japan) designed for their mobile phone network. i-mode provides features such as web browsing, e-mail, games, information management, calendars, and news services. i-mode is a proprietary system that uses cHTML, a customized version of HTML.

IMT-2000 or the International Mobile Telecommunications 2000—The IMT-2000 framework is for 3G systems and provides a seamless, global communications service that delivers high-speed multimedia data as well as voice through small, lightweight terminals. This specification was formerly referred to as the Future Public Land Mobile Telecommunications System (FPLMTS).

LBS or Location-Based Services—A class of services that have the capability to identify the location of a mobile phone or a vehicle. Location-based services are used both for emergency services as well as commercial applications such as location fleet and field force management, proximity-based services, routing and resource tracking.

M-Commerce—Mobile commerce.

MMS—Mobile Management Server such as Microsoft's Mobile Information Server.

Microbrowser—A browser is a software application that enables a user to view information placed on an Internet site or corporate intranet via an electronic device. The microbrowser is designed for small devices, such as mobile phones.

MVNO or Mobile Virtual Network Operator—A reseller of wireless services. These service providers do not own a license for spectrum and may or may not manage their own wireless network infrastructure. The mobile network operators provide service under their own brand name but use the facilities of existing carriers and network operators.

PAN or Personal Area Network—PAN technology was introduced at IBM's Almaden Research Center, San Jose, California, by Thomas Zimmerman. A PAN uses a tiny electrical current to transmit information from one person to another or from a person to an electronic device such as a PDA, a phone or a computer.

PDA—Personal Digital Assistant.

Portal—Service providing access to the World Wide Web. Portal services take the form of web pages that provide search engines or other directory services, and may also provide other types of information. Many portal services are funded by advertising messages.

SIM or Subscriber Identification Module—A smart card (i.e., a module) holding the user's identity and telephone account information. This device is a standard component of GSM-based phones.

SMS—Short Message Service (currently in use for most mobile subscribers in the form of text mail). The sending and receiving of short alphanumeric messages to and from mobile handsets on a cellular mobile network.

TDMA or Time Division Multiple Access—A cellular phone technology that weaves together multiple digital signals into a single communication channel. TDMA is the standard for GSM phones.

UMTS or Universal Mobile Telecommunications System—A digital packet-switching technology that has been adopted by the Europeans for implementation of 3G wireless phone networks. UMTS will offer higher bandwidth than GSM and will provide "always-on" access to the network and the Internet.

WAP or Wireless Application Protocol—A standard for providing phones, pagers, two-way radios, smart phones and other handheld devices with secure access to text-based e-mail and web pages. WAP standards are promulgated by the WAP Forum (www.wapforum.org) with the goal of enabling WAP to be useful on all network devices.

W-CDMA or Wideband-CDMA—A broadband 3G technology that uses CDMA rather than TDMA. W-CDMA was developed by NTT DoCoMo and has been adopted as a 3G standard by many carriers in Europe, Japan and North America.

WLAN or Wireless LAN—A local area network that is equipped with a wireless interface or wireless access point.

WML or Wireless Markup Language—A tag-based language similar to HTML that is used in the Wireless Application Protocol (WAP). It is essentially a streamlined version of HTML that can be used on small-screen displays.

ENDNOTE

1   Compiled by Brian E. Mennecke

About the Authors

Brian E. Mennecke is an Associate Professor of Management Information Systems in the College of Business at Iowa State University. Dr. Mennecke earned his PhD at Indiana University in Management Information Systems and also holds master's degrees in geology and business from Miami University. His research interests include mobile commerce, location-based services, the application of spatial technologies to business, technology-supported training, data visualization and virtual teams.

Troy J. Strader is Assistant Professor of Management Information Systems in the College of Business at Iowa State University. He received his PhD in Business Administration (Information Systems) from the University of Illinois at Urbana-Champaign in 1997. His research interests include mobile commerce, online investment banking, consumer behavior in online markets, and electronic commerce in the transportation and agribusiness industries.

* * *

Jukka Alanen is a Consultant at McKinsey and Company in Helsinki, Finland. He has experience advising leading wireless start-ups, venture capitalists and corporations in Europe, the USA and Israel in the areas of strategy, business building, organization and operations. Prior to McKinsey, he worked in the fields of venture creation, business building and investment banking. He holds an MSc (Technology) degree from Helsinki University of Technology and an MSc (Economics and Business Administration) from Helsinki School of Economics and Business Administration.

Formerly the Executive Vice President of Autodesk's GIS Solutions Division, **Joe Astroth** brings extensive location-based software experience to the Autodesk Location Services Division, also as Executive Vice President. A renowned industry expert, Dr. Astroth has worked in the GIS and computer mapping industry for more than 20 years and is responsible for the mobile and wireless solutions strategy at Autodesk. He has served as a member of the Autodesk executive management team since 1995. Dr. Astroth led the company's first effort to bring mobile and wireless technologies to Autodesk customers with the introduction of Autodesk OnSite. He is credited with leveraging Autodesk's knowledge and experience to expand the company's focus on, and investment in, location-based services. Dr. Astroth came to Autodesk from Convergent Group, where he was Vice President of Product Management and Engineering for the GDS Software Division. During his tenure, he helped the company to become a key worldwide provider of GIS software. Prior to joining Convergent Group, he was a Senior Principle Consultant at McDonnell-Douglas System Integration Company and EDS, Inc. Earlier in his career he was a part-time Technology Consultant and full-time Professor at the University of Missouri, where he co-founded a GIS research center for government and commercial projects. Dr. Astroth completed his doctoral work at the University of Chicago, and was awarded the Fulbright Hays Doctoral Dissertation Research Abroad Fellowship in 1982, and the Mellon Foundation Dissertation Award by the University of Chicago in 1983. He is the author of numerous articles, professional papers and presentations. His doctoral dissertation on Location Theory and Spatial Interaction Modeling was published by the University of Chicago Press.

Erkko Autio is Professor in Technology-Based Venturing at the Institute of Strategy and International Business of Helsinki University of Technology. Previously, he has held academic research and teaching positions at London Business School, University of Sussex and the Asian Institute of Technology. He has published, advised and consulted widely in the areas of venturing, innovation, mobile commerce, and technology and innovation policy, for public and private organizations alike. In all, he has authored or co-authored some 140 publications in his areas of research interest.

Stuart J. Barnes is Associate Professor of Electronic Commerce at the School of Information Management, Victoria University of Wellington, New Zealand. After starting as an economist, Dr. Stuart later completed a PhD in Business Administration at Manchester Business School, specializing in Information and Communications Technologies (ICTs). He spent the last six years working at the University of Bath before moving to Wellington in 2002. His current research interests include evaluating website and e-commerce quality, e-commerce strategy, information systems implementation, knowledge management systems and business applications of mobile technologies.

Greg Chiasson is a principal in the Worldwide Communications and Electronics Practice of PRTM, management consultants to technology-driven business. His expertise includes helping companies in the communications sector capitalize on opportunities afforded by the increasing availability of rich mobile communications. His experience includes working with traditional communication service and equipment providers as well as new value chain players, such as automotive OEMs and digital radio providers, on overall strategy definition and business planning, formation and launch. Mr. Chiasson has helped companies develop and evaluate their marketing, operational and business strategies for automotive and industrial telematics ventures, identify and recruit key value chain partners to reduce time to market and enable premium communications offerings, and launch new satellite radio, home networking and voice recognition software businesses. Prior to joining PRTM, Mr. Chiasson held general and product management positions with several telecommunications-related start-ups in Silicon Valley, and was previously part of Motorola's Corporate R&D group. Mr. Chiasson holds BS and MS degrees in Electrical Engineering from the University of Illinois at Urbana-Champaign, and an MBA from the Graduate School of Business at the University of Chicago. He holds seven U.S. patents, and has authored numerous journal, magazine, and conference articles on a range of wireless communications technology and mobile business issues. Greg Chiasson is based in PRTM's Chicago office and can be reached at gchiasson@prtm.com.

Irvine Clarke III is currently an Associate Professor of Marketing at James Madison University. Prior to joining JMU, he held the Freede Endowed Professorship of Teaching Excellence at Oklahoma City University. He has recently published in the *Journal of Business Strategy, Journal of International Marketing, International Marketing Review, Industrial Marketing Management, Central Business Review* and the *Journal of Marketing Education*. In the past five years Dr. Clarke has taught at locations in Canada, England, France, Germany, Malaysia, Mexico, Singapore and the Peoples Republic of China. He has 15 years of public- and private-sector organizational experience in various marketing areas.

Ioanna D. Constantiou is a PhD student at Athens University of Economics and Business in ELTRUN, where she also works as a research officer. In 1996 she graduated from the Department of International and European Economic Studies of Athens University of Economics and Business (AUEB). In 1997 she received her MSc from AUEB in International and European Economic Studies, majoring in International Banking and Finance. From 1997 to 1998 she worked at the Hellenic Centre for Investment (ELKE) in the Department of Research and Analysis. From October 1998 to March 1999 she worked as a researcher in Brussels in European

Commission, Directorate General III, in ESPRIT Programme. Since March 1999 she has been working in ELTRUN. She is Project Coordinator on MobiCom (IST project), which involves construction of evolution scenarios for emerging m-commerce services. In her PhD research, she is focusing on network economics and management. She is specializing on Internet and mobile packet networks services pricing as a tool for efficient resource allocation.

Jay Dominick is Assistant Vice President, Information Systems, Chief Information Officer, Wake Forest University. He earned his MBA from Wake Forest University in 1995; his MA in National Security Studies from Georgetown University in 1988; and his BS in Mathematical Sciences from UNC-CH in 1984. He is currently responsible for strategy, planning and operations for Wake Forest University's highly regarded Information Technology efforts. Wake Forest University is consistently ranked as a leader in the use of Information Technology in the teaching and learning process. As Chief Information Officer, he directs the efforts of the Information Systems Department, including networking, computer operations, help desk, telecommunications, programming and systems development. Mr. Dominick was responsible for the implementation and support of the ubiquitous laptop computing project at Wake Forest, which established a new model for technology deployment in higher education. He is active in statewide networking as Chairman of the North Carolina Research and Education Network (NCREN) Advisory Board and is a Co-Founder of WinstonNet—a community fiber optic network in Winston Salem. He worked as a Network Technician at the Research Triangle Institute in RTP, NC from 1988 through 1991 and supported Unix networking for in-house software development projects. He was also responsible for planning and budgeting for Unix computing support. Commissioned Lieutenant in the USAF in 1984. He served as a PC and Network Analyst at the Pentagon supporting Air Force Planning and Budget process. He also worked on the implementation team for the first TCP/IP network in the Pentagon, and was promoted to Captain before leaving the Service in 1988.

Theresa B. Flaherty is an Associate Professor of Marketing at James Madison University where she teaches strategic Internet marketing and integrated marketing communications. Prior to this position, she was a member of the marketing and e-commerce faculty at Old Dominion University where she taught graduate and undergraduate courses in various distance-learning environments. She also taught marketing management and marketing research classes at the University of Kentucky where she earned her PhD. She has published several research papers in outlets such as *Industrial Marketing Management, International Marketing Review, The Journal of Personal Selling and Sales Management, Central*

Business Review and *The Journal of Marketing Education.* Together Drs. Irvine Clarke and Flaherty are Research Fellows for the Commonwealth Information Security Center at James Madison University. They co-edited a special issue of *International Marketing Review* on the Impact of E-Commerce on International Distribution Strategy. Additionally, they were recent recipients of the Gene Teeple Outstanding Paper Award presented by the Atlantic Marketing Association, and have been recognized as outstanding teachers by the Society for Marketing Advances.

George M. Giaglis is Assistant Professor of Information Systems in the Department of Financial and Management Engineering at the University of Aegean, Greece. He has also held teaching and research positions at Brunel University (UK) and the Athens University of Economics and Business. His research interests include eBusiness, IS investment evaluation, business process modeling and simulation, and IS-enabled organizational change.

Susan H. Godar is Associate Professor and Chair in the Department of Marketing & Management Sciences at William Paterson University in Wayne, New Jersey. After a career in the aviation industry, she earned a PhD in International Business from Temple University. In addition to business ethics, her research interests are marketing professional services and virtual teams.

Before becoming an academic, **Debra Harker** worked as a Marketing Consultant with KPMG Peat Marwick Management Consultants in England and AGB McNair in Australia. Dr. Harker achieved a BA (Hons) in Business Studies at South Bank University, London, and evaluated the effectiveness of the advertising self-regulatory scheme in Australia for her doctorate, which was awarded by Griffith University in Brisbane. She is now a Senior Lecturer in Marketing at the University of the Sunshine Coast in Queensland, Australia, and publishes in quality journals around the world.

Anand Iyer is a Director in the Worldwide Communications and Electronics Practice of PRTM, management consultants to technology-driven business. He has 14 years of consulting, academic and industry experience spanning the communications, computer and software industries. Dr. Iyer is focused on helping companies in the wireless communications, life sciences and automotive sectors capitalize on opportunities created by the convergence of the Internet and wireless access. He has worked with OEMs, telecommunications service providers and telematics content providers to define collaborative telematics strategies, capital structure models, technology architectures and go-to-market plans. He has also worked with manufacturers of medical monitoring equipment and wireless data providers to define a role for wireless communications in the critical care unit. Dr. Iyer has helped to define and develop market- and technology-driven strategies, economic models and value chain positions for communications companies, and has helped to operationalize these strategies to launch viable businesses and significantly add to revenue and profitability. He holds a PhD and MS in Electrical and Computer Engineering and an MBA from Carnegie Mellon University, and a BE in Electrical and Computer Engineering from Carleton University. He is a member of the IEEE, SPIE and Sigma-Xi. He often speaks on convergence and telematics issues at national and international forums. Dr. Iyer is based in PRTM's Washington, DC, office and can be reached at aiyer@prtm.com.

Mirella Kleijnen is a PhD candidate with the Department of Marketing and Marketing Research, Maastricht University, Maastricht, The Netherlands. She is currently working on her dissertation research, which focuses on the adoption and diffusion of mobile services. Her main interests are electronic/mobile commerce, services marketing, adoption and diffusion theory, and consumer behavior. Her work has been published in the *International Journal of Service Industry Management* and *Total Quality Management*.

Panos Kourouthanassis holds an MSc in Decision Sciences (specialization in e-commerce) from the Athens University of Economics and Business (AUEB). Currently, he is a doctoral student at the same university and a researcher in ELTRUN, the eBusiness Centre of AUEB. His research is focused on emerging eCommerce technologies with emphasis on mobile commerce and its implementation on vertical sectors (retail, tourism and so on).

Mark S. Lee is a six-year veteran of The Coca-Cola Company, currently serving as Director of Strategic Marketing for Coca-Cola North America, with responsibility for projects around New Product and Packaging Commercialization and Acquisitions. Prior to joining the Strategic Marketing team, Mr. Lee served as Director of e-Business, with primary responsibility for B2E projects and all wireless activities. Before the e-Business assignment, he served as the Coca-Cola classic Ignition Leader for Fountain and as a Group Marketing Manager in National Accounts. Prior to joining Coca-Cola, he spent 12 ½ years at Coors Brewing Company. Mr. Lee worked in a variety of sales and marketing roles at Coors, including a post as Zima Brand Manager, leading the development and launch of Coors Brewing Company's most successful new product introduction.

David J. MacDonald is Project Coordinator of Strategic Alliances, i-mode, NTT DoCoMo, Inc. He has six years experience in the Japanese market, and is currently Project Coordinator within the i-mode Strategy Department, based in Tokyo, which has the mission of planning the development of NTT DoCoMo's i-mode service, and expanding the platform. He joined the i-mode team in the summer of 1999, shortly after service launch. In particular, Mr. MacDonald is responsible for new business development, focusing on strategic alliances. He manages NTT DoCoMo's relationships with international partners for the domestic service, including both content providers and technology vendors. He is also an adviser to NTT DoCoMo's overseas partners, as they study the implementation of i-mode-like services in their own markets. A frequent presenter at a variety of conferences and events on wireless Internet and mobile commerce, he has also contributed to several wireless professional journals. Mr. MacDonald is a graduate of Canada's Royal Military College. Upon graduation, he was commissioned as an officer in the Military Intelligence branch. He also holds the degree of Master of Science from the University of Sheffield's East Asian Business programme. He can be reached at macdonald@nttdocomo.co.jp.

Kirk Mitchell is the Asia Pacific Regional Sales Director of Webraska Mobile Technologies. Together with the rest of the local Webraska team, Mr. Mitchell is establishing Webraska as the leading regional provider of wireless location-based services. He has more than 12 years experience in Internet-based mapping and guidance, having previously served as a Business Development Manager with Pacific Access (the Australian publisher of fixed, wireless and in-car navigation services), as well as three years experience working in Europe with Tele Atlas (the world's largest digital map provider). He has a degree in Cartography and a Post-Graduate Diploma in Land Data Management (both undertaken at RMIT and completed with distinction) and is a member of the Mapping Science Institute of Australia (MSIA) and a board member of Intelligent Transport Systems (ITS) Australia.

Patricia J. O'Connor is an Associate Professor of Philosophy at Queens College, City University of New York. She earned a PhD from the University of Exeter, U.K. Her research interests include business ethics, the ethics of academic administration, and outcomes assessment.

George C. Polyzos is leading the Mobile Multimedia Lab at the Athens University of Economics and Business where he is a Professor of Computer Science. Previously, he was Professor of Computer Science and Engineering at the University of California, San Diego, where he was Co-Director of the Computer

Systems Lab, a member of the Steering Committee of the UCSD Center for Wireless Communications and Senior Fellow of the San Diego Supercomputer Center. He received his Dipl in EE from the National Technical University in Athens, Greece and his MASc in EE and PhD in Computer Science from the University of Toronto. His current research interests include mobile multimedia communications, ubiquitous computing, wireless networks, Internet protocols, distributed multimedia, telecommunications economics, and performance analysis of computer and communications systems. Professor Polyzos is on the editorial board of the journal, *Wireless Communications and Mobile Computing, and has been a guest editor for: IEEE Personal Communications, ACM/Springer Mobile Networking, IEEE JSAC* and *Computer Networks.* He has been on the program committees of many conferences and workshops, as well as reviewer for NSF, the California MICRO program, the European Commission and many scientific journals. He is a member of the ACM and the IEEE.

Minna Pura is a PhD candidate at the Department of Marketing and Corporate Geography at the Swedish School of Economics and Business Administration, Helsinki, Finland. Her research interests include: mobile branding and perceived service quality of and customer loyalty to mobile and Internet services. She has previously published in the *Yearbook on Services Management 2002—E-Services.* She is currently working as a Research Analyst at Eera Finland, which conducts strategic business consultancy combining management consulting, strategic communications and marketing and interactive solutions. Eera Finland closely cooperates with the Grey Group, which offers planning and implementation services in marketing communications. She can be reached at minna.pura@eerafinland.fi.

Andreas Rülke is a Principal in the Worldwide Communications and Electronics Practice of PRTM, management consultants to technology-driven business. He works closely with a broad range of start-ups and established companies in the communications and software industries. Mr. Rülke has led several projects involving the wireless Internet and next-generation wireless. He has worked with companies that cover many elements of the next-generation communications value chain, including telecom handset and network equipment providers, large software companies, telematics and Internet service providers, and telecom service providers. His expertise includes strategy, product and technology development, supply chain management, and customer service and support. Mr. Rülke has held interim operations management positions to help companies define, develop and integrate new operational and supply-chain-related functions across global business units. He currently is focused on working with telecommunications service providers to create standardized approaches in their supply chain operations to integrate customer-facing operations with their suppliers. Mr. Rülke holds an MBA from Carnegie Mellon University, and a Diplom Ingenieur (MS) degree in Mechanical Engineering from the Technical University of Munich, Germany. He is a member of Beta Gamma Sigma, a national scholastic honor society for business and management. Andreas Rülke is based in PRTM's Oxford, UK office and can be reached at arulke@prtm.com.

Ko de Ruyter is currently Professor of Interactive Marketing and Professor of International Service Research at Maastricht University, The Netherlands, and Director of the Maastricht Academic Center for Research in Services (MAXX). He has been a Visiting Professor at Purdue University. He holds master's degrees from the Free University Amsterdam and the University of Amsterdam. He received his PhD in Management Science from the University of Twente. He has published five books and over 150 refereed articles in, among others, the *Journal of Economic Psychology, International Journal of Research in Marketing, International Journal of Service Industry Management, Journal of Business Research, Journal of Service Research, European Journal of Marketing, Information and Management* and *Accounting, Organisation and Society.* He serves on the editorial boards of various international academic journals, including the *Journal of Service Research* and the *International Journal of Service Industry Management.* His research interests concern international service management, e-commerce and customer satisfaction and dissatisfaction.

Mark Schrauben has more than 18 years of experience in business process optimization, system integration, system development, management and consulting within the manufacturing industry. He has led multi-national teams providing value-added services for major manufacturing corporations. His experience includes the operational delivery of services, as well as strategic sales, business planning and execution. He has extensive experience in the automotive industry, as well as experience in process manufacturing, and the consumer and packaged goods industry. He has a BA in Accounting from Michigan State University.

Rick Solak has more than 13 years of experience in marketing, communications consulting and sales development within the manufacturing and high-tech industries. He managed teams in the analysis of client requirements and providing value-added marketing services for many successful projects that involved business and technology solutions for major corporations. He worked in the conceptual development and integration of market needs for service offerings, market trend and indicator analysis as well as providing analysis impacting corporate strategies for pricing, market competitiveness and business planning. Mr. Solak has developed and executed both regional and industry-level marketing plans focusing on a portfolio of applications/software and information technology outsourcing services. He has experience in the manufacturing, high-tech and Internet services industries. He has earned a BA in Marketing from Madonna University in 1989.

Mohan Tanniru is a Professor in MIS and the Director of the Applied Technology in Business (ATiB) Program at Oakland University in Rochester, Michigan. Under this program, he has directed over 200 projects in the areas of ERP, e-commerce, database migration, web benchmarking, intranet application development, network management, and business process analysis and redesign with over 40 participating firms such as DaimlerChrysler, EDS, Volkswagen of America, Lear, Eaton, Comerica, Compuware and Champion Enterprises. He has published over 40 articles in journals such as *MIS Quarterly, Communications of ACM, Information Systems Research, Journal of MIS, Decision Support Systems, Decision Sciences, IEEE Transactions in Engineering Management, Information and Management, International Journal of Human-Computer Studies* and *ES with Applications*, and made over 60 presentations at major national and international conferences. Dr. Tanniru was a consultant to Proctor & Gamble Pharmaceuticals, Carrier-UTC, Bristol Myers Squibb, Tata Consulting Services of India, and is currently a Research Consultant to Tata Infotech of India. He is a member of various professional organizations such as SIM, DSI, ACM and AIS. Dr. Tanniru has an MS in Engineering and MBA in Business Administration from the University of Wisconsin system. He earned his PhD in MIS from Northwestern University in 1978.

Peter Tarasewich is Assistant Professor of Information Science in the College of Computer Science at Northeastern University. His research interests include interface design and usability testing of mobile/wireless information systems. His publications appear in journals such as *Communications of the ACM, IEEE Transactions on Engineering Management, Quarterly Journal of Electronic Commerce, Internet Research, Journal of Computer Information Systems, Information Systems Management, Communications of the AIS, European Journal of Operational Research*, and the *Journal of the Operational Research Society*. Dr. Tarasewich holds a PhD in Operations and Information Management from the University of Connecticut, an MBA. from the University of Pittsburgh, and a BSE in Electrical Engineering with a second major in Computer Science from Duke University. He can be reached by e-mail at tarase@ccs.neu.edu and his home page can be found at: http://www.ccs.neu.edu/home/tarase/.

Argirios Tsamakos is a doctorate candidate and Research Assistant at the Department of Management Science and Technology of the Athens University of Economics and Business. He is currently working on two major European research projects related to mobile commerce. His main research interests include location-based services for mobile commerce, information technology applications in the tourism sector and indoor technologies for mobile location services.

Jeanette Van Akkeren is a Lecturer in Information Systems specializing in electronic commerce, data modelling and information systems for managers (MBA). Her main research interests include the adoption and diffusion of e-commerce technologies, and strategies for business use of mobile data technologies. Professor Van Akkeren has published extensively in the area of SME e-commerce adoption and diffusion in both journals and international conferences. In addition, she has worked as a consultant to small and medium-sized firms in the Australian states of Queensland, Victoria and NSW in the areas of end-user training and e-commerce strategy development.

Martin G.M. Wetzels is a Full Professor of Marketing with the Faculty of Technology Management, Technical University Eindhoven, Eindhoven, The Netherlands. His main research interests are: customer satisfaction/dissatisfaction, customer value, quality management in service organizations, services marketing, online marketing research, supply chain management, cross-functional cooperation and relationship marketing. His work has been published in the *International Journal of Research in Marketing*, the *Journal of Economic Psychology*, *Accounting, Organization and Society*, the *Journal of Business Research*, the *European Journal of Marketing*, *Advances in Services Marketing and Management*, *Total Quality Management*, the *Journal of Business and Industrial Marketing*, the *Journal of Management Studies*, the *Journal of Retailing and Consumer Services*, the *International Journal of Service Industry Management* and the *Journal of Service Research*. He has also contributed more than 30 papers to conference proceedings.

Mark Whitmore is the Asia Pacific Business Development Director for Webraska. Together with the rest of the local Webraska team, he is establishing Webraska as the leading regional provider of wireless location based mapping and guidance services. Mr. Whitmore has extensive experience in telecommunication and wireless data, having previously served as Senior Product Manager, Wireless Internet, Telstra OnAir and Commercial Manager, Cellular Networks, Telstra OnAir. He has also worked in the areas of Broadband Internet, interactive television and web production. He has a degree in Engineering and a master's of Project Management.

Thomas G. Zimmerman is a Research Staff Member in the USER Group of the IBM Almaden Research Center, working on new human/machine interface devices and paradigms. He received his BS in Humanities and Engineering and MS in Media Arts and Sciences from MIT. In the 1980s he invented the DataGlove, co-founding the field of virtual reality. At MIT he developed a personal area network to send data through the human body. His numerous patents cover position tracking, pen input, wireless communication, music training, biometrics and encryption. His interactive exhibits are installed at the Exploratorium in San Francisco, National Geographic Society in Washington, DC, and Science Center in Cincinnati, Ohio.

Index

Journal of Electronic Commerce in Organizations (JECO)

The International Journal of Electronic Commerce in Modern Organizations

ISSN: 1539-2937
eISSN: 1539-2929
Subscription: Annual fee per volume (4 issues):
Individual US $85
Institutional US $185

Editor: Mehdi Khosrow-Pour, D.B.A.
Information Resources
Management Association, USA

Mission

The Journal of Electronic Commerce in Organizations is designed to provide comprehensive coverage and understanding of the social, cultural, organizational, and cognitive impacts of e-commerce technologies and advances on organizations around the world. These impacts can be viewed from the impacts of electronic commerce on consumer behavior, as well as the impact of e-commerce on organizational behavior, development, and management in organizations. The secondary objective of this publication is to expand the overall body of knowledge regarding the human aspects of electronic commerce technologies and utilization in modern organizations, assisting researchers and practitioners to devise more effective systems for managing the human side of e-commerce.

Coverage

This publication includes topics related to electronic commerce as it relates to: Strategic Management, Management and Leadership, Organizational Behavior, Organizational Developement, Organizational Learning, Technologies and the Workplace, Employee Ethical Issues, Stress and Strain Impacts, Human Resources Management, Cultural Issues, Customer Behavior, Customer Relationships, National Work Force, Political Issues, and all other related issues that impact the overall utilization and management of electronic commerce technologies in modern organizations.

For subscription information, contact:

Idea Group Publishing
701 E Chocolate Ave., Ste 200
Hershey PA 17033-1240, USA
cust@idea-group.com

For paper submission information:

Dr. Mehdi Khosrow-Pour
Information Resources Management
Association
jeco@idea-group.com

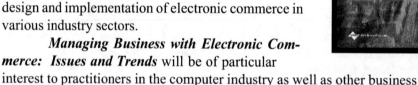